SELECTED SOLUTIONS MANUAL

Joseph Topich
Virginia Commonwealth University

CHEMISTRY
SECOND EDITION

John McMurry

Cornell University

Robert C. Fay

Cornell University

PRENTICE HALL, Upper Saddle River, NJ 07458

Associate Editor: Mary Hornby
Production Editor: Mindy De Palma
Supplement Cover Designer: Liz Nemeth
Special Projects Manager: Barbara A. Murray
Supplement Cover Manager: Paul Gourhan
Manufacturing Buyer: Ben Smith

Printed in the United States of America

10 9 8 7 6 5 4 3 2

ISBN 0-13-757527-0

Prentice-Hall International (UK) Limited, *London*
Prentice-Hall of Australia Pty. Limited, *Sydney*
Prentice-Hall Canada, Inc., *London*
Prentice-Hall Hispanoamericana, S.A., *Mexico*
Prentice-Hall of India Private Limited, *New Delhi*
Prentice-Hall of Japan, Inc., *Tokyo*
Simon & Schuster Asia Pte. Ltd., *Singapore*
Editora Prentice-Hall do Brazil, Ltda., *Rio de Janeiro*

Contents

Preface

The concepts and principles of chemistry are learned through a combination of different experiences. These experiences include reading about them in your text, hearing about them from your chemistry instructor, seeing them in laboratory experiments, and probably most importantly, using chemistry principles to solve chemistry problems.

The *Selected Solutions Manual* to accompany CHEMISTRY, 2/e by McMurry and Fay contains the solutions to all in-chapter, understanding key concept, and even numbered end-of-chapter problems. To be successful in chemistry you must be proficient in setting up and solving chemistry problems. In order to develop your chemistry problem solving skills, always attempt to set up and solve a problem before you consult the solutions manual. Read each problem carefully to see what it is asking you for and to see what information it contains. Then call upon your knowledge of chemistry principles to attempt a solution to the problem. Look at your answer and its units to see if they are reasonable. Only then should you consult the solutions manual to check your answer. You will not develop your own problem solving skills if you fall into the habit of consulting the solutions manual before you set up a problem.

I have worked to ensure that the solutions in this manual are as error free as possible. Solutions have been double-checked and in many cases triple-checked. Small differences in numerical answers between student results and those in the *Selected Solutions Manual* can result because of rounding and significant figure differences. Students should also be aware that there are, in many cases, more than one acceptable set up for a problem. Learn to use this *Selected Solutions Manual* to your advantage, not to your disadvantage.

I would like to thank John McMurry and Robert Fay for the opportunity to contribute to their CHEMISTRY package. I also want to thank them for their helpful comments as I worked on this solutions manual. I also want to acknowledge and thank Mary Hornby (Associate Editor, Science), Julie Grundman (accuracy checker), and Charlie Lebeda (line artist). Finally, I want to thank in a very special way my wife, Ruth, and our children, Joey and Judy, for their constant support and encouragement as I worked on this project.

Joseph Topich
Virginia Commonwealth University

1   Chemistry: Matter and Measurement

1.1   (a) Cu   (b) Pt   (c) Pu

1.2   (a) silver   (b) rhodium   (c) rhenium   (d) cesium   (e) argon   (f) arsenic

1.3   (a) Ti, metal   (b) Te, semimetal   (c) Se, nonmetal
       (d) Sc, metal   (e) At, semimetal   (f) Ar, nonmetal

1.4   (a) The decimal point must be shifted ten places to the right so the exponent is −10. The
           result is 3.72×10^{-10} m.
       (b) The decimal point must be shifted eleven places to the left so the exponent is 11. The
           result is 1.5×10^{11} m.

1.5   (a) microgram   (b) decimeter   (c) picosecond
       (d) kiloampere   (e) millimole

1.6 $°C = \dfrac{5}{9} \times (°F - 32) = \dfrac{5}{9} \times (98.6 - 32) = 37.0°C$

       $K = °C + 273.15 = 37.0 + 273.15 = 310.2\ K$

1.7   (a) $K = °C + 273.15 = -78 + 273.15 = 195.15\ K = 195\ K$

       (b) $°F = (\dfrac{9}{5} \times °C) + 32 = (\dfrac{9}{5} \times 158) + 32 = 316.4°F = 316°F$

       (c) $°C = K - 273.15 = 375 - 273.15 = 101.85°C = 102°C$

           $°F = (\dfrac{9}{5} \times °C) + 32 = (\dfrac{9}{5} \times 101.85) + 32 = 215.33°F = 215°F$

1.8   The actual mass of the bottle and the acetone = 38.0015 g + 0.7791 g = 38.7806 g. The
       measured values are 38.7798 g, 38.7795 g, and 38.7801 g. These values are both close to
       each other and close to the actual mass. Therefore the results are both precise and
       accurate.

1.9   (a) 76.600 kg has 5 significant figures because zeros at the end of a number and after the
       decimal point are always significant.
       (b) $4.502\ 00 \times 10^{3}$ g has 6 significant figures because zeros in the middle of a number are
       significant and zeros at the end of a number and after the decimal point are always
       significant.
       (c) 3000 nm has 1, 2, 3, or 4 significant figures because zeros at the end of a number and
       before the decimal point may or may not be significant.
       (d) 0.003 00 mL has 3 significant figures because zeros at the beginning of a number are

not significant and zeros at the end of a number and after the decimal point are always significant.

(e) 18 students has an infinite number of significant figures since this is an exact number.

1.10 (a) Since the digit to be dropped (the second 4) is less than 5, round down. The result is 3.774 L.

(b) Since the digit to be dropped (0) is less than 5, round down. The result is 255 K.

(c) Since the digit to be dropped is equal to 5 with nothing following and the preceding digit is even, round down. The result is 55.26 kg.

1.11 (a)

24.567	g	This result should be expressed with 3 decimal places.
+ 0.044 78	g	Since the digit to be dropped (7) is greater than 5, round up.
24.611 78	g	The result is 24.612 g (5 significant figures).

(b) 4.6742 g / 0.003 71 L = 1259.89 g/L

0.003 71 has only 3 significant figures so the result of the division should have only 3 significant figures. Since the digit to be dropped (first 9) is greater than 5, round up. The result is 1260 g/L (3 significant figures).

(c)

0.378	mL	This result should be expressed with 1 decimal place. Since
+ 42.3	mL	the digit to be dropped (9) is greater than 5, round up. The
− 1.5833	mL	result is 41.1 mL (3 significant figures).
41.0947 mL		

1.12 (a) Estimate: °F ≈ 2 x °C if °C is large. The melting point of gold ≈ 2000°F.

Calculation: $°F = (\frac{9}{5} \times °C) + 32 = (\frac{9}{5} \times 1064) + 32 = 1947°F$

(b) $r = d/2 = 3 \times 10^{-6}$ m $= 3 \times 10^{-4}$ cm; $h = 2 \times 10^{-6}$ m $= 2 \times 10^{-4}$ cm

Estimate: volume $= \pi r^2 h \approx 3r^2 h \approx 3(3 \times 10^{-4}$ cm$)^2(2 \times 10^{-4}$ cm$) \approx 5 \times 10^{-11}$ cm^3

Calculation: volume $= \pi r^2 h = (3.1416)(3 \times 10^{-4}$ cm$)^2(2 \times 10^{-4}$ cm$) = 6 \times 10^{-11}$ cm^3

1.13 1 mi = 1760 yd and 1 m = 1.0936 yd

$26 \text{ mi} \times \dfrac{1760 \text{ yd}}{1 \text{ mi}} = 45{,}760 \text{ yd}$

marathon in yards = 45,760 yd + 385 yd = 46,145 yd

$\text{marathon in meters} = 46{,}145 \text{ yd} \times \dfrac{1 \text{ m}}{1.0936 \text{ yd}} = 42{,}196 \text{ m}$

1.14 1 carat = 200 mg = 200 x 10^{-3} g = 0.200 g

$\text{Mass of Hope Diamond in grams} = 44.4 \text{ carats} \times \dfrac{0.200 \text{ g}}{1 \text{ carat}} = 8.88 \text{ g}$

1 ounce = 28.35 g

$\text{Mass of Hope Diamond in ounces} = 8.88 \text{ g} \times \dfrac{1 \text{ ounce}}{28.35 \text{ g}} = 0.313 \text{ ounces}$

1.15 $1100 \ \mu g = 1100 \times 10^{-6} \ g$

$$\frac{1100 \times 10^{-6} \ g}{2300 \ people} = 4.8 \times 10^{-7} \ g/person$$

1.16 $d = \dfrac{m}{V} = \dfrac{27.43 \ g}{12.40 \ cm^3} = 2.212 \ g/cm^3$

1.17 $volume = 9.37 \ g \times \dfrac{1 \ mL}{1.483 \ g} = 6.32 \ mL$

Understanding Key Concepts

1.18

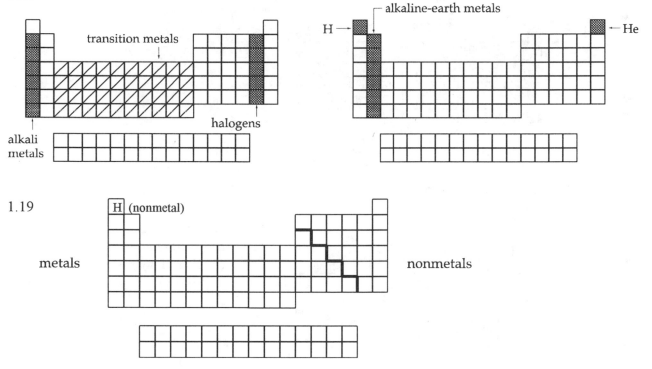

1.19

metals   nonmetals

1.20   red – gas; blue – 42; green – sodium

1.21   (a) Darts are scattered (poor precision) and are away from the bullseye (poor accuracy).
        (b) Darts are clustered together (good precision) and hit the bullseye (good accuracy).
        (c) Darts are clustered together (good precision) but are away from the bullseye (poor accuracy).

1.22   The density of solid lead is less than the density of liquid mercury so the lead will float in the mercury.

1.23   A hypothesis is an interpretation of experimental results. A theory is a consistent explanation of known observations.
A physical property is a characteristic that can be determined without changing the chemical makeup of a sample. A chemical property is a characteristic that results in a change in the chemical makeup of a sample.

Additional Problems
Elements and the Periodic Table

1.24   112 elements are presently known. About 90 elements occur naturally.

1.26   There are 18 groups in the periodic table. They are labeled as follows:
1A, 2A, 3B, 4B, 5B, 6B, 7B, 8B (3 groups), 1B, 2B, 3A, 4A, 5A, 6A, 7A, 8A

1.28

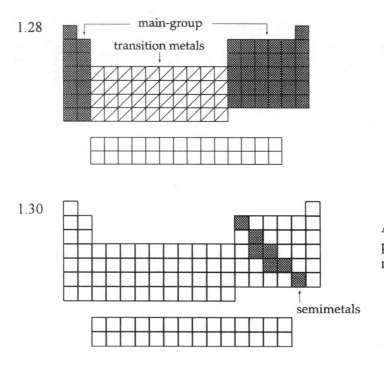

1.30

A semimetal is an element with properties that fall between those of metals and nonmetals.

1.32   Li, Na, K, Rb, and Cs

1.34   F, Cl, Br, and I

1.36   (a) americium, Am   (b) germanium, Ge   (c) technetium, Tc   (d) arsenic, As

1.38   (a) Te, tellurium   (b) Re, rhenium   (c) Be, beryllium
(d) Ar, argon   (e) Pu, plutonium

1.40   (a) Tin is Sn: Ti is titanium.   (b) Manganese is Mn: Mg is magnesium.
(c) Potassium is K: Po is polonium.   (d) The symbol for helium is He.

Units and Significant Figures

1.42   Accurate measurement is crucial in science because, if our experiments are to be reproducible, we must be able to describe fully the substances we are working with — their amounts, sizes, temperatures, etc.

1.44   Mass measures the amount of matter in an object, whereas weight measures the pull of gravity on an object by the earth or other celestial body.

1.46   (a) kilogram, kg   (b) meter, m   (c) kelvin, K   (d) cubic meter, m^3

1.48   A Celsius degree is larger than a Fahrenheit degree by a factor of $\dfrac{9}{5}$.

1.50   The volume of a cubic decimeter (dm^3) and a liter (L) are the same.

1.52   (a) and (b) are exact because they are obtained by counting. (c) is not exact because it results from a measurement.

1.54   cL is centiliter (10^{-2} L)

1.56 $1\ mg = 1 \times 10^{-3}$ g and $1\ pg = 1 \times 10^{-12}$ g

$$\frac{1 \times 10^{-3}\ g}{1\ mg} \times \frac{1\ pg}{1 \times 10^{-12}\ g} = 1 \times 10^{9}\ pg/mg$$

$35\ ng = 35 \times 10^{-9}$ g $\dfrac{35 \times 10^{-9}\ g}{35\ ng} \times \dfrac{1\ pg}{1 \times 10^{-12}\ g} = 3.5 \times 10^{4}\ pg/35\ ng$

1.58   (a) $5\ pm = 5 \times 10^{-12}$ m

$5 \times 10^{-12}\ m \times \dfrac{100\ cm}{1\ m} = 5 \times 10^{-10}\ cm$

$5 \times 10^{12}\ m \times \dfrac{1\ nm}{1 \times 10^{-9}\ m} = 5 \times 10^{-3}\ nm$

(b) $8.5\ cm^3 \times \left(\dfrac{1\ m}{100\ cm}\right)^3 = 8.5 \times 10^{-6}\ m^3$

$8.5\ cm^3 \times \left(\dfrac{10\ mm}{1\ cm}\right)^3 = 8.5 \times 10^{3}\ mm^3$

(c) $65.2\ mg \times \dfrac{1 \times 10^{-3}\ g}{1\ mg} = 0.0652\ g$

$65.2\ mg \times \dfrac{1 \times 10^{-3}\ g}{1\ mg} \times \dfrac{1\ pg}{1 \times 10^{-12}\ g} = 6.52 \times 10^{10}\ pg$

1.60   (a) 35.0445 g has 6 significant figures because zeros in the middle of a number are significant.
(b) 59.0001 cm has 6 significant figures because zeros in the middle of a number are significant.
(c) 0.030 03 kg has 4 significant figures because zeros at the beginning of a number are not significant and zeros in the middle of a number are significant.
(d) 0.004 50 m has 3 significant figures because zeros at the beginning of a number are not significant and zeros at the end of a number and after the decimal point are always significant.
(e) 67,000 m^2 has 2, 3, 4, or 5 significant figures because zeros at the end of a number and before the decimal point may or may not be significant.
(f) 3.8200 x 10^3 L has 5 significant figures because zeros at the end of a number and after the decimal point are always significant.

1.62   To convert 3,666,500 m^3 to scientific notation, move the decimal point 6 places to the left and include an exponent of 10^6. The result is 3.6665 x 10^6.

1.64   (a) To convert 453.32 mg to scientific notation, move the decimal point 2 places to the left and include an exponent of 10^2. The result is 4.5332 x 10^2 mg.
(b) To convert 0.000 042 1 mL to scientific notation, move the decimal point 5 places to the right and include an exponent of 10^{-5}. The result is 4.21 x 10^{-5} mL.
(c) To convert 667,000 g to scientific notation, move the decimal point 5 places to the left and include an exponent of 10^5. The result is 6.67 x 10^5 g.

1.66   (a) Since the digit to be dropped (0) is less than 5, round down. The result is 3.567 x 10^4 or 35,670 m (4 significant figures).
Since the digit to be dropped (the second 6) is greater than 5, round up. The result is 35,670.1 m (6 significant figures).
(b) Since the digit to be dropped is 5 with nonzero digits following, round up. The result is 69 g (2 significant figures).
Since the digit to be dropped (0) is less than 5, round down. The result is 68.5 g (3 significant figures).
(c) Since the digit to be dropped is 5 with nothing following and the preceding digit is odd, round up. The result is 5.00 x 10^3 cm (3 significant figures).
(d) Since the digit to be dropped is 5 with nothing following and the preceding digit is even, round down. The result is 2.3098 x 10^{-4} kg (5 significant figures).

1.68   (a) 4.884 x 2.05 = 10.012
The result should contain only 3 significant figures because 2.05 contains 3 significant figures (the smaller number of significant figures of the two). Since the digit to be dropped (1) is less than 5, round down. The result is 10.0.
(b) 94.61 / 3.7 = 25.57
The result should contain only 2 significant figures because 3.7 contains 2 significant figures (the smaller number of significant figures of the two). Since the digit to be dropped (second 5) is 5 with nonzero digits following, round up. The result is 26.

(c) $3.7 / 94.61 = 0.0391$

The result should contain only 2 significant figures because 3.7 contains 2 significant figures (the smaller number of significant figures of the two). Since the digit to be dropped (1) is less than 5, round down. The result is 0.039.

(d)

$$
\begin{array}{r}
5502.3 \\
24 \\
+ \quad 0.01 \\
\hline
5526.31
\end{array}
$$

This result should be expressed with no decimal places. Since the digit to be dropped (3) is less than 5, round down. The result is 5526.

(e)

$$
\begin{array}{r}
86.3 \\
+ \quad 1.42 \\
- \quad 0.09 \\
\hline
87.63
\end{array}
$$

This result should be expressed with only 1 decimal place. Since the digit to be dropped (3) is less than 5, round down. The result is 87.6.

(f) $5.7 \times 2.31 = 13.167$

The result should contain only 2 significant figures because 5.7 contains 2 significant figures (the smaller number of significant figures of the two). Since the digit to be dropped (second 1) is less than 5, round down. The result is 13.

Unit Conversions

1.70 (a) $0.25 \text{ lb} \times \dfrac{453.59 \text{ g}}{1 \text{ lb}} = 113.4 \text{ g} = 110 \text{ g}$

(b) $1454 \text{ ft} \times \dfrac{12 \text{ in}}{1 \text{ ft}} \times \dfrac{2.54 \text{ cm}}{1 \text{ in}} \times \dfrac{1 \text{ m}}{100 \text{ cm}} = 443.2 \text{ m}$

(c) $2,941,526 \text{ mi}^2 \times \left(\dfrac{1.6093 \text{ km}}{1 \text{ mi}}\right)^2 \times \left(\dfrac{1000 \text{ m}}{1 \text{ km}}\right)^2 = 7.6181 \times 10^{12} \text{ m}^2$

1.72 (a) $1 \text{ acre–ft} \times \dfrac{1 \text{ mi}^2}{640 \text{ acres}} \times \left(\dfrac{5280 \text{ ft}}{1 \text{ mi}}\right)^2 = 43,560 \text{ ft}^3$

(b) $116 \text{ mi}^3 \times \left(\dfrac{5280 \text{ ft}}{1 \text{ mi}}\right)^3 \times \dfrac{1 \text{ acre–ft}}{43,560 \text{ ft}^3} - 3.92 \times 10^8 \text{ acre–ft}$

1.74 (a) $\dfrac{200 \text{ mg}}{100 \text{ mL}} \times \dfrac{1000 \text{ mL}}{1 \text{ L}} = 2000 \text{ mg/L}$

(b) $\dfrac{200 \text{ mg}}{100 \text{ mL}} \times \dfrac{1 \times 10^{-3} \text{ g}}{1 \text{ mg}} \times \dfrac{1 \text{ } \mu g}{1 \times 10^{-6} \text{ g}} = 2000 \text{ } \mu g/mL$

(c) $\dfrac{200 \text{ mg}}{100 \text{ mL}} \times \dfrac{1 \times 10^{-3} \text{ g}}{1 \text{ mg}} \times \dfrac{1000 \text{ mL}}{1 \text{ L}} = 2 \text{ g/L}$

(d) $\dfrac{200 \text{ mg}}{100 \text{ mL}} \times \dfrac{1 \times 10^{-3} \text{ g}}{1 \text{ mg}} \times \dfrac{1000 \text{ mL}}{1 \text{ L}} \times \dfrac{1 \text{ ng}}{1 \times 10^{-9} \text{ g}} \times \dfrac{1 \times 10^{-6} \text{ L}}{1 \text{ } \mu L} = 2000 \text{ ng/} \mu L$

(e) $2 \text{ g/L} \times 5 \text{ L} = 10 \text{ g}$

1.76 $\quad 55 \dfrac{\text{mi}}{\text{h}} \times \dfrac{5280 \text{ ft}}{1 \text{ mi}} \times \dfrac{12 \text{ in}}{1 \text{ ft}} \times \dfrac{2.54 \text{ cm}}{1 \text{ in}} \times \dfrac{1 \text{ h}}{3600 \text{ s}} \times \dfrac{2.5 \times 10^{-4} \text{ s}}{1 \text{ shake}} = 0.61 \dfrac{\text{cm}}{\text{shake}}$

Temperature

1.78 $\quad {}^\circ\text{F} = (\dfrac{9}{5} \times {}^\circ\text{C}) + 32$

$\quad\quad {}^\circ\text{F} = (\dfrac{9}{5} \times 39.9{}^\circ\text{C}) + 32 = 103.8{}^\circ\text{F}$

$\quad\quad {}^\circ\text{F} = (\dfrac{9}{5} \times 22.2{}^\circ\text{C}) + 32 = 72.0{}^\circ\text{F}$

1.80 $\quad {}^\circ\text{C} = \dfrac{5}{9} \times ({}^\circ\text{F} - 32) = \dfrac{5}{9} \times (6170 - 32) = 3410{}^\circ\text{C}$

$\quad\quad \text{K} = {}^\circ\text{C} + 273.15 = 3410 + 273.15 = 3683.15 \text{ K} = 3683 \text{ K}$

1.82 $\quad$ Ethanol boiling point $\quad$ 78.5°C $\quad\quad$ 173.3°F $\quad\quad$ 200°E
$\quad\quad$ Ethanol melting point $\;-117.3$°C $\quad\;\; -179.1$°F $\quad\quad$ 0°E

$\quad\quad$ (a) $\quad \dfrac{200{}^\circ\text{E}}{[78.5{}^\circ\text{C} - (-117.3{}^\circ\text{C})]} = \dfrac{200{}^\circ\text{E}}{195.8{}^\circ\text{C}} = 1.021 \text{ {}°E/°C}$

$\quad\quad$ (b) $\quad \dfrac{200{}^\circ\text{E}}{[173.3{}^\circ\text{F} - (-179.1{}^\circ\text{F})]} = \dfrac{200{}^\circ\text{E}}{352.4{}^\circ\text{F}} = 0.5675 \text{ {}°E/°F}$

$\quad\quad$ (c) $\quad {}^\circ\text{E} = \dfrac{200}{195.8} \times ({}^\circ\text{C} + 117.3)$

$\quad\quad$ H_2O melting point $= 0{}^\circ\text{C}; \quad {}^\circ\text{E} = \dfrac{200}{195.8} \times (0 + 117.3) = 119.8{}^\circ\text{E}$

$\quad\quad$ H_2O boiling point $= 100{}^\circ\text{C}; \quad {}^\circ\text{E} = \dfrac{200}{195.8} \times (100 + 117.3) = 220.0{}^\circ\text{E}$

$\quad\quad$ (d) $\quad {}^\circ\text{E} = \dfrac{200}{352.4} \times ({}^\circ\text{F} + 179.1) = \dfrac{200}{352.4} \times (98.6 + 179.1) = 157.6{}^\circ\text{E}$

$\quad\quad$ (e) $\quad {}^\circ\text{F} = \left({}^\circ\text{E} \times \dfrac{352.4}{200}\right) - 179.1 = \left(130 \times \dfrac{352.4}{200}\right) - 179.1 = 50.0{}^\circ\text{F}$

$\quad\quad$ Since the outside temperature is 50.0°F, I would wear a sweater or light jacket.

Density

1.84 $\quad 250 \text{ mg} \times \dfrac{1 \times 10^{-3} \text{ g}}{1 \text{ mg}} = 0.25 \text{ g}; \quad\quad V = 0.25 \text{ g} \times \dfrac{1 \text{ cm}^3}{1.40 \text{ g}} = 0.18 \text{ cm}^3$

$\quad\quad 500 \text{ lb} \times \dfrac{453.59 \text{ g}}{1 \text{ lb}} = 226{,}795 \text{ g}; \quad V = 226{,}795 \text{ g} \times \dfrac{1 \text{ cm}^3}{1.40 \text{ g}} = 161{,}996 \text{ cm}^3 = 162{,}000 \text{ cm}^3$

1.86 $d = \dfrac{m}{V} = \dfrac{220.9 \text{ g}}{(0.50 \times 1.55 \times 25.00) \text{ cm}^3} = 11.4 \dfrac{\text{g}}{\text{cm}^3} = 11 \dfrac{\text{g}}{\text{cm}^3}$

1.88 $d = \dfrac{m}{V} = \dfrac{8.763 \text{ g}}{(28.76 - 25.00) \text{ mL}} = \dfrac{8.763 \text{ g}}{3.76 \text{ mL}} = 2.331 \dfrac{\text{g}}{\text{cm}^3} = 2.33 \dfrac{\text{g}}{\text{cm}^3}$

General Problems

1.90   Physical properties: color, melting point, mass, and solubility.
Chemical properties: rusting (of iron), combustion (of methane), hardening (of cement), explosion (of nitroglycerine).

1.92   (a) Se, selenium   (b) Sb, antimony   (c) Ba, barium
       (d) Cs, cesium   (e) Np, neptunium

1.94   (a) cL, centiliter   (b) m^2, square meter   (c) K, kelvin   (d) nm, nanometer

1.96   NaCl melting point = 1074 K
$^\circ C = K - 273.15 = 1074 - 273.15 = 800.85^\circ C = 801^\circ C$

$^\circ F = (\dfrac{9}{5} \times {}^\circ C) + 32 = (\dfrac{9}{5} \times 800.85) + 32 = 1473.53^\circ F = 1474^\circ F$

NaCl boiling point = 1686 K
$^\circ C = K - 273.15 = 1686 - 273.15 = 1412.85^\circ C = 1413^\circ C$

$^\circ F = (\dfrac{9}{5} \times {}^\circ C) + 32 = (\dfrac{9}{5} \times 1412.85) + 32 = 2575.13^\circ F = 2575^\circ F$

1.98 $V = 112.5 \text{ g} \times \dfrac{1 \text{ mL}}{1.4832 \text{ g}} = 75.85 \text{ mL}$

1.100 $V = 9.536 \times 10^{10} \text{ lb} \times \dfrac{453.59 \text{ g}}{1 \text{ lb}} \times \dfrac{1 \text{ mL}}{1.831 \text{ g}} \times \dfrac{1 \text{ L}}{1000 \text{ mL}} = 2.362 \times 10^{10} \text{ L}$

1.102 $^\circ C = \dfrac{5}{9} \times ({}^\circ F - 32)$;   Set $^\circ C = {}^\circ F$: $^\circ C = \dfrac{5}{9} \times ({}^\circ C - 32)$

Solve for $^\circ C$: $^\circ C \times \dfrac{9}{5} = {}^\circ C - 32$

$({}^\circ C \times \dfrac{9}{5}) - {}^\circ C = -32$

$^\circ C \times \dfrac{4}{5} = -32$

$^\circ C = \dfrac{5}{4}(-32) = -40^\circ C$

The Celsius and Fahrenheit scales "cross" at $-40^\circ C$ ($-40^\circ F$).

1.104   Convert 8 min, 25 s to s. $8 \text{ min} \times \dfrac{60 \text{ s}}{1 \text{ min}} + 25 \text{ s} = 505 \text{ s}$

Convert 293.2 K to °F $293.2 - 273.15 = 20.05°C$

$$°F = (\dfrac{9}{5} \times 20.05) + 32 = 68.09°F$$

Final temperature $= 68.09°F + 505 \text{ s} \times \dfrac{3.0°F}{60 \text{ s}} = 93.34°F$

$$°C = \dfrac{5}{9} \times (93.34 - 32) = 34.1°C$$

1.106   Average brass density $= (0.670)(8.92 \text{ g/cm}^3) + (0.330)(7.14 \text{ g/cm}^3) = 8.333 \text{ g/cm}^3$

length $= 1.62 \text{ in} \times \dfrac{2.54 \text{ cm}}{1 \text{ in.}} = 4.115 \text{ cm}$

diameter $= 0.514 \text{ in} \times \dfrac{2.54 \text{ cm}}{1 \text{ in.}} = 1.306 \text{ cm}$

volume $= \pi r^2 h = (3.1416)[(1.306 \text{ cm})/2]^2(4.115 \text{ cm}) = 5.512 \text{ cm}^3$

mass $= 5.512 \text{ cm}^3 \times \dfrac{8.333 \text{ g}}{1 \text{ cm}^3} = 45.9 \text{ g}$

Atoms, Molecules, and Ions

2.1   First, find the S:O ratio in each compound.
Substance 1: S:O mass ratio = (6.00 g S) / (5.99 g O) = 1.00
Substance 2: S:O mass ratio = (8.60 g S) / (12.88 g O) = 0.668

$$\frac{\text{S:O mass ratio in substance 1}}{\text{S:O mass ratio in substance 2}} = \frac{1.00}{0.668} = 1.50 = \frac{3}{2}$$

2.2 $0.0002 \text{ in} \times \dfrac{2.54 \text{ cm}}{1 \text{ in}} \times \dfrac{1 \text{ Au atom}}{2.9 \times 10^{-8} \text{ cm}} = 2 \times 10^4 \text{ Au atoms}$

2.3 $1 \times 10^{19} \text{ C atoms} \times \dfrac{1.5 \times 10^{-10} \text{ m}}{\text{C atom}} \times \dfrac{1 \text{ km}}{1000 \text{ m}} \times \dfrac{1 \text{ time}}{40{,}075 \text{ km}} = 37.4 \text{ times} \approx 40 \text{ times}$

2.4 $^{75}_{34}\text{Se}$ has 34 protons, 34 electrons, and (75 – 34) = 41 neutrons.

2.5 $^{35}_{17}\text{Cl}$ has (35 – 17) = 18 neutrons. $^{37}_{17}\text{Cl}$ has (37 – 17) = 20 neutrons.

2.6   The element with 47 protons is Ag. The mass number is the sum of the protons and the neutrons, 47 + 62 = 109. The isotope symbol is $^{109}_{47}\text{Ag}$.

2.7   atomic weight = (0.6917 x 62.94 amu) + (0.3083 x 64.93 amu) = 63.55 amu

2.8 $2.15 \text{ g} \times \dfrac{1 \text{ amu}}{1.6605 \times 10^{-24} \text{ g}} \times \dfrac{1 \text{ Cu}}{63.55 \text{ amu}} = 2.04 \times 10^{22} \text{ Cu atoms}$

2.9
```
        H   H
        |   |
   H — C — N — H
        |
        H
```

2.10 (a) LiBr is composed of a metal (Li) and nonmetal (Br) and is ionic.
(b) $SiCl_4$ is composed of only nonmetals and is molecular.
(c) BF_3 is composed of only nonmetals and is molecular.
(d) CaO is composed of a metal (Ca) and nonmetal (O) and is ionic.

2.11 (a) HF is an acid. In water, HF dissociates to produce $H^+(aq)$.
(b) $Ca(OH)_2$ is a base. In water, $Ca(OH)_2$ dissociates to produce $OH^-(aq)$.
(c) LiOH is a base. In water, LiOH dissociates to produce $OH^-(aq)$.
(d) HCN is an acid. In water, HCN dissociates to produce $H^+(aq)$.

2.12   (a) CsF, cesium fluoride   (b) K_2O, potassium oxide   (c) CuO, copper(II) oxide
        (d) BaS, barium sulfide   (e) $BeBr_2$, beryllium bromide

2.13   (a) vanadium(III) chloride, VCl_3 (b) manganese(IV) oxide, MnO_2
        (c) copper(II) sulfide, CuS   (d) aluminum oxide, Al_2O_3

2.14   (a) NCl_3, nitrogen trichloride   (b) P_4O_6, tetraphosphorus hexoxide
        (c) S_2F_2, disulfur difluoride

2.15   (a) disulfur dichloride, S_2Cl_2 (b) iodine monochloride, ICl
        (c) nitrogen triiodide, NI_3

2.16   (a) $Ca(ClO)_2$, calcium hypochlorite
        (b) $Ag_2S_2O_3$, silver(I) thiosulfate or silver thiosulfate
        (c) NaH_2PO_4, sodium dihydrogen phosphate
        (d) $Sn(NO_3)_2$, tin(II) nitrate   (e) $Pb(C_2H_3O_2)_4$, lead(IV) acetate

2.17   (a) lithium phosphate, Li_3PO_4 (b) magnesium hydrogen sulfate, $Mg(HSO_4)_2$
        (c) manganese(II) nitrate, $Mn(NO_3)_2$ (d) chromium(III) sulfate, $Cr_2(SO_4)_3$

2.18   (a) HIO_4, periodic acid   (b) $HBrO_2$, bromous acid   (c) H_2CrO_4, chromic acid

Understanding Key Concepts

2.19   Drawing (a) depicts a collection of SO_2 molecules.

2.20

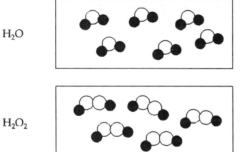

2.21   A Na atom has 11 protons and 11 electrons [drawing (b)].
        A Ca^{2+} ion has 20 protons and 18 electrons [drawing (c)].
        A F^- ion has 9 protons and 10 electrons [drawing (a)].

2.22

     Ne   (10+)   10– Br^- (35+)   36– Au^{3+} (79+)   76–

2.23   In order to obey the law of mass conservation the correct drawing must have the same number of red and yellow spheres as in drawing (a). The correct drawing is (d).

2.24   Figures (b) and (d) illustrate the law of multiple proportions. The ∘∘•/∘• mass ratio is 2.

2.25   (a) gallium, Ga   (b) chromium, Cr   (c) aluminum, Al   (d) aluminum-27, ^{27}Al

2.26.   alanine, $C_3H_7NO_2$;   ethylene glycol, $C_2H_6O_2$;   acetic acid, $C_2H_4O_2$

2.27

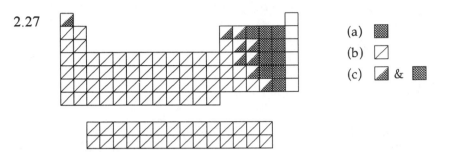

(a) ▨
(b) ▨
(c) ◩ & ▨

Additional Problems
Atomic Theory

2.28   Different samples of the same compound always contain the same mass ratio of elements.

2.30   The law of mass conservation in terms of Dalton's atomic theory states that chemical reactions only rearrange the way that atoms are combined; the atoms themselves are not changed.
The law of definite proportions in terms of Dalton's atomic theory states that the chemical combination of elements to make different substances occurs when atoms join together in small, whole-number ratios.

2.32   First, find the C:H ratio in each compound.
Benzene: C:H mass ratio = (4.61 g C) / (0.39 g H) = 12
Ethane: C:H mass ratio (4.00 g C) / (1.00 g H) = 4.0
Ethylene: C:H mass ratio = (4.29 g C) / (0.71 g H) = 6.0

$$\frac{\text{C:H mass ratio in benzene}}{\text{C:H mass ratio in ethane}} = \frac{12}{4.0} = \frac{3}{1}$$

$$\frac{\text{C:H mass ratio in benzene}}{\text{C:H mass ratio in ethylene}} = \frac{12}{6.0} = \frac{2}{1}$$

$$\frac{\text{C:H mass ratio in ethylene}}{\text{C:H mass ratio in ethane}} = \frac{6.0}{4.0} = \frac{3}{2}$$

2.34   (a) For benzene:

$$4.61 \text{ g} \times \frac{1 \text{ amu}}{1.6605 \times 10^{-24} \text{ g}} \times \frac{1 \text{ C atom}}{12.011 \text{ amu}} = 2.31 \times 10^{23} \text{ C atoms}$$

$$0.39 \text{ g} \times \frac{1 \text{ amu}}{1.6605 \times 10^{-24} \text{ g}} \times \frac{1 \text{ H atom}}{1.008 \text{ amu}} = 2.3 \times 10^{23} \text{ H atoms}$$

$$\frac{C}{H} = \frac{2.31 \times 10^{23} \text{ C atoms}}{2.3 \times 10^{23} \text{ H atoms}} = \frac{1 \text{ C}}{1 \text{ H}}$$

A possible formula for benzene is CH.

For ethane:

$$4.00 \text{ g} \times \frac{1 \text{ amu}}{1.6605 \times 10^{-24} \text{ g}} \times \frac{1 \text{ C atom}}{12.011 \text{ amu}} = 2.01 \times 10^{23} \text{ C atoms}$$

$$1.00 \text{ g} \times \frac{1 \text{ amu}}{1.6605 \times 10^{-24} \text{ g}} \times \frac{1 \text{ H atom}}{1.008 \text{ amu}} = 5.97 \times 10^{23} \text{ H atoms}$$

$$\frac{C}{H} = \frac{2.01 \times 10^{23} \text{ C atoms}}{5.97 \times 10^{23} \text{ H atoms}} = \frac{1 \text{ C}}{3 \text{ H}}$$

A possible formula for ethane is CH_3.

For ethylene:

$$4.29 \text{ g} \times \frac{1 \text{ amu}}{1.6605 \times 10^{-24} \text{ g}} \times \frac{1 \text{ C atom}}{12.011 \text{ amu}} = 2.15 \times 10^{23} \text{ C atoms}$$

$$0.71 \text{ g} \times \frac{1 \text{ amu}}{1.6605 \times 10^{-24} \text{ g}} \times \frac{1 \text{ H atom}}{1.008 \text{ amu}} = 4.2 \times 10^{23} \text{ H atoms}$$

$$\frac{C}{H} = \frac{2.15 \times 10^{23} \text{ C atoms}}{4.2 \times 10^{23} \text{ H atoms}} = \frac{1 \text{ C}}{2 \text{ H}}$$

A possible formula for ethylene is CH_2.

(b) The results in part (a) give the smallest whole-number ratio of C to H for benzene, ethane, and ethylene, and these ratios are consistent with their modern formulas.

2.36   (a) $(1.67 \times 10^{-24} \frac{g}{\text{H atom}})(6.02 \times 10^{23} \text{ H atoms}) = 1.01 \text{ g}$

This result is numerically equal to the atomic weight of H in grams.

(b) $(26.558 \times 10^{-24} \frac{g}{\text{O atom}})(6.02 \times 10^{23} \text{ O atoms}) = 16.0 \text{ g}$

This result is numerically equal to the atomic weight of O in grams.

2.38   Assume a 1.00 g sample of the binary compound of zinc and sulfur.
$0.671 \times 1.00 \text{ g} = 0.671 \text{ g Zn};$ $0.329 \times 1.00 \text{ g} = 0.329 \text{ g S}$

$$0.671 \text{ g} \times \frac{1 \text{ amu}}{1.6605 \times 10^{-24} \text{ g}} \times \frac{1 \text{ Zn atom}}{65.39 \text{ amu}} = 6.18 \times 10^{21} \text{ Zn atoms}$$

$$0.329 \text{ g} \times \frac{1 \text{ amu}}{1.6605 \times 10^{-24} \text{ g}} \times \frac{1 \text{ S atom}}{32.066 \text{ amu}} = 6.18 \times 10^{21} \text{ S atoms}$$

$$\frac{Zn}{S} = \frac{6.18 \times 10^{21} \text{ Zn atoms}}{6.18 \times 10^{21} \text{ S atoms}} = \frac{1 \text{ Zn}}{1 \text{ S}}; \text{ therefore the formula is ZnS.}$$

Elements and Atoms

2.40   A cathode–ray tube is a glass tube from which the air has been removed and in which two thin pieces of metal called electrodes have been sealed. When a sufficiently high voltage is applied across the electrodes, an electric current flows through the tube from the negatively charged electrode (the cathode) to the positively charged electrode (the anode). If the tube is not fully evacuated but still contains a small amount of air or other gas, the flowing current is visible as a glow called a cathode ray.

2.42   The atomic number is equal to the number of protons.
The mass number is equal to the sum of the number of protons and the number of neutrons.

2.44   Atoms of the same element that have different numbers of neutrons are called isotopes.

2.46   The subscript giving the atomic number of an atom is often left off of an isotope symbol because one can readily look up the atomic number in the periodic table.

2.48   (a) carbon, C   (b) argon, Ar   (c) vanadium, V

2.50   (a) $^{220}_{86}\text{Rn}$ (b) $^{210}_{84}\text{Po}$ (c) $^{197}_{79}\text{Au}$

2.52   (a) $^{15}_{7}\text{N}$, 7 protons, 7 electrons, $(15 - 7) = 8$ neutrons

   (b) $^{60}_{27}\text{Co}$, 27 protons, 27 electrons, $(60 - 27) = 33$ neutrons

   (c) $^{131}_{53}\text{I}$, 53 protons, 53 electrons, $(131 - 53) = 78$ neutrons

2.54   (a) $^{24}_{12}\text{Mg}$, magnesium   (b) $^{58}_{28}\text{Ni}$, nickel

2.56 $(0.199 \times 10.0129 \text{ amu}) + (0.801 \times 11.009\,31 \text{ amu}) = 10.8 \text{ amu for B}$

2.58 $24.305 \text{ amu} = (0.7899 \times 23.985 \text{ amu}) + (0.1000 \times 24.986 \text{ amu}) + (0.1101 \times Z)$
Solve for Z. $Z = 25.982$ (^{26}Mg)

Compounds and Mixtures, Molecules and Ions

2.60   A compound is a pure substance with constant composition. A mixture is formed when two or more substances are mixed together in some random proportion without chemically changing the individual substances in the mixture. Sucrose ($C_{12}H_{22}O_{11}$) is a compound. Sucrose dissolved in water is a mixture.

2.62   (a) muddy water, heterogeneous mixture
(b) concrete, heterogeneous mixture
(c) house paint, homogeneous mixture
(d) a soft drink, homogeneous mixture (heterogeneous mixture if it contains CO_2 bubbles)

2.64   An atom is the smallest particle that retains the chemical properties of an element. A molecule is matter that results when two or more atoms are joined by covalent bonds. H and O are atoms, H_2O is a water molecule.

2.66   A covalent bond results when two atoms share several (usually two) of their electrons. An ionic bond results from a complete transfer of one or more electrons from one atom to another. The C—H bonds in methane (CH_4) are covalent bonds. The bond in NaCl (Na^+Cl^-) is an ionic bond.

2.68   Element symbols are composed of one or two letters. If the element symbol is two letters, the first letter is uppercase and the second is lowercase. CO stands for carbon and oxygen in carbon monoxide.

2.70   (a) Be^{2+}, 4 protons and 2 electrons   (b) Rb^+, 37 protons and 36 electrons
(c) Se^{2-}, 34 protons and 36 electrons   (d) Au^{3+}, 79 protons and 76 electrons

2.72 C_3H_8O

2.74

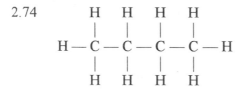

Acids and Bases

2.76   (a) HI, acid   (b) CsOH, base   (c) H_3PO_4, acid
(d) $Ba(OH)_2$, base   (e) H_2CO_3, acid

2.78 $HI(aq) \rightarrow H^+(aq) + I^-(aq)$; the anion is I^-
$H_3PO_4(aq) \rightarrow H^+(aq) + H_2PO_4^-(aq)$; the predominant anion is $H_2PO_4^-$
$H_2CO_3(aq) \rightarrow H^+(aq) + HCO_3^-(aq)$; the predominant anion is HCO_3^-

Naming Compounds

2.80   (a) KCl   (b) $SnBr_2$ (c) CaO   (d) $BaCl_2$ (e) AlH_3

2.82   (a) barium ion   (b) cesium ion   (c) vanadium(III) ion
(d) hydrogen carbonate ion   (e) ammonium ion   (f) nickel(II) ion
(g) nitrite ion   (h) chlorite ion   (i) manganese(II) ion   (j) perchlorate ion

2.84 (a) SO_3^{2-} (b) PO_4^{3-} (c) Zr^{4+} (d) CrO_4^{2-} (e) $CH_3CO_2^-$ (f) $S_2O_3^{2-}$

2.86 (a) zinc(II) cyanide   (b) iron(III) nitrite   (c) titanium(IV) sulfate
    (d) tin(II) phosphate   (e) mercury(I) sulfide   (f) manganese(IV) oxide
    (g) potassium periodate   (h) copper(II) acetate

2.88 (a) Na^+ and SO_4^{2-}; therefore the formula is Na_2SO_4
    (b) Ba^{2+} and PO_4^{3-}; therefore the formula is $Ba_3(PO_4)_2$
    (c) Ga^{3+} and SO_4^{2-}; therefore the formula is $Ga_2(SO_4)_3$

General Problems

2.90   atomic weight $= (0.205 \times 69.924 \text{ amu}) + (0.274 \times 71.922 \text{ amu})$
                $+ (0.078 \times 72.923 \text{ amu}) + (0.365 \times 73.921 \text{ amu})$
                $+ (0.078 \times 75.921 \text{ amu}) = 72.6 \text{ amu}$

2.92 (a) sodium bromate   (b) phosphoric acid
    (c) phosphorous acid   (d) vanadium(V) oxide

2.94   For NH_3, $(2.34 \text{ g N})\left(\dfrac{3 \times 1.0079 \text{ amu H}}{14.0067 \text{ amu N}}\right) = 0.505 \text{ g H}$

    For N_2H_4, $(2.34 \text{ g N})\left(\dfrac{4 \times 1.0079 \text{ amu H}}{2 \times 14.0067 \text{ amu N}}\right) = 0.337 \text{ g H}$

2.96 TeO_4^{2-}, tellurate; TeO_3^{2-}, tellurite.
    TeO_4^{2-} and TeO_3^{2-} are analogous to SO_4^{2-} and SO_3^{2-}.

2.98 (a) I^- (b) Au^{3+} (c) Kr

2.100 $\dfrac{39.9626 \text{ amu}}{15.9994 \text{ amu}} = \dfrac{X}{16.0000 \text{ amu}}$; $X = 39.9641$ amu for ^{40}Ca prior to 1961.

2.102 (a) calcium-40, ^{40}Ca
    (b) Not enough information, several different isotopes can have 63 neutrons.
    (c) The neutral atom contains 26 electrons. The ion is iron-56, $^{56}Fe^{3+}$.
    (d) Se^{2-}

2.104 $^1H^{35}Cl$ has 18 protons, 18 neutrons, and 18 electrons.
    $^1H^{37}Cl$ has 18 protons, 20 neutrons, and 18 electrons.
    $^2H^{35}Cl$ has 18 protons, 19 neutrons, and 18 electrons.
    $^2H^{37}Cl$ has 18 protons, 21 neutrons, and 18 electrons.
    $^3H^{35}Cl$ has 18 protons, 20 neutrons, and 18 electrons.
    $^3H^{37}Cl$ has 18 protons, 22 neutrons, and 18 electrons.

2.106

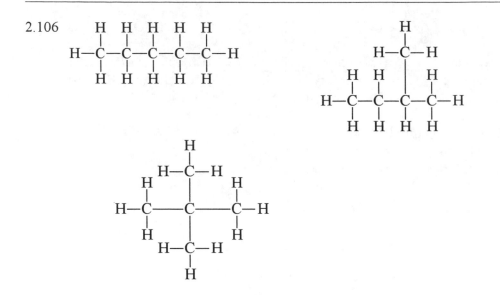

2.108 Molecular weight = (8 x 12.011 amu) + (9 x 1.0079 amu) + (1 x 14.0067 amu)
+ (2 x 15.9994 amu) = 151.165 amu

Formulas, Equations, and Moles

3.1 $2 \text{ KClO}_3 \rightarrow 2 \text{ KCl} + 3 \text{ O}_2$

3.2 (a) $C_6H_{12}O_6 \rightarrow 2 \text{ C}_2H_6O + 2 \text{ CO}_2$
(b) $4 \text{ Fe} + 3 \text{ O}_2 \rightarrow 2 \text{ Fe}_2O_3$
(c) $4 \text{ NH}_3 + \text{Cl}_2 \rightarrow \text{N}_2\text{H}_4 + 2 \text{ NH}_4\text{Cl}$

3.3 (a) Fe_2O_3: $2(55.85) + 3(16.00) = 159.7$ amu
(b) H_2SO_4: $2(1.01) + 1(32.07) + 4(16.00) = 98.1$ amu
(c) $C_6H_8O_7$: $6(12.01) + 8(1.01) + 7(16.00) = 192.1$ amu
(d) $C_{16}H_{18}N_2O_4S$: $16(12.01) + 18(1.01) + 2(14.01) + 4(16.00) + 1(32.07) = 334.4$ amu

3.4 $\text{Fe}_2\text{O}_3(s) + 3 \text{ CO}(g) \rightarrow 2 \text{ Fe}(s) + 3 \text{ CO}_2(g)$

$0.500 \text{ mol Fe}_2\text{O}_3 \times \dfrac{3 \text{ mol CO}}{1 \text{ mol Fe}_2\text{O}_3} = 1.50 \text{ mol CO}$

3.5 $C_9H_8O_4$, 180.2 amu; $500 \text{ mg} = 500 \times 10^{-3} \text{ g} = 0.500 \text{ g}$

$0.500 \text{ g} \times \dfrac{1 \text{ mol}}{180.2 \text{ g}} = 2.77 \times 10^{-3}$ mol aspirin

$2.77 \times 10^{-3} \text{ mol} \times \dfrac{6.02 \times 10^{23} \text{ molecules}}{1 \text{ mol}} = 1.67 \times 10^{21}$ aspirin molecules

3.6 salicylic acid, $C_7H_6O_3$, 138.1 amu;   acetic anhydride, $C_4H_6O_3$, 102.1 amu
aspirin, $C_9H_8O_4$, 180.2 amu;   acetic acid, $C_2H_4O_2$, 60.1 amu

$4.50 \text{ g } C_7H_6O_3 \times \dfrac{1 \text{ mol } C_7H_6O_3}{138.1 \text{ g } C_7H_6O_3} \times \dfrac{1 \text{ mol } C_4H_6O_3}{1 \text{ mol } C_7H_6O_3} \times \dfrac{102.1 \text{ g } C_4H_6O_3}{1 \text{ mol } C_4H_6O_3} = 3.33 \text{ g } C_4H_6O_3$

$4.50 \text{ g } C_7H_6O_3 \times \dfrac{1 \text{ mol } C_7H_6O_3}{138.1 \text{ g } C_7H_6O_3} \times \dfrac{1 \text{ mol } C_9H_8O_4}{1 \text{ mol } C_7H_6O_3} \times \dfrac{180.2 \text{ g } C_9H_8O_4}{1 \text{ mol } C_9H_8O_4} = 5.87 \text{ g } C_9H_8O_4$

$4.50 \text{ g } C_7H_6O_3 \times \dfrac{1 \text{ mol } C_7H_6O_3}{138.1 \text{ g } C_7H_6O_3} \times \dfrac{1 \text{ mol } C_2H_4O_2}{1 \text{ mol } C_7H_6O_3} \times \dfrac{60.1 \text{ g } C_2H_4O_2}{1 \text{ mol } C_2H_4O_2} = 1.96 \text{ g } C_2H_4O_2$

3.7 C_2H_4, 28.1 amu; C_2H_6O, 46.1 amu

$4.6 \text{ g } C_2H_4 \times \dfrac{1 \text{ mol } C_2H_4}{28.1 \text{ g } C_2H_4} \times \dfrac{1 \text{ mol } C_2H_6O}{1 \text{ mol } C_2H_4} \times \dfrac{46.1 \text{ g } C_2H_6O}{1 \text{ mol } C_2H_6O} = 7.5 \text{ g } C_2H_6O$ (theoretical yield)

$\text{Percent yield} = \dfrac{\text{Actual yield}}{\text{Theoretical yield}} \times 100\% = \dfrac{4.7 \text{ g}}{7.5 \text{ g}} \times 100\% = 63\%$

3.8 CH_4, 16.04 amu; CH_2Cl_2, 84.93 amu;   1.85 kg = 1850 g

$$1850g \; CH_4 \times \frac{1 \; mol \; CH_4}{16.04 \; g \; CH_4} \times \frac{1 \; mol \; CH_2Cl_2}{1 \; mol \; CH_4} \times \frac{84.93 \; g \; CH_2Cl_2}{1 \; mol \; CH_2Cl_2} = 9800g \; CH_2Cl_2 \; (\text{theoretical yield})$$

Actual yield = (9800 g)(0.431) = 4220 g CH_2Cl_2

3.9 $C_3H_8 + 5 \; O_2 \rightarrow 3 \; CO_2 + 4 \; H_2O$; C_3H_8, 44.1 amu; O_2, 32.0 amu

$$65 \; g \; O_2 \times \frac{1 \; mol \; O_2}{32.0 \; g \; O_2} \times \frac{1 \; mol \; C_3H_8}{5 \; mol \; O_2} \times \frac{44.1 \; g \; C_3H_8}{1 \; mol \; C_3H_8} = 18 \; g \; C_3H_8 \; \text{reacted}$$

mass C_3H_8 remaining = 700.0 g – 18 g = 682 g C_3H_8 remaining

3.10 Li_2O, 29.9 amu: 65 kg = 65,000 g; H_2O, 18.0 amu: 80.0 kg = 80,000 g

$$65,000 \; g \; Li_2O \times \frac{1 \; mol \; Li_2O}{29.9 \; g \; Li_2O} = 2.17 \times 10^3 \; mol \; Li_2O$$

$$80,000 \; g \; H_2O \times \frac{1 \; mol \; H_2O}{18.0 \; g \; H_2O} = 4.44 \times 10^3 \; mol \; H_2O$$

The reaction stoichiometry between Li_2O and H_2O is one to one. There are twice as many moles of H_2O as there are moles of Li_2O. Therefore, Li_2O is the limiting reactant.
$(4.44 \times 10^3 \; mol - 2.17 \times 10^3 \; mol) = 2.27 \times 10^3 \; mol \; H_2O$ remaining

$$2.27 \times 10^3 \; mol \; H_2O \times \frac{18.0 \; g \; H_2O}{1 \; mol \; H_2O} = 40,860 \; g \; H_2O = 40.9 \; kg = 41 \; kg \; H_2O$$

3.11   LiOH, 23.9 amu; CO_2, 44.0 amu

$$500.0 \; g \; LiOH \times \frac{1 \; mol \; LiOH}{23.9 \; g \; LiOH} \times \frac{1 \; mol \; CO_2}{1 \; mol \; LiOH} \times \frac{44.0 \; g \; CO_2}{1 \; mol \; CO_2} = 921 \; g \; CO_2$$

3.12   (a)   125 mL = 0.125 L;   (0.20 mol/L)(0.125 L) = 0.025 mol $NaHCO_3$
       (b)   650.0 mL = 0.6500 L;   (2.50 mol/L)(0.650 L) = 1.62 mol H_2SO_4

3.13   (a)   NaOH, 40.0 amu; 500.0 mL = 0.5000 L

$$1.25 \; \frac{mol \; NaOH}{L} \times 0.500 \; L \times \frac{40.0 \; g \; NaOH}{1 \; mol \; NaOH} = 25.0 \; g \; NaOH$$

       (b) $C_6H_{12}O_6$, 180.2 amu

$$0.250 \; \frac{mol \; C_6H_{12}O_6}{L} \times 1.50 \; L \times \frac{180.2 \; g \; C_6H_{12}O_6}{1 \; mol \; C_6H_{12}O_6} = 67.6 \; g \; C_6H_{12}O_6$$

3.14 $C_6H_{12}O_6$, 180.2 amu;

$$25.0 \; g \; C_6H_{12}O_6 \times \frac{1 \; mol \; C_6H_{12}O_6}{180.2 \; g \; C_6H_{12}O_6} = 0.1387 \; mol \; C_6H_{12}O_6$$

$$0.1387 \; mol \times \frac{1 \; L}{0.20 \; mol} = 0.69 \; L; \; 0.69 \; L = 690 \; mL$$

3.15 $C_{27}H_{46}O$, 386.7 amu; 750 mL = 0.750 L

$$0.005 \frac{\text{mol } C_{27}H_{46}O}{L} \times 0.750 \text{ L} \times \frac{386.7 \text{ g } C_{27}H_{46}O}{1 \text{ mol } C_{27}H_{46}O} = 1 \text{ g } C_{27}H_{46}O$$

3.16 $M_i \times V_i = M_f \times V_f$; $M_f = \dfrac{M_i \times V_i}{V_f} = \dfrac{3.50 \text{ M} \times 75.0 \text{ mL}}{400.0 \text{ mL}} = 0.656 \text{ M}$

3.17 $M_i \times V_i = M_f \times V_f$; $V_i = \dfrac{M_f \times V_f}{M_i} = \dfrac{0.500 \text{ M} \times 250.0 \text{ mL}}{18.0 \text{ M}} = 6.94 \text{ mL}$

Dilute 6.94 mL of 18.0 M H_2SO_4 with enough water to make 250.0 mL of solution. The resulting solution will be 0.500 M H_2SO_4.

3.18 50.0 mL = 0.0500 L; (0.100 mol/L)(0.0500 L) = 5.00×10^{-3} mol NaOH

$$5.00 \times 10^{-3} \text{ mol NaOH} \times \frac{1 \text{ mol } H_2SO_4}{2 \text{ mol NaOH}} = 2.50 \times 10^{-3} \text{ mol } H_2SO_4$$

volume = 2.50×10^{-3} mol $\times \dfrac{1 \text{ L}}{0.250 \text{ mol}} = 0.0100$ L; 0.0100 L = 10.0 mL H_2SO_4

3.19 $HNO_3(aq) + KOH(aq) \rightarrow KNO_3(aq) + H_2O(l)$
25.0 mL = 0.0250 L and 68.5 mL = 0.0685 L

$$0.150 \frac{\text{mol KOH}}{L} \times 0.0250 \text{ L} \times \frac{1 \text{ mol } HNO_3}{1 \text{ mol KOH}} = 3.75 \times 10^{-3} \text{ mol } HNO_3$$

HNO_3 molarity = $\dfrac{3.75 \times 10^{-3} \text{ mol}}{0.0685 \text{ L}} = 5.47 \times 10^{-2}$ M

3.20 From the reaction stoichiometry, moles NaOH = moles CH_3COOH
(0.200 mol/L)(0.0947 L) = 0.018 94 mol NaOH = 0.018 94 mol CH_3COOH

molarity = $\dfrac{0.018 \; 94 \text{ mol}}{0.0250 \text{ L}} = 0.758$ M

3.21 For dimethylhydrazine, $C_2H_8N_2$, divide each subscript by 2 to obtain the empirical formula. The empirical formula is CH_4N. $C_2H_8N_2$, 60.1 amu or 60.1 g/mol

% C = $\dfrac{2 \times 12.0 \text{ g}}{60.1 \text{ g}} \times 100\% = 39.9\%$

% H = $\dfrac{8 \times 1.01 \text{ g}}{60.1 \text{ g}} \times 100\% = 13.4\%$

% N = $\dfrac{2 \times 14.0 \text{ g}}{60.1 \text{ g}} \times 100\% = 46.6\%$

3.22 Assume a 100.0 g sample. From the percent composition data, a 100.0 g sample contains 14.25 g C, 56.93 g O, and 28.83 g Mg.

$$14.25 \text{ g C} \times \frac{1 \text{ mol C}}{12.0 \text{ g C}} = 1.19 \text{ mol C}$$

$$56.93 \text{ g O} \times \frac{1 \text{ mol O}}{16.0 \text{ g O}} = 3.56 \text{ mol O}$$

$$28.83 \text{ g Mg} \times \frac{1 \text{ mol Mg}}{24.3 \text{ g Mg}} = 1.19 \text{ mol Mg}$$

$Mg_{1.19}C_{1.19}O_{3.56}$; divide each subscript by the smallest, 1.19.
$Mg_{1.19/1.19}C_{1.19/1.19}O_{3.56/1.19}$
The empirical formula is $MgCO_3$.

3.23 $\quad 1.161 \text{ g H}_2\text{O} \times \dfrac{1 \text{ mol H}_2\text{O}}{18.0 \text{ g H}_2\text{O}} \times \dfrac{2 \text{ mol H}}{1 \text{ mol H}_2\text{O}} = 0.129 \text{ mol H}$

$$2.818 \text{ g CO}_2 \times \frac{1 \text{ mol CO}_2}{44.0 \text{ g CO}_2} \times \frac{1 \text{ mol C}}{1 \text{ mol CO}_2} = 0.0640 \text{ mol C}$$

$$0.129 \text{ mol H} \times \frac{1.01 \text{ g H}}{1 \text{ mol H}} = 0.130 \text{ g H}$$

$$0.0640 \text{ mol C} \times \frac{12.0 \text{ g C}}{1 \text{ mol C}} = 0.768 \text{ g C}$$

$1.00 \text{ g total} - (0.130 \text{ g H} + 0.768 \text{ g C}) = 0.102 \text{ g O}$

$$0.102 \text{ g O} \times \frac{1 \text{ mol O}}{16.0 \text{ g O}} = 0.006 \, 38 \text{ mol O}$$

$C_{0.0640}H_{0.129}O_{0.006\,38}$; divide each subscript by the smallest, 0.006 38.
$C_{0.0640/0.006\,38}H_{0.129/0.006\,38}O_{0.006\,38/0.006\,38}$
$C_{10.03}H_{20.22}O_1 \qquad$ The empirical formula is $C_{10}H_{20}O$.

3.24 $\quad$ The empirical formula is CH_2O, 30 amu: molecular weight = 150 amu.
$$\frac{\text{molecular weight}}{\text{empirical formula weight}} = \frac{150 \text{ amu}}{30 \text{ amu}} = 5; \text{ therefore}$$
molecular formula = 5 x empirical formula = $C_{(5 \times 1)}H_{(5 \times 2)}O_{(5 \times 1)} = C_5H_{10}O_5$

3.25 $\quad$ (a) Assume a 100.0 g sample. From the percent composition data, a 100.0 g sample contains 21.86 g H and 78.14 g B.

$$21.86 \text{ g H} \times \frac{1 \text{ mol H}}{1.01 \text{ g H}} = 21.6 \text{ mol H}$$

$$78.14 \text{ g B} \times \frac{1 \text{ mol B}}{10.8 \text{ g B}} = 7.24 \text{ mol B}$$

$B_{7.24}H_{21.6}$; divide each subscript by the smaller, 7.24.
$B_{7.24/7.24}H_{21.6/7.24} \qquad$ The empirical formula is BH_3, 13.8 amu.
27.7 amu / 13.8 amu = 2; molecular formula = $B_{(2 \times 1)}H_{(2 \times 3)} = B_2H_6$.
(b) Assume a 100.0 g sample. From the percent composition data, a 100.0 g sample contains 6.71 g H, 40.00 g C, and 53.28 g O.

$$6.71 \text{ g H} \times \frac{1 \text{ mol H}}{1.01 \text{ g H}} = 6.64 \text{ mol H}$$

$$40.00 \text{ g C} \times \frac{1 \text{ mol C}}{12.0 \text{ g C}} = 3.33 \text{ mol C}$$

$$53.28 \text{ g O} \times \frac{1 \text{ mol O}}{16.0 \text{ g O}} = 3.33 \text{ mol O}$$

$C_{3.33} H_{6.64} O_{3.33}$; divide each subscript by the smallest, 3.33.

$C_{3.33/3.33} H_{6.64/3.33} O_{3.33/3.33}$ The empirical formula is CH_2O, 30.0 amu.

90.08 amu / 30.0 amu = 3; molecular formula = $C_{(3 \times 1)}H_{(3 \times 2)}O_{(3 \times 1)} = C_3H_6O_3$

Understanding Key Concepts

3.26   The concentration of a solution is cut in half when the volume is doubled. This is best represented by box (b).

3.27   (c) $2 A + B_2 \rightarrow A_2B_2$

3.28 $C_2H_4 + 3 O_2 \rightarrow 2 CO_2 + 2 H_2O$

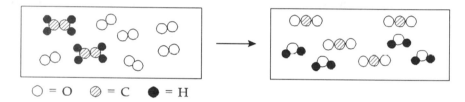

$\bigcirc = O$ $\oslash = C$ $\bullet = H$

3.29   reactants, box (d), and products, box (c)

3.30 $C_5H_{11}NO_2S$ 5(12.01) + 11(1.01) + 1(14.01) + 2(16.00) + 1(32.07) = 149.2 amu

3.31   Because the two volumes are equal (let the volume = y L), the concentrations are proportional to the number of solute ions.

$$\text{OH}^- \text{ concentration} = 1.00 \text{ M} \times \frac{y \text{ L}}{12 \text{ H}^+} \times \frac{8 \text{ OH}^-}{y \text{ L}} = 0.67 \text{ M}$$

3.32   (a) B_2 is the limiting reactant because it is completely consumed.
(b) $A_2 + 3 B_2 \rightarrow 2 AB_3$
(c) For 1.0 mol of A_2, 3.0 mol of B_2 are required. Because only 1.0 mol of B_2 is available, B_2 is the limiting reactant.

$$1 \text{ mol } B_2 \times \frac{2 \text{ mol } AB_3}{3 \text{ mol } B_2} = 2/3 \text{ mol } AB_3$$

3.33   Adding a subscript of "2" to the oxygen in N_2O would change nitrous oxide (N_2O) into a different compound. This is not allowed to balance an equation.

Chapter 3 – Formulas, Equations, and Moles

Additional Problems
Balancing Equations

3.34   Equation (b) is balanced, (a) is not balanced .

3.36   (a) $Mg + 2 HNO_3 \rightarrow H_2 + Mg(NO_3)_2$
      (b) $CaC_2 + 2 H_2O \rightarrow Ca(OH)_2 + C_2H_2$
      (c) $3 O_2 + 2 S \rightarrow 2 SO_3$
      (d) $UO_2 + 4 HF \rightarrow UF_4 + 2 H_2O$

Molecular Weights and Moles

3.38 Hg_2Cl_2: $2(200.59) + 2(35.45) = 472.1$ amu
      $C_4H_8O_2$: $4(12.01) + 8(1.01) + 2(16.00) = 88.1$ amu
      CF_2Cl_2: $1(12.01) + 2(19.00) + 2(35.45) = 120.9$ amu

3.40   One mole equals the atomic weight or molecular weight in grams.
      (a) Ti, 47.88 g   (b) Br_2, 159.81 g   (c) Hg, 200.59 g   (d) H_2O, 18.02 g

3.42   There are 2 ions per each formula unit of NaCl. (2.5 mol)(2 mol ions/mol) = 5.0 mol ions

3.44   There are 3 ions (one Mg^{2+} and 2 Cl^-) per each formula unit of $MgCl_2$.
      $MgCl_2$, 95.2 amu

$$27.5 \text{ g } MgCl_2 \times \frac{1 \text{ mol } MgCl_2}{95.2 \text{ g } MgCl_2} \times \frac{3 \text{ mol ions}}{1 \text{ mol } MgCl_2} = 0.867 \text{ mol ions}$$

3.46   Molar mass $= \dfrac{3.28 \text{ g}}{0.0275 \text{ mol}} = 119 \text{ g/mol}$;   molecular weight = 119 amu.

3.48 $FeSO_4$, 151.9 amu; 300 mg = 0.300 g

$$0.300 \text{ g } FeSO_4 \times \frac{1 \text{ mol } FeSO_4}{151.9 \text{ g } FeSO_4} = 1.97 \times 10^{-3} \text{ mol } FeSO_4$$

$$1.97 \times 10^{-3} \text{ mol } FeSO_4 \times \frac{6.02 \times 10^{23} \text{ Fe(II) atoms}}{1 \text{ mol } FeSO_4} = 1.19 \times 10^{21} \text{ Fe(II) atoms}$$

3.50 $C_8H_{10}N_4O_2$, 194.2 amu; 125 mg = 0.125 g

$$0.125 \text{ g caffeine} \times \frac{1 \text{ mol caffeine}}{194.2 \text{ g caffeine}} = 6.44 \times 10^{-4} \text{ mol caffeine}$$

$$0.125 \text{ g caffeine} \times \frac{1 \text{ mol caffeine}}{194.2 \text{ g caffeine}} \times \frac{6.022 \times 10^{23} \text{ molecules}}{1 \text{ mol}} = 3.88 \times 10^{20} \text{ caffeine molecules}$$

3.52 (a) $1.0 \text{ g Li } \times \dfrac{1 \text{ mol Li}}{6.94 \text{ g Li}} = 0.14 \text{ mol Li}$

(b) $1.0 \text{ g Au } \times \dfrac{1 \text{ mol Au}}{197.0 \text{ g Au}} = 0.0051 \text{ mol Au}$

(c) penicillin G: $C_{16}H_{17}N_2O_4SK$, 372.5 amu

$1.0 \text{ g } \times \dfrac{1 \text{ mol penicillin G}}{372.5 \text{ g penicillin G}} = 2.7 \times 10^{-3} \text{ mol penicillin G}$

Stoichiometry Calculations

3.54 TiO_2, 79.88 amu; $100 \text{ kg Ti } \times \dfrac{79.88 \text{ kg TiO}_2}{47.88 \text{ kg Ti}} = 167 \text{ kg TiO}_2$

3.56 (a) $2 \text{ Fe}_2O_3 + 3 \text{ C} \rightarrow 4 \text{ Fe} + 3 \text{ CO}_2$

(b) Fe_2O_3, 159.7 amu; $525 \text{ g Fe}_2O_3 \times \dfrac{1 \text{ mol Fe}_2O_3}{159.7 \text{ g Fe}_2O_3} \times \dfrac{3 \text{ mol C}}{2 \text{ mol Fe}_2O_3} = 4.93 \text{ mol C}$

(c) $4.93 \text{ mol C} \times \dfrac{12.01 \text{ g C}}{1 \text{ mol C}} = 59.2 \text{ g C}$

3.58 (a) $2 \text{ Mg} + \text{O}_2 \rightarrow 2 \text{ MgO}$

(b) Mg, 24.30 amu; O_2, 32.00 amu; MgO, 40.30 amu

$25.0 \text{ g Mg } \times \dfrac{1 \text{ mol Mg}}{24.30 \text{ g Mg}} \times \dfrac{1 \text{ mol O}_2}{2 \text{ mol Mg}} \times \dfrac{32.00 \text{ g O}_2}{1 \text{ mol O}_2} = 16.5 \text{ g O}_2$

$25.0 \text{ g Mg } \times \dfrac{1 \text{ mol Mg}}{24.30 \text{ g Mg}} \times \dfrac{2 \text{ mol MgO}}{2 \text{ mol Mg}} \times \dfrac{40.30 \text{ g MgO}}{1 \text{ mol MgO}} = 41.5 \text{ g MgO}$

(c) $25.0 \text{ g O}_2 \times \dfrac{1 \text{ mol O}_2}{32.00 \text{ g O}_2} \times \dfrac{2 \text{ mol Mg}}{1 \text{ mol O}_2} \times \dfrac{24.30 \text{ g Mg}}{1 \text{ mol Mg}} = 38.0 \text{ g Mg}$

$25.0 \text{ g O}_2 \times \dfrac{1 \text{ mol O}_2}{32.00 \text{ g O}_2} \times \dfrac{2 \text{ mol MgO}}{1 \text{ mol O}_2} \times \dfrac{40.30 \text{ g MgO}}{1 \text{ mol MgO}} = 63.0 \text{ g MgO}$

3.60 (a) $2 \text{ HgO} \rightarrow 2 \text{ Hg} + \text{O}_2$

(b) HgO, 216.6 amu; Hg, 200.6 amu; O_2, 32.0 amu

$45.5 \text{ g HgO } \times \dfrac{1 \text{ mol HgO}}{216.6 \text{ g HgO}} \times \dfrac{2 \text{ mol Hg}}{2 \text{ mol HgO}} \times \dfrac{200.6 \text{ g Hg}}{1 \text{ mol Hg}} = 42.1 \text{ g Hg}$

$45.5 \text{ g HgO } \times \dfrac{1 \text{ mol HgO}}{216.6 \text{ g HgO}} \times \dfrac{1 \text{ mol O}_2}{2 \text{ mol HgO}} \times \dfrac{32.00 \text{ g O}_2}{1 \text{ mol O}_2} = 3.36 \text{ g O}_2$

(c) $33.3 \text{ g O}_2 \times \dfrac{1 \text{ mol O}_2}{32.00 \text{ g O}_2} \times \dfrac{2 \text{ mol HgO}}{1 \text{ mol O}_2} \times \dfrac{216.6 \text{ g HgO}}{1 \text{ mol HgO}} = 451 \text{ g HgO}$

3.62 $2.00 \text{ g Ag} \times \dfrac{1 \text{ mol Ag}}{107.9 \text{ g Ag}} = 0.0185 \text{ mol Ag}$

$0.657 \text{ g Cl} \times \dfrac{1 \text{ mol Cl}}{35.45 \text{ g Cl}} = 0.0185 \text{ mol Cl}$

$Ag_{0.0185}Cl_{0.0185}$ Both subscripts are the same. The empirical formula is AgCl.

Limiting Reactants and Reaction Yield

3.64 $3.44 \text{ mol N}_2 \times \dfrac{3 \text{ mol H}_2}{1 \text{ mol N}_2} = 10.3 \text{ mol H}_2 \text{ required.}$

Since there is only 1.39 mol H_2, H_2 is the limiting reactant.

$1.39 \text{ mol H}_2 \times \dfrac{2 \text{ mol NH}_3}{3 \text{ mol H}_2} \times \dfrac{17.03 \text{ g NH}_3}{1 \text{ mol NH}_3} = 15.8 \text{ g NH}_3$

$1.39 \text{ mol H}_2 \times \dfrac{1 \text{ mol N}_2}{3 \text{ mol H}_2} \times \dfrac{28.01 \text{ g N}_2}{1 \text{ mol N}_2} = 13.0 \text{ g N}_2 \text{ reacted}$

$3.44 \text{ mol N}_2 \times \dfrac{28.01 \text{ g N}_2}{1 \text{ mol N}_2} = 96.3 \text{ g N}_2 \text{ initially}$

$(96.3 \text{ g} - 13.0 \text{ g}) = 83.3 \text{ g N}_2 \text{ left over}$

3.66 C_2H_4, 28.05 amu; Cl_2, 70.91 amu; $C_2H_4Cl_2$, 98.96 amu

$15.4 \text{ g C}_2\text{H}_4 \times \dfrac{1 \text{ mol C}_2\text{H}_4}{28.05 \text{ g C}_2\text{H}_4} = 0.549 \text{ mol C}_2\text{H}_4$

$3.74 \text{ g Cl}_2 \times \dfrac{1 \text{ mol Cl}_2}{70.91 \text{ g Cl}_2} = 0.0527 \text{ mol Cl}_2$

Since the reaction stoichiometry between C_2H_4 and Cl_2 is one to one, Cl_2 is the limiting reactant.

$0.0527 \text{ mol Cl}_2 \times \dfrac{1 \text{ mol C}_2\text{H}_4\text{Cl}_2}{1 \text{ mol Cl}_2} \times \dfrac{98.96 \text{ g C}_2\text{H}_4\text{Cl}_2}{1 \text{ mol C}_2\text{H}_4\text{Cl}_2} = 5.22 \text{ g C}_2\text{H}_4\text{Cl}_2$

3.68 $CaCO_3$, 100.1 amu; HCl, 36.46 amu

$CaCO_3 + 2 HCl \rightarrow CaCl_2 + H_2O + CO_2$

$2.35 \text{ g CaCO}_3 \times \dfrac{1 \text{ mol CaCO}_3}{100.1 \text{ g CaCO}_3} = 0.0235 \text{ mol CaCO}_3$

$2.35 \text{ g HCl} \times \dfrac{1 \text{ mol HCl}}{36.46 \text{ g HCl}} = 0.0645 \text{ mol HCl}$

The reaction stoichiometry is 1 mole of $CaCO_3$ for every 2 moles of HCl. For 0.0235 mol $CaCO_3$, we only need 2(0.0235 mol) = 0.0470 mol HCl. We have 0.0645 mol HCl, therefore $CaCO_3$ is the limiting reactant.

$$0.0235 \text{ mol } CaCO_3 \ \times \ \frac{1 \text{ mol } CO_2}{1 \text{ mol } CaCO_3} \ \times \ \frac{22.4 \text{ L}}{1 \text{ mol } CO_2} \ = \ 0.526 \text{ L } CO_2$$

3.70 $CH_3COOH \ + \ C_5H_{12}O \ \rightarrow \ C_7H_{14}O_2 \ + \ H_2O$
   CH_3COOH, 60.05 amu; $C_5H_{12}O$, 88.15 amu; $C_7H_{14}O_2$, 130.19 amu

$$3.58 \text{ g } CH_3COOH \ \times \ \frac{1 \text{ mol } CH_3COOH}{60.05 \text{ g } CH_3COOH} \ = \ 0.0596 \text{ mol } CH_3COOH$$

$$4.75 \text{ g } C_5H_{12}O \ \times \ \frac{1 \text{ mol } C_5H_{12}O}{88.15 \text{ g } C_5H_{12}O} \ = \ 0.0539 \text{ mol } C_5H_{12}O$$

Since the reaction stoichiometry between CH_3COOH and $C_5H_{12}O$ is one to one, isopentyl alcohol ($C_5H_{12}O$) is the limiting reactant.

$$0.0539 \text{ mol } C_5H_{12}O \ \times \ \frac{1 \text{ mol } C_7H_{14}O_2}{1 \text{ mol } C_5H_{12}O} \ \times \ \frac{130.19 \text{ g } C_7H_{14}O_2}{1 \text{ mol } C_7H_{14}O_2} \ = \ 7.02 \text{ g } C_7H_{14}O_2$$

7.02 g $C_7H_{14}O_2$ is the theoretical yield. Actual yield = (7.02 g)(0.45) = 3.2 g.

3.72 $CH_3COOH \ + \ C_5H_{12}O \ \rightarrow \ C_7H_{14}O_2 \ + \ H_2O$
   CH_3COOH, 60.05 amu; $C_5H_{12}O$, 88.15 amu; $C_7H_{14}O_2$, 130.19 amu

$$1.87 \text{ g } CH_3COOH \ \times \ \frac{1 \text{ mol } CH_3COOH}{60.05 \text{ g } CH_3COOH} \ = \ 0.0311 \text{ mol } CH_3COOH$$

$$2.31 \text{ g } C_5H_{12}O \ \times \ \frac{1 \text{ mol } C_5H_{12}O}{88.15 \text{ g } C_5H_{12}O} \ = \ 0.0262 \text{ mol } C_5H_{12}O$$

Since the reaction stoichiometry between CH_3COOH and $C_5H_{12}O$ is one to one, isopentyl alcohol ($C_5H_{12}O$) is the limiting reactant.

$$0.0262 \text{ mol } C_5H_{12}O \ \times \ \frac{1 \text{ mol } C_7H_{14}O_2}{1 \text{ mol } C_5H_{12}O} \ \times \ \frac{130.19 \text{ g } C_7H_{14}O_2}{1 \text{ mol } C_7H_{14}O_2} \ = \ 3.41 \text{ g } C_7H_{14}O_2$$

3.41 g $C_7H_{14}O_2$ is the theoretical yield.

$$\% \text{ Yield} = \frac{\text{Actual yield}}{\text{Theoretical yield}} \times 100\% = \frac{2.96 \text{ g}}{3.41 \text{ g}} \times 100\% = 86.8\%$$

Molarity, Solution Stoichiometry, Dilution, and Titration

3.74   (a) 35.0 mL = 0.0350 L; $1.200 \ \dfrac{\text{mol } HNO_3}{\text{L}} \ \times \ 0.0350 \text{ L} \ = \ 0.0420 \text{ mol } HNO_3$

   (b) 175 mL = 0.175 L; $0.67 \ \dfrac{\text{mol } C_6H_{12}O_6}{\text{L}} \ \times \ 0.175 \text{ L} \ = \ 0.12 \text{ mol } C_6H_{12}O_6$

3.76 $BaCl_2$, 208.2 amu

$$15.0 \text{ g } BaCl_2 \ \times \ \frac{1 \text{ mol } BaCl_2}{208.2 \text{ g } BaCl_2} \ = \ 0.0720 \text{ mol } BaCl_2$$

$$0.0720 \text{ mol} \times \frac{1.0 \text{ L}}{0.45 \text{ mol}} = 0.16 \text{ L}; \quad 0.16 \text{ L} = 160 \text{ mL}$$

3.78   NaCl, 58.4 amu;   400 mg = 0.400 g;   100 mL = 0.100 L

$$0.400 \text{ g NaCl} \times \frac{1 \text{ mol NaCl}}{58.4 \text{ g NaCl}} = 0.006\ 85 \text{ mol NaCl}$$

$$\text{molarity} = \frac{0.006\ 85 \text{ mol}}{0.100 \text{ L}} = 0.0685 \text{ M}$$

3.80   NaCl, 58.4 amu;   KCl, 74.6 amu; $CaCl_2$, 111.0 amu;   500 mL = 0.500 L

$$4.30 \text{ g NaCl} \times \frac{1 \text{ mol NaCl}}{58.4 \text{ g NaCl}} = 0.0736 \text{ mol NaCl}$$

$$0.150 \text{ g KCl} \times \frac{1 \text{ mol KCl}}{74.6 \text{ g KCl}} = 0.002\ 01 \text{ mol KCl}$$

$$0.165 \text{ g CaCl}_2 \times \frac{1 \text{ mol CaCl}_2}{111.0 \text{ g CaCl}_2} = 0.001\ 49 \text{ mol CaCl}_2$$

$$0.0736 \text{ mol} + 0.002\ 01 \text{ mol} + 2(0.001\ 49 \text{ mol}) = 0.0786 \text{ mol Cl}^-$$

$$Na^+ \text{ molarity} = \frac{0.0736 \text{ mol}}{0.500 \text{ L}} = 0.147 \text{ M}$$

$$Ca^{2+} \text{ molarity} = \frac{0.001\ 49 \text{ mol}}{0.500 \text{ L}} = 0.002\ 98 \text{ M}$$

$$K^+ \text{ molarity} = \frac{0.002\ 01 \text{ mol}}{0.500 \text{ L}} = 0.004\ 02 \text{ M}$$

$$Cl^- \text{ molarity} = \frac{0.0786 \text{ mol}}{0.500 \text{ L}} = 0.157 \text{ M}$$

3.82 $M_f \times V_f = M_i \times V_i$; $M_f = \dfrac{M_i \times V_i}{V_f} = \dfrac{12.0 \text{ M} \times 35.7 \text{ mL}}{250.0 \text{ mL}} = 1.71 \text{ M HCl}$

3.84 $2 \text{ HBr(aq)} + K_2CO_3\text{(aq)} \rightarrow 2 \text{ KBr(aq)} + CO_2\text{(g)} + H_2O\text{(l)}$
     K_2CO_3, 138.2 amu;   450 mL = 0.450 L

$$0.500 \text{ } \frac{\text{mol HBr}}{\text{L}} \times 0.450 \text{ L} = 0.225 \text{ mol HBr}$$

$$0.225 \text{ mol HBr} \times \frac{1 \text{ mol K}_2CO_3}{2 \text{ mol HBr}} \times \frac{138.2 \text{ g K}_2CO_3}{1 \text{ mol K}_2CO_3} = 15.5 \text{ g K}_2CO_3$$

3.86 $H_2C_2O_4$, 90.04 amu

$$3.225 \text{ g H}_2C_2O_4 \times \frac{1 \text{ mol H}_2C_2O_4}{90.04 \text{ g H}_2C_2O_4} \times \frac{2 \text{ mol KMnO}_4}{5 \text{ mol H}_2C_2O_4} = 0.0143 \text{ mol KMnO}_4$$

$$0.0143 \text{ mol} \times \frac{1 \text{ L}}{0.250 \text{ mol}} = 0.0572 \text{ L} = 57.2 \text{ mL}$$

Formulas and Elemental Analysis

3.88 CH_4N_2O, 60.1 amu

$$\% \text{ C} = \frac{12.0 \text{ g C}}{60.1 \text{ g}} \times 100\% = 20.0\%$$

$$\% \text{ H} = \frac{4 \times 1.01 \text{ g H}}{60.1 \text{ g}} \times 100\% = 6.72\%$$

$$\% \text{ N} = \frac{2 \times 14.0 \text{ g N}}{60.1 \text{ g}} \times 100\% = 46.6\%$$

$$\% \text{ O} = \frac{16.0 \text{ g O}}{60.1 \text{ g}} \times 100\% = 26.6\%$$

3.90   Assume a 100.0 g sample. From the percent composition data, a 100.0 g sample contains 24.25 g F and 75.75 g Sn.

$$24.25 \text{ g F} \times \frac{1 \text{ mol F}}{19.00 \text{ g F}} = 1.276 \text{ mol F}$$

$$75.75 \text{ g Sn} \times \frac{1 \text{ mol Sn}}{118.7 \text{ g Sn}} = 0.6382 \text{ mol Sn}$$

$Sn_{0.6382}F_{1.276}$; divide each subscript by the smaller, 0.6382.
$Sn_{0.6382 / 0.6382}F_{1.276 / 0.6382}$ The empirical formula is SnF_2.

3.92   Toluene contains only C and H.
152.5 mg = 0.1525 g and 35.67 mg = 0.035 67 g

$$0.1525 \text{ g CO}_2 \times \frac{1 \text{ mol CO}_2}{44.01 \text{ g CO}_2} \times \frac{1 \text{ mol C}}{1 \text{ mol CO}_2} = 0.003 \ 465 \text{ mol C}$$

$$0.035 \ 67 \text{ g H}_2\text{O} \times \frac{1 \text{ mol H}_2\text{O}}{18.02 \text{ g H}_2\text{O}} \times \frac{2 \text{ mol H}}{1 \text{ mol H}_2\text{O}} = 0.003 \ 959 \text{ mol H}$$

$C_{0.003 \ 465}H_{0.003 \ 959}$; divide each subscript by the smaller, 0.003 465.
$C_{0.003 \ 465 / 0.003 \ 465}H_{0.003 \ 959 / 0.003 \ 465}$
$CH_{1.14}$; multiply each subscript by 7 to obtain integers.
The empirical formula is C_7H_8.

3.94   Let X equal the molecular weight of cytochrome c.

$$0.0043 = \frac{55.847 \text{ amu}}{X}; \qquad X = \frac{55.847 \text{ amu}}{0.0043} = 13{,}000 \text{ amu}$$

3.96   Let X equal the molecular weight of disilane.

$$0.9028 = \frac{2 \times 28.09 \text{ amu}}{X}; \qquad X = \frac{2 \times 28.09 \text{ amu}}{0.9028} = 62.23 \text{ amu}$$

62.23 amu – 2(Si atomic wt) = 62.23 amu – 2(28.09 amu) = 6.05 amu

6.05 amu is the total mass of H atoms.

6.05 amu x $\dfrac{1\text{ H atom}}{1.01\text{ amu}}$ = 6 H atoms;　　　Disilane is Si_2H_6.

General Problems

3.98　(a) $C_6H_{12}O_6$, 180.2 amu

$\%\,C = \dfrac{6 \times 12.01\text{ g C}}{180.2\text{ g}} \times 100\% = 39.99\%$

$\%\,H = \dfrac{12 \times 1.008\text{ g H}}{180.2\text{ g}} \times 100\% = 6.71\%$

$\%\,O = \dfrac{6 \times 16.00\text{ g O}}{180.2\text{ g}} \times 100\% = 53.27\%$

(b) H_2SO_4, 98.08 amu

$\%\,H = \dfrac{2 \times 1.008\text{ g H}}{98.08\text{ g}} \times 100\% = 2.06\%$

$\%\,S = \dfrac{32.07\text{ g S}}{98.08\text{ g}} \times 100\% = 32.70\%$

$\%\,O = \dfrac{4 \times 16.00\text{ g O}}{98.08\text{ g}} \times 100\% = 65.25\%$

(c) $KMnO_4$, 158.0 amu

$\%\,K = \dfrac{39.10\text{ g K}}{158.0\text{ g}} \times 100\% = 24.75\%$

$\%\,Mn = \dfrac{54.94\text{ g Mn}}{158.0\text{ g}} \times 100\% = 34.77\%$

$\%\,O = \dfrac{4 \times 16.00\text{ g O}}{158.0\text{ g}} \times 100\% = 40.51\%$

(d) $C_7H_5NO_3S$, 183.2 amu

$\%\,C = \dfrac{7 \times 12.01\text{ g C}}{183.2\text{ g}} \times 100\% = 45.89\%$

$\%\,H = \dfrac{5 \times 1.008\text{ g H}}{183.2\text{ g}} \times 100\% = 2.75\%$

$\%\,N = \dfrac{14.01\text{ g N}}{183.2\text{ g}} \times 100\% = 7.65\%$

$\%\,O = \dfrac{3 \times 16.00\text{ g O}}{183.2\text{ g}} \times 100\% = 26.20\%$

$\%\,S = \dfrac{32.07\text{ g S}}{183.2\text{ g}} \times 100\% = 17.51\%$

3.100 (a) $SiCl_4 + 2 H_2O \rightarrow SiO_2 + 4 HCl$
(b) $P_4O_{10} + 6 H_2O \rightarrow 4 H_3PO_4$
(c) $CaCN_2 + 3 H_2O \rightarrow CaCO_3 + 2 NH_3$
(d) $3 NO_2 + H_2O \rightarrow 2 HNO_3 + NO$

3.102 NaH, 24.00 amu; B_2H_6, 27.67 amu; $NaBH_4$, 37.83 amu
$2 NaH + B_2H_6 \rightarrow 2 NaBH_4$

$$8.55 \text{ g NaH} \times \frac{1 \text{ mol NaH}}{24.00 \text{ g NaH}} = 0.356 \text{ mol NaH}$$

$$6.75 \text{ g } B_2H_6 \times \frac{1 \text{ mol } B_2H_6}{27.67 \text{ g } B_2H_6} = 0.244 \text{ mol } B_2H_6$$

For 0.244 mol B_2H_6, 2 x (0.244) = 0.488 mol NaH are needed. Because only 0.356 mol of NaH is available, NaH is the limiting reactant.

$$0.356 \text{ mol NaH} \times \frac{2 \text{ mol } NaBH_4}{2 \text{ mol NaH}} \times \frac{37.83 \text{ g } NaBH_4}{1 \text{ mol } NaBH_4} = 13.5 \text{ g } NaBH_4 \text{ produced}$$

$$0.356 \text{ mol NaH} \times \frac{1 \text{ mol } B_2H_6}{2 \text{ mol NaH}} \times \frac{27.67 \text{ g } B_2H_6}{1 \text{ mol } B_2H_6} = 4.93 \text{ g } B_2H_6 \text{ reacted}$$

B_2H_6 left over = 6.75 g – 4.93 g = 1.82 g B_2H_6

3.104 Assume a 100.0 g sample of ferrocene. From the percent composition data, a 100.0 g sample contains 5.42 g H, 64.56 g C, and 30.02 g Fe.

$$5.42 \text{ g H} \times \frac{1 \text{ mol H}}{1.01 \text{ g H}} = 5.37 \text{ mol H}$$

$$64.56 \text{ g C} \times \frac{1 \text{ mol C}}{12.01 \text{ g C}} = 5.38 \text{ mol C}$$

$$30.02 \text{ g Fe} \times \frac{1 \text{ mol Fe}}{55.85 \text{ g Fe}} = 0.538 \text{ mol Fe}$$

$C_{5.38}H_{5.37}Fe_{0.538}$; divide each subscript by the smallest, 0.538.
$C_{5.38/0.538}H_{5.37/0.538}Fe_{0.538/0.538}$ The empirical formula is $C_{10}H_{10}Fe$.

3.106 Na_2SO_4, 142.04 amu; Na_3PO_4, 163.94 amu; Li_2SO_4, 109.95 amu; 100.00 mL = 0.10000 L

$$0.550 \text{ g } Na_2SO_4 \times \frac{1 \text{ mol } Na_2SO_4}{142.04 \text{ g } Na_2SO_4} = 0.003\ 872 \text{ mol } Na_2SO_4$$

$$1.188 \text{ g } Na_3PO_4 \times \frac{1 \text{ mol } Na_3PO_4}{163.94 \text{ g } Na_3PO_4} = 0.007\ 247 \text{ mol } Na_3PO_4$$

$$0.223 \text{ g } Li_2SO_4 \times \frac{1 \text{ mol } Li_2SO_4}{109.95 \text{ g } Li_2SO_4} = 0.002\ 028 \text{ mol } Li_2SO_4$$

$$Na^+ \text{ molarity} = \frac{(2 \times 0.003\ 872 \text{ mol}) + (3 \times 0.007\ 247 \text{ mol})}{0.100\ 00 \text{ L}} = 0.295 \text{ M}$$

$$\text{Li}^+ \text{ molarity} = \frac{2 \times 0.002\ 028 \text{ mol}}{0.100\ 00 \text{ L}} = 0.0406 \text{ M}$$

$$\text{SO}_4^{2-} \text{ molarity} = \frac{(1 \times 0.003\ 872 \text{ mol}) + (1 \times 0.002\ 028 \text{ mol})}{0.100\ 00 \text{ L}} = 0.0590 \text{ M}$$

$$\text{PO}_4^{3-} \text{ molarity} = \frac{1 \times 0.007\ 247 \text{ mol}}{0.100\ 00 \text{ L}} = 0.0725 \text{ M}$$

3.108 High resolution mass spectrometry is capable of measuring the mass of molecules with a particular isotopic composition.

3.110 $C_6H_{12}O_6 + 6\ O_2 \rightarrow 6\ CO_2 + 6\ H_2O$; $C_6H_{12}O_6$, 180.16 amu; CO_2, 44.01 amu

$$66.3 \text{ g } C_6H_{12}O_6 \times \frac{1 \text{ mol } C_6H_{12}O_6}{180.16 \text{ g } C_6H_{12}O_6} \times \frac{6 \text{ mol } CO_2}{1 \text{ mol } C_6H_{12}O_6} \times \frac{44.01 \text{ g } CO_2}{1 \text{ mol } CO_2} = 97.2 \text{ g } CO_2$$

$$66.3 \text{ g } C_6H_{12}O_6 \times \frac{1 \text{ mol } C_6H_{12}O_6}{180.16 \text{ g } C_6H_{12}O_6} \times \frac{6 \text{ mol } CO_2}{1 \text{ mol } C_6H_{12}O_6} \times \frac{25.4 \text{ L } CO_2}{1 \text{ mol } CO_2} = 56.1 \text{ L } CO_2$$

3.112 Mass of Cu = 2.196 g; mass of S = 2.748 g – 2.196 g = 0.552 g S

(a) $\%\text{Cu} = \dfrac{2.196 \text{ g}}{2.748 \text{ g}} \times 100\% = 79.9\%$

$\%\text{S} = \dfrac{0.552 \text{ g}}{2.748 \text{ g}} \times 100\% = 20.1\%$

(b) $2.196 \text{ g Cu} \times \dfrac{1 \text{ mol Cu}}{63.55 \text{ g Cu}} = 0.0346 \text{ mol Cu}$

$0.552 \text{ g S} \times \dfrac{1 \text{ mol S}}{32.07 \text{ g S}} = 0.0172 \text{ mol S}$

$Cu_{0.0346}S_{0.0172}$; divide each subscript by the smaller, 0.0172.
$Cu_{0.0346\ /\ 0.0172}S_{0.0172\ /\ 0.0172}$ The empirical formula is Cu_2S.

(c) Cu_2S, 159.16 amu

$$\frac{5.6 \text{ g } Cu_2S}{1 \text{ cm}^3} \times \frac{1 \text{ mol } Cu_2S}{159.16 \text{ g } Cu_2S} \times \frac{2 \text{ mol } Cu^+ \text{ ions}}{1 \text{ mol } Cu_2S} \times \frac{6.022 \times 10^{23} \text{ } Cu^+ \text{ ions}}{1 \text{ mol } Cu^+ \text{ ions}}$$
$$= 4.2 \times 10^{22} \text{ } Cu^+ \text{ ions/cm}^3$$

3.114 PCl_3, 137.33 amu; PCl_5, 208.24 amu
Let Y = mass of PCl_3 in the mixture, and (10.00 – Y) = mass of PCl_5 in the mixture.

$$\text{fraction Cl in } PCl_3 = \frac{(3)(35.453 \text{ g/mol})}{137.33 \text{ g/mol}} = 0.774\ 48$$

$$\text{fraction Cl in } PCl_5 = \frac{(5)(35.453 \text{ g/mol})}{208.24 \text{ g/mol}} = 0.851\ 25$$

(mass of Cl in PCl_3) + (mass of Cl in PCl_5) = mass of Cl in the mixture
0.774 48Y + 0.851 25(10.00 g – Y) = (0.8104)(10.00 g)
Y = 5.32 g PCl_3 and 10.00 – Y = 4.68 g PCl_5

Reactions in Aqueous Solution

4.1   (a) precipitation   (b) redox   (c) acid-base neutralization

4.2 $FeBr_3$ contains 3 Br^- ions.
       The molar concentration of Br^- ions = 3 x 0.225 M = 0.675 M

4.3   (a) Ionic equation:
       $2\,Ag^+(aq) + 2\,NO_3^-(aq) + 2\,Na^+(aq) + CrO_4^{2-}(aq) \rightarrow Ag_2CrO_4(s) + 2\,Na^+(aq) + 2\,NO_3^-(aq)$
       Delete spectator ions from the ionic equation to get the net ionic equation.
       Net ionic equation: $2\,Ag^+(aq) + CrO_4^{2-}(aq) \rightarrow Ag_2CrO_4(s)$
       (b) Ionic equation:
       $2\,H^+(aq) + SO_4^{2-}(aq) + MgCO_3(s) \rightarrow H_2O(l) + CO_2(g) + Mg^{2+}(aq) + SO_4^{2-}(aq)$
       Delete spectator ions from the ionic equation to get the net ionic equation.
       Net ionic equation: $2\,H^+(aq) + MgCO_3(s) \rightarrow H_2O(l) + CO_2(g) + Mg^{2+}(aq)$

4.4   (a) $CdCO_3$, insoluble   (b) MgO, insoluble   (c) Na_2S, soluble
       (d) $PbSO_4$, insoluble   (e) $(NH_4)_3PO_4$, soluble   (f) $HgCl_2$, soluble

4.5   (a) Ionic equation:
       $Ni^{2+}(aq) + 2\,Cl^-(aq) + 2\,NH_4^+(aq) + S^{2-}(aq) \rightarrow NiS(s) + 2\,NH_4^+(aq) + 2\,Cl^-(aq)$
       Delete spectator ions from the ionic equation to get the net ionic equation.
       Net ionic equation: $Ni^{2+}(aq) + S^{2-}(aq) \rightarrow NiS(s)$
       (b) Ionic equation:
       $2\,Na^+(aq) + CrO_4^{2-}(aq) + Pb^{2+}(aq) + 2\,NO_3^-(aq) \rightarrow PbCrO_4(s) + 2\,Na^+(aq) + 2\,NO_3^-(aq)$
       Delete spectator ions from the ionic equation to get the net ionic equation.
       Net ionic equation: $Pb^{2+}(aq) + CrO_4^{2-}(aq) \rightarrow PbCrO_4(s)$
       (c) Ionic equation:
       $2\,Ag^+(aq) + 2\,ClO_4^-(aq) + Ca^{2+}(aq) + 2\,Br^-(aq) \rightarrow 2\,AgBr(s) + Ca^{2+}(aq) + 2\,ClO_4^-(aq)$
       Delete spectator ions from the ionic equation, and reduce coefficients to get the net ionic
       equation.   Net ionic equation: $Ag^+(aq) + Br^-(aq) \rightarrow AgBr(s)$

4.6 $3\,CaCl_2(aq) + 2\,Na_3PO_4(aq) \rightarrow Ca_3(PO_4)_2(s) + 6\,NaCl(aq)$
       Ionic equation:
       $3\,Ca^{2+}(aq) + 6\,Cl^-(aq) + 6\,Na^+(aq) + 2\,PO_4^{3-}(aq) \rightarrow Ca_3(PO_4)_2(s) + 6\,Na^+(aq) + 6\,Cl^-(aq)$
       Delete spectator ions from the ionic equation to get the net ionic equation.
       Net ionic equation: $3\,Ca^{2+}(aq) + 2\,PO_4^{3-}(aq) \rightarrow Ca_3(PO_4)_2(s)$

4.7   (a) Ionic equation:
       $2\,Cs^+(aq) + 2\,OH^-(aq) + 2\,H^+(aq) + SO_4^{2-}(aq) \rightarrow 2\,Cs^+(aq) + SO_4^{2-}(aq) + 2\,H_2O(l)$
       Delete spectator ions from the ionic equation, and reduce coefficients to get the net ionic
       equation.   Net ionic equation: $H^+(aq) + OH^-(aq) \rightarrow H_2O(l)$

(b) Ionic equation:

$Ca^{2+}(aq) + 2\ OH^-(aq) + 2\ CH_3COOH(aq) \rightarrow Ca^{2+}(aq) + 2\ CH_3CO_2^-(aq) + 2\ H_2O(l)$

Delete spectator ions from the ionic equation, and reduce coefficients to get the net ionic equation.   Net ionic equation: $CH_3COOH(aq) + OH^-(aq) \rightarrow CH_3CO_2^-(aq) + H_2O(l)$

4.8 (a) $SnCl_4$:   Cl –1, Sn +4   (b) CrO_3:   O –2, Cr +6
(c) $VOCl_3$:   O –2, Cl –1, V +5   (d) V_2O_3:   O –2, V +3
(e) HNO_3:   O –2, H +1, N +5   (f) $FeSO_4$:   O –2, S +6, Fe +2

4.9 $2\ Cu^{2+}(aq) + 4\ I^-(aq) \rightarrow 2\ CuI(s) + I_2(aq)$
oxidation numbers: Cu^{2+} +2; I^- –1;   CuI: Cu +1, I –1; I_2: 0
oxidizing agent (oxidation number decreases), Cu^{2+}
reducing agent (oxidation number decreases) , I^-

4.10 (a) $SnO_2(s) + 2\ C(s) \rightarrow Sn(s) + 2\ CO(g)$
C is oxidized (its oxidation number increases from 0 to +2). C is the reducing agent.
The Sn in SnO_2 is reduced (its oxidation number decreases from +4 to 0). SnO_2 is the oxidizing agent.
(b) $Sn^{2+}(aq) + 2\ Fe^{3+}(aq) \rightarrow Sn^{4+}(aq) + 2\ Fe^{2+}(aq)$
Sn^{2+} is oxidized (its oxidation number increases from +2 to +4). Sn^{2+} is the reducing agent.
Fe^{3+} is reduced (its oxidation number decreases from +3 to +2). Fe^{3+} is the oxidizing agent.

4.11 (a) Pt is below H in the activity series; therefore NO REACTION.
(b) Mg is below Ca in the activity series; therefore NO REACTION.

4.12 $Cr_2O_7{}^{2-}(aq) + I^-(aq) \rightarrow Cr^{3+}(aq) + IO_3{}^-(aq)$

$Cr_2O_7{}^{2-}(aq) + I^-(aq) \rightarrow 2\ Cr^{3+}(aq) + IO_3{}^-(aq)$
+6   –1   +3   +5
2(+6)= +12   2(+3)= +6
lose 6e$^-$
gain 6e$^-$

$8\ H^+(aq) + Cr_2O_7{}^{2-}(aq) + I^-(aq) \rightarrow 2\ Cr^{3+}(aq) + IO_3^-(aq) + 4\ H_2O(l)$

4.13 $MnO_4{}^-(aq) + Br^-(aq) \rightarrow MnO_2(s) + BrO_3{}^-(aq)$
+7   –1   +4   +5
gain 3e$^-$
lose 6e$^-$

$2\ MnO_4^-(aq) + Br^-(aq) \rightarrow 2\ MnO_2(s) + BrO_3^-(aq)$
$2\ H^+(aq) + 2\ MnO_4^-(aq) + Br^-(aq) \rightarrow 2\ MnO_2(s) + BrO_3^-(aq) + H_2O(l)$

$$2\ H^+(aq) + 2\ OH^-(aq) + 2\ MnO_4^-(aq) + Br^-(aq) \rightarrow$$
$$2\ MnO_2(s) + BrO_3^-(aq) + H_2O(l)\ +\ 2\ OH^-(aq)$$
$$H_2O(l)\ +\ 2\ MnO_4^-(aq)\ +\ Br^-(aq)\ \rightarrow\ 2\ MnO_2(s)\ +\ BrO_3^-(aq)\ +\ 2\ OH^-(aq)$$

4.14　(a) $MnO_4^-(aq)\ \rightarrow\ MnO_2(s)$　　　(reduction)
　　　　　$IO_3^-(aq)\ \rightarrow\ IO_4^-(aq)$　　　　(oxidation)
　　　(b) $NO_3^-(aq)\ \rightarrow\ NO_2(g)$　　　　(reduction)
　　　　　$SO_2(aq)\ \rightarrow\ SO_4^{2-}(aq)$　　　(oxidation)

4.15　$NO_3^-(aq)\ +\ Cu(s)\ \rightarrow\ NO(g)\ +\ Cu^{2+}(aq)$
　　　$[Cu(s)\ \rightarrow\ Cu^{2+}(aq)\ +\ 2\ e^-]\ x\ 3$　　　　(oxidation half reaction)

　　　$NO_3^-(aq)\ \rightarrow\ NO(g)$
　　　$NO_3^-(aq)\ \rightarrow\ NO(g)\ +\ 2\ H_2O(l)$
　　　$4\ H^+(aq)\ +\ NO_3^-(aq)\ \rightarrow\ NO(g)\ +\ 2\ H_2O(l)$
　　　$[3\ e^-\ +\ 4\ H^+(aq)\ +\ NO_3^-(aq)\ \rightarrow\ NO(g)\ +\ 2\ H_2O(l)]\ x\ 2$　　　(reduction half reaction)

　　　Combine the two half reactions.
　　　$2\ NO_3^-(aq) + 8\ H^+(aq) + 3\ Cu(s) \rightarrow 3\ Cu^{2+}(aq) + 2\ NO(g) + 4\ H_2O(l)$

4.16　$Fe(OH)_2(s)\ +\ O_2(g)\ \rightarrow\ Fe(OH)_3(s)$
　　　$[Fe(OH)_2(s)\ +\ OH^-(aq)\ \rightarrow\ Fe(OH)_3(s) + e^-]\ x\ 4$　(oxidation half reaction)

　　　$O_2(g)\ \rightarrow\ 2\ H_2O(l)$
　　　$4\ H^+(aq)\ +\ O_2(g)\ \rightarrow\ 2\ H_2O(l)$
　　　$4\ e^-\ +\ 4\ H^+(aq)\ +\ O_2(g)\ \rightarrow\ 2\ H_2O(l)$
　　　$4\ e^-\ +\ 4\ H^+(aq)\ +\ 4\ OH^-(aq)\ +\ O_2(g)\ \rightarrow\ 2\ H_2O(l)\ +\ 4\ OH^-(aq)$
　　　$4\ e^-\ +\ 4\ H_2O(l)\ +\ O_2(g)\ \rightarrow\ 2\ H_2O(l) + 4\ OH^-(aq)$
　　　$4\ e^-\ +\ 2\ H_2O(l) + O_2(g)\ \rightarrow\ 4\ OH^-(aq)$　　　(reduction half reaction)

　　　Combine the two half reactions.
　　　$4\ Fe(OH)_2(s) + 4\ OH^-(aq)\ +\ 2\ H_2O(l)\ +\ O_2(g)\ \rightarrow\ 4\ Fe(OH)_3(s)\ +\ 4\ OH^-(aq)$
　　　$4\ Fe(OH)_2(s) + 2\ H_2O(l)\ +\ O_2(g)\ \rightarrow\ 4\ Fe(OH)_3(s)$

4.17　$31.50\ mL = 0.031\ 50\ L;\ \ 10.00\ mL = 0.010\ 00\ L$

$$0.031\ 50\ L \times \frac{0.105\ mol\ BrO_3^-}{1\ L}\ \times\ \frac{6\ mol\ Fe^{2+}}{1\ mol\ BrO_3^-} = 1.98 \times 10^{-2}\ mol\ Fe^{2+}$$

$$molarity = \frac{1.98 \times 10^{-2}\ mol\ Fe^{2+}}{0.010\ 00\ L} = 1.98\ M\ Fe^{2+}\ solution$$

Understanding Key Concepts

4.18　(a) $Ba^{2+}(aq)\ +\ SO_4^{2-}(aq)\ \rightarrow\ BaSO_4(s)$
　　　(b) $2\ H^+(aq)\ +\ CO_3^{2-}(aq)\ \rightarrow\ CO_2(g)\ +\ H_2O(l)$

4.19　"Any element higher in the activity series will react with the ion of any element lower in the activity series."

$C + B^+ \rightarrow C^+ + B$; therefore C is higher than B.

$A^+ + D \rightarrow$ no reaction; therefore A is higher than D.

$C^+ + A \rightarrow$ no reaction; therefore C is higher than A.

$D + B^+ \rightarrow D^+ + B$; therefore D is higher than B.

The net result is $C > A > D > B$.

4.20　(a)　The reaction, $A^+ + C \rightarrow A + C^+$, will occur since C is above A in the activity series.

(b)　The reaction, $A^+ + B \rightarrow A + B^+$, will not occur since B is below A in the activity series.

4.21　(a)　$2\,Na^+(aq) + CO_3^{2-}(aq)$　does not form a precipitate.　This is represented by box (1).

(b)　$Ba^{2+}(aq) + CrO_4^{2-}(aq) \rightarrow BaCrO_4(s)$.　This is represented by box (2).

(c)　$2\,Ag^+(aq) + SO_4^{2-}(aq) \rightarrow Ag_2SO_4(s)$.　This is represented by box (3).

4.22　In the precipitate there are two cations (blue) for each anion (green).　Looking at the ions in the list, the anion must have a 2– charge and the cation a 1+ charge for charge neutrality of the precipitate.　The cation must be Ag^+ because all Na^+ salts are soluble.　Ag_2CrO_4 and Ag_2CO_3 are insoluble and consistent with the observed result.

4.23　One OH^- will react with each available H^+ on the acid forming H_2O.　The acid is identified by how many of the 12 OH^- react with three molecules of each acid.

(a)　Three HF's react with three OH^-, leaving nine OH^- unreacted (box 2).

(b)　Three H_2SO_3's react with six OH^-, leaving six OH^- unreacted (box 3).

(c)　Three H_3PO_4's react with nine OH^-, leaving three OH^- unreacted (box 1).

Additional Problems
Aqueous Reactions and Net Ionic Equations

4.24　(a) precipitation　　　　　(b) redox　　　　　(c) acid-base neutralization

4.26　(a) Ionic equation:

$Hg^{2+}(aq) + 2\,NO_3^-(aq) + 2\,Na^+(aq) + 2\,I^-(aq) \rightarrow 2\,Na^+(aq) + 2\,NO_3^-(aq) + HgI_2(s)$

Delete spectator ions from the ionic equation to get the net ionic equation.

Net ionic equation: $Hg^{2+}(aq) + 2\,I^-(aq) \rightarrow HgI_2(s)$

(b) $2\,HgO(s) \xrightarrow{\text{Heat}} 2\,Hg(l) + O_2(g)$

(c) Ionic equation:

$H_3PO_4(aq) + 3\,K^+(aq) + 3\,OH^-(aq) \rightarrow 3\,K^+(aq) + PO_4^{3-}(aq) + 3\,H_2O(l)$

Delete spectator ions from the ionic equation to get the net ionic equation.

Net ionic equation: $H_3PO_4(aq) + 3\,OH^-(aq) \rightarrow PO_4^{3-}(aq) + 3\,H_2O(l)$

4.28　A strong electrolyte is a compound that completely dissociates into its ions when placed in water.　When a weak electrolyte is placed in water it dissociates only slightly into its ions.

4.30   (a) HBr, strong electrolyte   (b) HF, weak electrolyte
        (c) $NaClO_4$, strong electrolyte   (d) $(NH_4)_2CO_3$, strong electrolyte
        (e) NH_3, weak electrolyte

4.32   (a) K_2CO_3 contains 3 ions (2 K^+ and 1 CO_3^{2-}).
        The molar concentration of ions = 3 x 0.750 M = 2.25 M.
        (b) $AlCl_3$ contains 4 ions (1 Al^{3+} and 3 Cl^-).
        The molar concentration of ions = 4 x 0.355 M = 1.42 M.

Precipitation Reactions and Solubility Rules

4.34   (a) Ag_2O, insoluble   (b) $Ba(NO_3)_2$, soluble
        (c) $SnCO_3$, insoluble   (d) Fe_2O_3, insoluble

4.36   (a) No precipitate will form.
        (b) $FeCl_2(aq)$ + 2 $KOH(aq)$ → $Fe(OH)_2(s)$ + 2 $KCl(aq)$
        (c) No precipitate will form.

4.38   (a) $Pb(NO_3)_2(aq)$ + $Na_2SO_4(aq)$ → $PbSO_4(s)$ + 2 $NaNO_3(aq)$
        (b) 3 $MgCl_2(aq)$ + 2 $K_3PO_4(aq)$ → $Mg_3(PO_4)_2(s)$ + 6 $KCl(aq)$
        (c) $ZnSO_4(aq)$ + $Na_2CrO_4(aq)$ → $ZnCrO_4(s)$ + $Na_2SO_4(aq)$

4.40   Add $HCl(aq)$; it will selectively precipitate $AgCl(s)$.

4.42 Ag^+ is eliminated because it would have precipitated as $AgCl(s)$; Ba^{2+} is eliminated
        because it would have precipitated as $BaSO_4(s)$. The solution might contain Cs^+ and/or
        NH_4^+. Neither of these will precipitate with OH^-, SO_4^{2-}, or Cl^-.

Acids, Bases, and Neutralization Reactions

4.44   Add the solution to an active metal, such as magnesium. Bubbles of H_2 gas indicate the
        presence of an acid.

4.46   (a) 2 $H^+(aq)$ + 2 $ClO_4^-(aq)$ + $Ca^{2+}(aq)$ + 2 $OH^-(aq)$ → $Ca^{2+}(aq)$ + 2 $ClO_4^-(aq)$ + 2 $H_2O(l)$
        (b) $CH_3COOH(aq)$ + $Na^+(aq)$ + $OH^-(aq)$ → $CH_3CO_2^-(aq)$ + $Na^+(aq)$ + $H_2O(l)$

4.48   (a) $LiOH(aq)$ + $HI(aq)$ → $LiI(aq)$ + $H_2O(l)$
        Ionic equation: $Li^+(aq)$ + $OH^-(aq)$ + $H^+(aq)$ + $I^-(aq)$ → $Li^+(aq)$ + $I^-(aq)$ + $H_2O(l)$
        Delete spectator ions from the ionic equation to get the net ionic equation.
        Net ionic equation: $H^+(aq)$ + $OH^-(aq)$ → $H_2O(l)$

        (b) 2 $HBr(aq)$ + $Ca(OH)_2(aq)$ → $CaBr_2(aq)$ + 2 $H_2O(l)$
        Ionic equation:
        2 $H^+(aq)$ + 2 $Br^-(aq)$ + $Ca^{2+}(aq)$ + 2 $OH^-(aq)$ → $Ca^{2+}(aq)$ + 2 $Br^-(aq)$ + 2 $H_2O(l)$
        Delete spectator ions from the ionic equation to get the net ionic equation.
        Net ionic equation: $H^+(aq)$ + $OH^-(aq)$ → $H_2O(l)$

Redox Reactions and Oxidation Numbers

4.50 The best reducing agents are at the bottom left of the periodic table. The best oxidizing agents are at the top right of the periodic table (excluding inert gases).

4.52 (a) An oxidizing agent gains electrons.
(b) A reducing agent loses electrons.
(c) A substance undergoing oxidation loses electrons.
(d) A substance undergoing reduction gains electrons.

4.54 (a) NO_2 O –2, N +4   (b) SO_3 O –2, S +6
(c) $COCl_2$ O –2, Cl –1, C +4   (d) CH_2Cl_2 Cl –1, H +1, C 0
(e) $KClO_3$ O –2, K +1, Cl +5   (f) HNO_3 O –2, H +1, N +5

4.56 (a) ClO_3^- O –2, Cl +5   (b) SO_3^{2-} O –2, S +4
(c) $C_2O_4^{2-}$ O –2, C +3   (d) NO_2^- O –2, N +3
(e) BrO^- O –2, Br +1

4.58 (a) $Ca(s) + Sn^{2+}(aq) \rightarrow Ca^{2+}(aq) + Sn(s)$.
$Ca(s)$ is oxidized (oxidation number increases from 0 to +2).
$Sn^{2+}(aq)$ is reduced (oxidation number decreases from +2 to 0).
(b) $ICl(s) + H_2O(l) \rightarrow HCl(aq) + HOI(aq)$
No oxidation numbers change. The reaction is not a redox reaction.

4.60 (a) Zn is below Na^+; therefore no reaction.
(b) Pt is below H^+; therefore no reaction.
(c) Au is below Ag^+; therefore no reaction.
(d) Ag is above Au^{3+}; the reaction is $Au^{3+}(aq) + 3\ Ag(s) \rightarrow 3\ Ag^+(aq) + Au(s)$.

4.62 "Any element higher in the activity series will react with the ion of any element lower in the activity series."
$A + B^+ \rightarrow A^+ + B$; therefore A is higher than B.
$C^+ + D \rightarrow$ no reaction; therefore C is higher than D.
$B + D^+ \rightarrow B^+ + D$; therefore B is higher than D.
$B + C^+ \rightarrow B^+ + C$; therefore B is higher than C.
The net result is A > B > C > D.

4.64 (a) C is below A^+; therefore no reaction.
(b) D is below A^+; therefore no reaction.

Balancing Redox Reactions

4.66 (a) N oxidation number decreases from +5 to +2; reduction.
(b) Zn oxidation number increases from 0 to +2; oxidation.
(c) Ti oxidation number increases from +3 to +4; oxidation.
(d) Sn oxidation number decreases from +4 to +2; reduction.

4.68 (a) $NO_3^-(aq) \rightarrow NO(g)$
$NO_3^-(aq) \rightarrow NO(g) + 2\ H_2O(l)$
$4\ H^+(aq) + NO_3^-(aq) \rightarrow NO(g) + 2\ H_2O(l)$
$3\ e^- + 4\ H^+(aq) + NO_3^-(aq) \rightarrow NO(g) + 2\ H_2O(l)$

(b) $Zn(s) \rightarrow Zn^{2+}(aq) + 2\ e^-$

(c) $Ti^{3+}(aq) \rightarrow TiO_2(s)$
$Ti^{3+}(aq) + 2\ H_2O(l) \rightarrow TiO_2(s)$
$Ti^{3+}(aq) + 2\ H_2O(l) \rightarrow TiO_2(s) + 4\ H^+(aq)$
$Ti^{3+}(aq) + 2\ H_2O(l) \rightarrow TiO_2(s) + 4\ H^+(aq) + e^-$

(d) $Sn^{4+}(aq) + 2\ e^- \rightarrow Sn^{2+}(aq)$

4.70 (a) $Te(s) + NO_3^-(aq) \rightarrow TeO_2(s) + NO(g)$
oxidation: $Te(s) \rightarrow TeO_2(s)$
reduction: $NO_3^-(aq) \rightarrow NO(g)$

(b) $H_2O_2(aq) + Fe^{2+}(aq) \rightarrow Fe^{3+}(aq) + H_2O(l)$
oxidation: $Fe^{2+}(aq) \rightarrow Fe^{3+}(aq)$
reduction: $H_2O_2(aq) \rightarrow H_2O(l)$

4.72 (a) $Cr_2O_7^{2-}(aq) \rightarrow Cr^{3+}(aq)$
$Cr_2O_7^{2-}(aq) \rightarrow 2\ Cr^{3+}(aq)$
$Cr_2O_7^{2-}(aq) \rightarrow 2\ Cr^{3+}(aq) + 7\ H_2O(l)$
$14\ H^+(aq) + Cr_2O_7^{2-}(aq) \rightarrow 2\ Cr^{3+}(aq) + 7\ H_2O(l)$
$14\ H^+(aq) + Cr_2O_7^{2-}(aq) + 6\ e^- \rightarrow 2\ Cr^{3+}(aq) + 7\ H_2O(l)$

(b) $CrO_4^{2-}(aq) \rightarrow Cr(OH)_4^-(aq)$
$4\ H^+(aq) + CrO_4^{2-}(aq) \rightarrow Cr(OH)_4^-(aq)$
$4\ H^+(aq) + 4\ OH^-(aq) + CrO_4^{2-}(aq) \rightarrow Cr(OH)_4^-(aq) + 4\ OH^-(aq)$
$4\ H_2O(l) + CrO_4^{2-}(aq) \rightarrow Cr(OH)_4^-(aq) + 4\ OH^-(aq)$
$4\ H_2O(l) + CrO_4^{2-}(aq) + 3\ e^- \rightarrow Cr(OH)_4^-(aq) + 4\ OH^-(aq)$

(c) $Bi^{3+}(aq) \rightarrow BiO_3^-(aq)$
$Bi^{3+}(aq) + 3\ H_2O(l) \rightarrow BiO_3^-(aq)$
$Bi^{3+}(aq) + 3\ H_2O(l) \rightarrow BiO_3^-(aq) + 6\ H^+(aq)$
$Bi^{3+}(aq) + 3\ H_2O(l) + 6\ OH^-(aq) \rightarrow BiO_3^-(aq) + 6\ H^+(aq) + 6\ OH^-(aq)$
$Bi^{3+}(aq) + 3\ H_2O(l) + 6\ OH^-(aq) \rightarrow BiO_3^-(aq) + 6\ H_2O(l)$
$Bi^{3+}(aq) + 6\ OH^-(aq) \rightarrow BiO_3^-(aq) + 3\ H_2O(l)$
$Bi^{3+}(aq) + 6\ OH^-(aq) \rightarrow BiO_3^-(aq) + 3\ H_2O(l) + 2\ e^-$

(d) $ClO^-(aq) \rightarrow Cl^-(aq)$
$ClO^-(aq) \rightarrow Cl^-(aq) + H_2O(l)$
$2\ H^+(aq) + ClO^-(aq) \rightarrow Cl^-(aq) + H_2O(l)$
$2\ H^+(aq) + 2\ OH^-(aq) + ClO^-(aq) \rightarrow Cl^-(aq) + H_2O(l) + 2\ OH^-(aq)$

$2 H_2O(l) + ClO^-(aq) \rightarrow Cl^-(aq) + H_2O(l) + 2 OH^-(aq)$

$H_2O(l) + ClO^-(aq) \rightarrow Cl^-(aq) + 2 OH^-(aq)$

$H_2O(l) + ClO^-(aq) + 2 e^- \rightarrow Cl^-(aq) + 2 OH^-(aq)$

4.74   (a) $MnO_4^-(aq) \rightarrow MnO_2(s)$

$MnO_4^-(aq) \rightarrow MnO_2(s) + 2 H_2O(l)$

$4 H^+(aq) + MnO_4^-(aq) \rightarrow MnO_2(s) + 2 H_2O(l)$

$[4 H^+(aq) + MnO_4^-(aq) + 3 e^- \rightarrow MnO_2(s) + 2 H_2O(l)] \times 2$ (reduction half reaction)

$IO_3^-(aq) \rightarrow IO_4^-(aq)$

$H_2O(l) + IO_3^-(aq) \rightarrow IO_4^-(aq)$

$H_2O(l) + IO_3^-(aq) \rightarrow IO_4^-(aq) + 2 H^+(aq)$

$[H_2O(l) + IO_3^-(aq) \rightarrow IO_4^-(aq) + 2 H^+(aq) + 2 e^-] \times 3$ (oxidation half reaction)

Combine the two half reactions.

$8 H^+(aq) + 3 H_2O(l) + 2 MnO_4^-(aq) + 3 IO_3^-(aq) \rightarrow$
$$6 H^+(aq) + 4 H_2O(l) + 2 MnO_2(s) + 3 IO_4^-(aq)$$

$2 H^+(aq) + 2 MnO_4^-(aq) + 3 IO_3^-(aq) \rightarrow 2 MnO_2(s) + 3 IO_4^-(aq) + H_2O(l)$

$2 H^+(aq) + 2 OH^-(aq) + 2 MnO_4^-(aq) + 3 IO_3^-(aq) \rightarrow$
$$2 MnO_2(s) + 3 IO_4^-(aq) + H_2O(l) + 2 OH^-(aq)$$

$2 H_2O(l) + 2 MnO_4^-(aq) + 3 IO_3^-(aq) \rightarrow$
$$2 MnO_2(s) + 3 IO_4^-(aq) + H_2O(l) + 2 OH^-(aq)$$

$H_2O(l) + 2 MnO_4^-(aq) + 3 IO_3^-(aq) \rightarrow 2 MnO_2(s) + 3 IO_4^-(aq) + 2 OH^-(aq)$

   (b) $Cu(OH)_2(s) \rightarrow Cu(s)$

$Cu(OH)_2(s) \rightarrow Cu(s) + 2 H_2O(l)$

$2 H^+(aq) + Cu(OH)_2(s) \rightarrow Cu(s) + 2 H_2O(l)$

$[2 H^+(aq) + Cu(OH)_2(s) + 2 e^- \rightarrow Cu(s) + 2 H_2O(l)] \times 2$ (reduction half reaction)

$N_2H_4(aq) \rightarrow N_2(g)$

$N_2H_4(aq) \rightarrow N_2(g) + 4 H^+(aq)$

$N_2H_4(aq) \rightarrow N_2(g) + 4 H^+(aq) + 4 e^-$ (oxidation half reaction)

Combine the two half reactions.

$4 H^+(aq) + 2 Cu(OH)_2(s) + N_2H_4(aq) \rightarrow 2 Cu(s) + 4 H_2O(l) + N_2(g) + 4 H^+(aq)$

$2 Cu(OH)_2(s) + N_2H_4(aq) \rightarrow 2 Cu(s) + 4 H_2O(l) + N_2(g)$

   (c) $Fe(OH)_2(s) \rightarrow Fe(OH)_3(s)$

$Fe(OH)_2(s) + H_2O(l) \rightarrow Fe(OH)_3(s)$

$Fe(OH)_2(s) + H_2O(l) \rightarrow Fe(OH)_3(s) + H^+(aq)$

$[Fe(OH)_2(s) + H_2O(l) \rightarrow Fe(OH)_3(s) + H^+(aq) + e^-] \times 3$ (oxidation half reaction)

$CrO_4^{2-}(aq) \rightarrow Cr(OH)_4^-(aq)$

$4 H^+(aq) + CrO_4^{2-}(aq) \rightarrow Cr(OH)_4^-(aq)$

$4 H^+(aq) + CrO_4^{2-}(aq) + 3 e^- \rightarrow Cr(OH)_4^-(aq)$ (reduction half reaction)

Combine the two half reactions.

$3 \text{ Fe(OH)}_2(s) + 3 \text{ H}_2\text{O}(l) + 4 \text{ H}^+(aq) + \text{CrO}_4^{2-}(aq) \rightarrow$
$\qquad\qquad\qquad 3 \text{ Fe(OH)}_3(s) + 3 \text{ H}^+(aq) + \text{Cr(OH)}_4^-(aq)$

$3 \text{ Fe(OH)}_2(s) + 3 \text{ H}_2\text{O}(l) + \text{H}^+(aq) + \text{CrO}_4^{2-}(aq) \rightarrow 3 \text{ Fe(OH)}_3(s) + \text{Cr(OH)}_4^-(aq)$

$3 \text{ Fe(OH)}_2(s) + 3 \text{ H}_2\text{O}(l) + \text{H}^+(aq) + \text{OH}^-(aq) + \text{CrO}_4^{2-}(aq) \rightarrow$
$\qquad\qquad\qquad 3 \text{ Fe(OH)}_3(s) + \text{Cr(OH)}_4^-(aq) + \text{OH}^-(aq)$

$3 \text{ Fe(OH)}_2(s) + 4 \text{ H}_2\text{O}(l) + \text{CrO}_4^{2-}(aq) \rightarrow 3 \text{ Fe(OH)}_3(s) + \text{Cr(OH)}_4^-(aq) + \text{OH}^-(aq)$

(d) $\text{ClO}_4^-(aq) \rightarrow \text{ClO}_2^-(aq)$
$\text{ClO}_4^-(aq) \rightarrow \text{ClO}_2^-(aq) + 2 \text{ H}_2\text{O}(l)$
$4 \text{ H}^+(aq) + \text{ClO}_4^-(aq) \rightarrow \text{ClO}_2^-(aq) + 2 \text{ H}_2\text{O}(l)$
$4 \text{ H}^+(aq) + \text{ClO}_4^-(aq) + 4 \text{ e}^- \rightarrow \text{ClO}_2^-(aq) + 2 \text{ H}_2\text{O}(l)$ (reduction half reaction)

$\text{H}_2\text{O}_2(aq) \rightarrow \text{O}_2(g)$
$\text{H}_2\text{O}_2(aq) \rightarrow \text{O}_2(g) + 2 \text{ H}^+(aq)$
$[\text{H}_2\text{O}_2(aq) \rightarrow \text{O}_2(g) + 2 \text{ H}^+(aq) + 2 \text{ e}^-] \times 2$ (oxidation half reaction)

Combine the two half reactions.
$4 \text{ H}^+(aq) + \text{ClO}_4^-(aq) + 2 \text{ H}_2\text{O}_2(aq) \rightarrow \text{ClO}_2^-(aq) + 2 \text{ H}_2\text{O}(l) + 2 \text{ O}_2(g) + 4 \text{ H}^+(aq)$
$\text{ClO}_4^-(aq) + 2 \text{ H}_2\text{O}_2(aq) \rightarrow \text{ClO}_2^-(aq) + 2 \text{ H}_2\text{O}(l) + 2 \text{ O}_2(g)$

4.76 (a) $\text{Zn}(s) \rightarrow \text{Zn}^{2+}(aq)$
$\text{Zn}(s) \rightarrow \text{Zn}^{2+}(aq) + 2 \text{ e}^-$ (oxidation half reaction)

$\text{VO}^{2+}(aq) \rightarrow \text{V}^{3+}(aq)$
$\text{VO}^{2+}(aq) \rightarrow \text{V}^{3+}(aq) + \text{H}_2\text{O}(l)$
$2 \text{ H}^+(aq) + \text{VO}^{2+}(aq) \rightarrow \text{V}^{3+}(aq) + \text{H}_2\text{O}(l)$
$[2 \text{ H}^+(aq) + \text{VO}^{2+}(aq) + \text{e}^- \rightarrow \text{V}^{3+}(aq) + \text{H}_2\text{O}(l)] \times 2$ (reduction half reaction)

Combine the two half reactions.
$\text{Zn}(s) + 2 \text{ VO}^{2+}(aq) + 4 \text{ H}^+(aq) \rightarrow \text{Zn}^{2+}(aq) + 2 \text{ V}^{3+}(aq) + 2 \text{ H}_2\text{O}(l)$

(b) $\text{Ag}(s) \rightarrow \text{Ag}^+(aq)$
$\text{Ag}(s) \rightarrow \text{Ag}^+(aq) + \text{e}^-$ (oxidation half reaction)

$\text{NO}_3^-(aq) \rightarrow \text{NO}_2(g)$
$\text{NO}_3^-(aq) \rightarrow \text{NO}_2(g) + \text{H}_2\text{O}(l)$
$2 \text{ H}^+(aq) + \text{NO}_3^-(aq) \rightarrow \text{NO}_2(g) + \text{H}_2\text{O}(l)$
$2 \text{ H}^+(aq) + \text{NO}_3^-(aq) + \text{e}^- \rightarrow \text{NO}_2(g) + \text{H}_2\text{O}(l)$ (reduction half reaction)

Combine the two half reactions.
$2 \text{ H}^+(aq) + \text{Ag}(s) + \text{NO}_3^-(aq) \rightarrow \text{Ag}^+(aq) + \text{NO}_2(g) + \text{H}_2\text{O}(l)$

(c) $\text{Mg}(s) \rightarrow \text{Mg}^{2+}(aq)$
$[\text{Mg}(s) \rightarrow \text{Mg}^{2+}(aq) + 2 \text{ e}^-] \times 3$ (oxidation half reaction)

$$VO_4^{3-}(aq) \rightarrow V^{2+}(aq)$$
$$VO_4^{3-}(aq) \rightarrow V^{2+}(aq) + 4 H_2O(l)$$
$$8 H^+(aq) + VO_4^{3-}(aq) \rightarrow V^{2+}(aq) + 4 H_2O(l)$$
$$[8 H^+(aq) + VO_4^{3-}(aq) + 3 e^- \rightarrow V^{2+}(aq) + 4 H_2O(l)] \times 2 \quad \text{(reduction half reaction)}$$

Combine the two half reactions.
$$3 Mg(s) + 16 H^+(aq) + 2 VO_4^{3-}(aq) \rightarrow 3 Mg^{2+}(aq) + 2 V^{2+}(aq) + 8 H_2O(l)$$

(d) $I^-(aq) \rightarrow I_3^-(aq)$
$$3 I^-(aq) \rightarrow I_3^-(aq)$$
$$[3 I^-(aq) \rightarrow I_3^-(aq) + 2 e^-] \times 8 \qquad \text{(oxidation half reaction)}$$

$$IO_3^-(aq) \rightarrow I_3^-(aq)$$
$$3 IO_3^-(aq) \rightarrow I_3^-(aq)$$
$$3 IO_3^-(aq) \rightarrow I_3^-(aq) + 9 H_2O(l)$$
$$18 H^+(aq) + 3 IO_3^-(aq) \rightarrow I_3^-(aq) + 9 H_2O(l)$$
$$18 H^+(aq) + 3 IO_3^-(aq) + 16 e^- \rightarrow I_3^-(aq) + 9 H_2O(l) \quad \text{(reduction half reaction)}$$

Combine the two half reactions.
$$18 H^+(aq) + 3 IO_3^-(aq) + 24 I^-(aq) \rightarrow 9 I_3^-(aq) + 9 H_2O(l)$$
Divide each coefficient by 3.
$$6 H^+(aq) + IO_3^-(aq) + 8 I^-(aq) \rightarrow 3 I_3^-(aq) + 3 H_2O(l)$$

Redox Titrations

4.78 $I_2(aq) + 2 S_2O_3^{2-}(aq) \rightarrow S_4O_6^{2-}(aq) + 2 I^-(aq);$ $35.20 \text{ mL} = 0.032 \ 50 \text{ L}$

$$0.035 \ 20 \text{ L} \times \frac{0.150 \text{ mol } S_2O_3^{2-}}{\text{L}} \times \frac{1 \text{ mol } I_2}{2 \text{ mol } S_2O_3^{2-}} \times \frac{253.8 \text{ g } I_2}{1 \text{ mol } I_2} = 0.670 \text{ g } I_2$$

4.80 $3 H_3AsO_3(aq) + BrO_3^-(aq) \rightarrow Br^-(aq) + 3 H_3AsO_4(aq)$
22.35 mL = 0.022 35 L and 50.00 mL = 0.050 00 L

$$0.022 \ 35 \text{ L} \times \frac{0.100 \text{ mol } BrO_3^-}{\text{L}} \times \frac{3 \text{ mol } H_3AsO_3}{1 \text{ mol } BrO_3^-} = 6.70 \times 10^{-3} \text{ mol } H_3AsO_3$$

$$\text{molarity} = \frac{6.70 \times 10^{-3} \text{ mol}}{0.050 \ 00 \text{ L}} = 0.134 \text{ M As(III)}$$

4.82 $2 Fe^{3+}(aq) + Sn^{2+}(aq) \rightarrow 2 Fe^{2+}(aq) + Sn^{4+}(aq);$ $13.28 \text{ mL} = 0.013 \ 28 \text{ L}$

$$0.013 \ 28 \text{ L} \times \frac{0.1015 \text{ mol } Sn^{2+}}{\text{L}} \times \frac{2 \text{ mol } Fe^{3+}}{1 \text{ mol } Sn^{2+}} \times \frac{55.847 \text{ g } Fe^{3+}}{1 \text{ mol } Fe^{3+}} = 0.1506 \text{ g } Fe^{3+}$$

$$\text{mass \% Fe} = \frac{0.1506 \text{ g}}{0.1875 \text{ g}} \times 100\% = 80.32\%$$

Chapter 4 – Reactions in Aqueous Solutions

4.84 $C_2H_5OH(aq) + 2\ Cr_2O_7^{2-}(aq) + 16\ H^+(aq) \rightarrow 2\ CO_2(g) + 4\ Cr^{3+}(aq) + 11\ H_2O(l)$
C_2H_5OH, 46.07 amu; 8.76 mL = 0.008 76 L

$$0.008\ 76\ L \times \frac{0.049\ 88\ mol\ Cr_2O_7^{2-}}{L} \times \frac{1\ mol\ C_2H_5OH}{2\ mol\ Cr_2O_7^{2-}} \times \frac{46.07\ g\ C_2H_5OH}{1\ mol\ C_2H_5OH}$$

= 0.010 07 g C_2H_5OH

mass % C_2H_5OH = $\dfrac{0.010\ 07\ g}{10.002\ g}$ × 100% = 0.101%

General Problems

4.86 (a) $[Fe(CN)_6]^{3-}(aq) \rightarrow Fe(CN)_6]^{4-}(aq)$
$([Fe(CN)_6]^{3-}(aq) + e^- \rightarrow [Fe(CN)_6]^{4-}(aq)) \times 4$ (reduction half reaction)

$N_2H_4(aq) \rightarrow N_2(g)$
$N_2H_4(aq) \rightarrow N_2(g) + 4\ H^+(aq)$
$N_2H_4(aq) \rightarrow N_2(g) + 4\ H^+(aq) + 4\ e^-$
$N_2H_4(aq) + 4\ OH^-(aq) \rightarrow N_2(g) + 4\ H^+(aq) + 4\ OH^-(aq) + 4\ e^-$
$N_2H_4(aq) + 4\ OH^-(aq) \rightarrow N_2(g) + 4\ H_2O(l) + 4\ e^-$ (oxidation half reaction)

Combine the two half reactions.
$4\ [Fe(CN)_6]^{3-}(aq) + N_2H_4(aq) + 4\ OH^-(aq) \rightarrow$
$\qquad\qquad 4\ [Fe(CN)_6]^{4-}(aq) + N_2(g) + 4\ H_2O(l)$

(b) $Cl_2(g) \rightarrow Cl^-(aq)$
$Cl_2(g) \rightarrow 2\ Cl^-(aq)$
$Cl_2(g) + 2\ e^- \rightarrow 2\ Cl^-(aq)$ (reduction half reaction)

$SeO_3^{2-}(aq) \rightarrow SeO_4^{2-}(aq)$
$SeO_3^{2-}(aq) + H_2O(l) \rightarrow SeO_4^{2-}(aq)$
$SeO_3^{2-}(aq) + H_2O(l) \rightarrow SeO_4^{2-}(aq) + 2\ H^+(aq)$
$SeO_3^{2-}(aq) + H_2O(l) \rightarrow SeO_4^{2-}(aq) + 2\ H^+(aq) + 2\ e^-$
$SeO_3^{2-}(aq) + H_2O(l) + 2\ OH^-(aq) \rightarrow SeO_4^{2-}(aq) + 2\ H^+(aq) + 2\ OH^-(aq) + 2\ e^-$
$SeO_3^{2-}(aq) + H_2O(l) + 2\ OH^-(aq) \rightarrow SeO_4^{2-}(aq) + 2\ H_2O(l) + 2\ e^-$
$SeO_3^{2-}(aq) + 2\ OH^-(aq) \rightarrow SeO_4^{2-}(aq) + H_2O(l) + 2\ e^-$ (oxidation half reaction)

Combine the two half reactions.
$SeO_3^{2-}(aq) + Cl_2(g) + 2\ OH^-(aq) \rightarrow SeO_4^{2-}(aq) + 2\ Cl^-(aq) + H_2O(l)$

(c) $CoCl_2(aq) \rightarrow Co(OH)_3(s) + Cl^-(aq)$
$CoCl_2(aq) \rightarrow Co(OH)_3(s) + 2\ Cl^-(aq)$
$CoCl_2(aq) + 3\ H_2O(l) \rightarrow Co(OH)_3(s) + 2\ Cl^-(aq)$
$CoCl_2(aq) + 3\ H_2O(l) \rightarrow Co(OH)_3(s) + 2\ Cl^-(aq) + 3\ H^+(aq)$
$[CoCl_2(aq) + 3\ H_2O(l) \rightarrow Co(OH)_3(s) + 2\ Cl^-(aq) + 3\ H^+(aq) + e^-] \times 2$
$\qquad\qquad\qquad\qquad\qquad\qquad\qquad\qquad$ (oxidation half reaction)

43

$HO_2^-(aq) \rightarrow H_2O(l)$

$HO_2^-(aq) \rightarrow 2 H_2O(l)$

$3 H^+(aq) + HO_2^-(aq) \rightarrow 2 H_2O(l)$

$3 H^+(aq) + HO_2^-(aq) + 2 e^- \rightarrow 2 H_2O(l)$ (reduction half reaction)

Combine the two half reactions.

$2 CoCl_2(aq) + 6 H_2O(l) + 3 H^+(aq) + HO_2^-(aq) \rightarrow$
$\qquad\qquad 2 Co(OH)_3(s) + 4 Cl^-(aq) + 6 H^+(aq) + 2 H_2O(l)$

$2 CoCl_2(aq) + 4 H_2O(l) + HO_2^-(aq) \rightarrow 2 Co(OH)_3(s) + 4 Cl^-(aq) + 3 H^+(aq)$

$2 CoCl_2(aq) + 4 H_2O(l) + HO_2^-(aq) + 3 OH^-(aq) \rightarrow$
$\qquad\qquad 2 Co(OH)_3(s) + 4 Cl^-(aq) + 3 H^+(aq) + 3 OH^-(aq)$

$2 CoCl_2(aq) + 4 H_2O(l) + HO_2^-(aq) + 3 OH^-(aq) \rightarrow$
$\qquad\qquad 2 Co(OH)_3(s) + 4 Cl^-(aq) + 3 H_2O(l)$

$2 CoCl_2(aq) + H_2O(l) + HO_2^-(aq) + 3 OH^-(aq) \rightarrow 2 Co(OH)_3(s) + 4 Cl^-(aq)$

4.88   (a) TiO_2, insoluble   (b) $SnCl_4$, soluble   (c) Ag_2S, insoluble   (d) $Pd(NO_3)_2$, soluble

4.90   (a) C_2H_6 H +1,   C –3
       (b) $Na_2B_4O_7$ O –2,   Na +1,   B +3
       (c) Mg_2SiO_4 O –2,   Mg +2,   Si +4

4.92   Ions below Fe in the activity series (Table 4.3) can be reduced to their elemental forms by Fe. Of Ni^{2+}, Au^{3+}, Zn^{2+}, and Ba^{2+}, only Ni^{2+} and Au^{3+} can be reduced to their elemental forms by Fe.

4.94 $MgF_2(s) \rightleftarrows Mg^{2+}(aq) + 2 F^-(aq)$
                      x   2x

$[Mg^{2+}] = x = 2.6 \times 10^{-4}$ M and $[F^-] = 2x = 2(2.6 \times 10^{-4}$ M$) = 5.2 \times 10^{-4}$ M in a saturated solution.

$K_{sp} = [Mg^{2+}][F^-]^2 = (2.6 \times 10^{-4}$ M$)(5.2 \times 10^{-4}$ M$)^2 = 7.0 \times 10^{-11}$

4.96   (a) (1) $I^-(aq) \rightarrow I_3^-(aq)$

$\qquad\qquad 3 I^-(aq) \rightarrow I_3^-(aq)$

$\qquad\qquad 3 I^-(aq) \rightarrow I_3^-(aq) + 2 e^-$ (oxidation half reaction)

$\qquad HNO_2(aq) \rightarrow NO(g)$

$\qquad HNO_2(aq) \rightarrow NO(g) + H_2O(l)$

$\qquad H^+(aq) + HNO_2(aq) \rightarrow NO(g) + H_2O(l)$

$\qquad [e^- + H^+(aq) + HNO_2(aq) \rightarrow NO(g) + H_2O(l)] \times 2$ (reduction half reaction)

Combine the two half reactions.

$3 I^-(aq) + 2 H^+(aq) + 2 HNO_2(aq) \rightarrow I_3^-(aq) + 2 NO(g) + 2 H_2O(l)$

(2) $S_2O_3^{2-}(aq) \rightarrow S_4O_6^{2-}(aq)$
      $2\ S_2O_3^{2-}(aq) \rightarrow S_4O_6^{2-}(aq)$
      $2\ S_2O_3^{2-}(aq) \rightarrow S_4O_6^{2-}(aq) + 2\ e^-$ (oxidation half reaction)

      $I_3^-(aq) \rightarrow I^-(aq)$
      $I_3^-(aq) \rightarrow 3\ I^-(aq)$
      $2\ e^- + I_3^-(aq) \rightarrow 3\ I^-(aq)$ (reduction half reaction)

      Combine the two half reactions.
      $2\ S_2O_3^{2-}(aq) + I_3^-(aq) \rightarrow S_4O_6^{2-}(aq) + 3\ I^-(aq)$

(b)   18.77 mL = 0.018 77 L; NO_2^-, 46.01 amu

      $0.1500\ \dfrac{mol\ S_2O_3^{2-}}{1\ L} \times 0.018\ 77\ L = 0.002\ 815\ 5\ mol\ S_2O_3^{2-}$

      mass NO_2^- =

      $0.002\ 815\ 5\ mol\ S_2O_3^{2-} \times \dfrac{1\ mol\ I_3^-}{2\ mol\ S_2O_3^{2-}} \times \dfrac{2\ mol\ NO_2^-}{1\ mol\ I_3^-} \times \dfrac{46.01\ g\ NO_2^-}{1\ mol\ NO_2^-} =$

      $0.1295\ g\ NO_2^-$

      mass % NO_2^- = $\dfrac{0.1295\ g}{2.935\ g} \times 100\% = 4.412\%$

4.98   (a) Add HCl to precipitate Hg_2Cl_2. $Hg_2^{2+}(aq) + 2Cl^-(aq) \rightarrow Hg_2Cl_2(s)$
       (b) Add H_2SO_4 to precipitate $PbSO_4$. $Pb^{2+}(aq) + SO_4^{2-}(aq) \rightarrow PbSO_4(s)$
       (c) Add Na_2CO_3 to precipitate $CaCO_3$. $Ca^{2+}(aq) + CO_3^{2-}(aq) \rightarrow CaCO_3(s)$
       (d) Add Na_2SO_4 to precipitate $BaSO_4$. $Ba^{2+}(aq) + SO_4^{2-}(aq) \rightarrow BaSO_4(s)$

4.100 All four reactions are redox reactions.
       (a) $Mn(OH)_2(s) \rightarrow Mn(OH)_3(s)$
             $Mn(OH)_2(s) + OH^-(aq) \rightarrow Mn(OH)_3(s)$
             $[Mn(OH)_2(s) + OH^-(aq) \rightarrow Mn(OH)_3(s) + e^-] \times 2$ (oxidation half reaction)

             $H_2O_2(aq) \rightarrow 2\ H_2O(l)$
             $2\ H^+(aq) + H_2O_2(aq) \rightarrow 2\ H_2O(l)$
             $2\ e^- + 2\ H^+(aq) + H_2O_2(aq) \rightarrow 2\ H_2O(l)$
             $2\ e^- + 2\ OH^-(aq) + 2\ H^+(aq) + H_2O_2(aq) \rightarrow 2\ H_2O(l) + 2\ OH^-(aq)$
             $2\ e^- + 2\ H_2O(l) + H_2O_2(aq) \rightarrow 2\ H_2O(l) + 2\ OH^-(aq)$
             $2\ e^- + H_2O_2(aq) \rightarrow 2\ OH^-(aq)$ (reduction half reaction)

             Combine the two half reactions.
             $2\ Mn(OH)_2(s) + 2\ OH^-(aq) + H_2O_2(aq) \rightarrow 2\ Mn(OH)_3(s) + 2\ OH^-(aq)$
             $2\ Mn(OH)_2(s) + H_2O_2(aq) \rightarrow 2\ Mn(OH)_3(s)$

(b) $[MnO_4^{2-}(aq) \rightarrow MnO_4^-(aq) + e^-] \times 2$ (oxidation half reaction)

$MnO_4^{2-}(aq) \rightarrow MnO_2(s)$

$MnO_4^{2-}(aq) \rightarrow MnO_2(s) + 2 H_2O(l)$

$4 H^+(aq) + MnO_4^{2-}(aq) \rightarrow MnO_2(s) + 2 H_2O(l)$

$2 e^- + 4 H^+(aq) + MnO_4^{2-}(aq) \rightarrow MnO_2(s) + 2 H_2O(l)$ (reduction half reaction)

Combine the two half reactions.

$4 H^+(aq) + 3 MnO_4^{2-}(aq) \rightarrow MnO_2(s) + 2 MnO_4^-(aq) + 2 H_2O(l)$

(c) $I^-(aq) \rightarrow I_3^-(aq)$

$3 I^-(aq) \rightarrow I_3^-(aq)$

$[3 I^-(aq) \rightarrow I_3^-(aq) + 2 e^-] \times 8$ (oxidation half reaction)

$IO_3^-(aq) \rightarrow I_3^-(aq)$

$3 IO_3^-(aq) \rightarrow I_3^-(aq)$

$3 IO_3^-(aq) \rightarrow I_3^-(aq) + 9 H_2O(l)$

$18 H^+(aq) + 3 IO_3^-(aq) \rightarrow I_3^-(aq) + 9 H_2O(l)$

$16 e^- + 18 H^+(aq) + 3 IO_3^-(aq) \rightarrow I_3^-(aq) + 9 H_2O(l)$ (reduction half reaction)

Combine the two half reactions.

$24 I^-(aq) + 3 IO_3^-(aq) + 18 H^+(aq) \rightarrow 9 I_3^-(aq) + 9 H_2O(l)$

Divide all coefficients by 3.

$8 I^-(aq) + IO_3^-(aq) + 6 H^+(aq) \rightarrow 3 I_3^-(aq) + 3 H_2O(l)$

(d) $P(s) \rightarrow HPO_3^{2-}(aq)$

$3 H_2O(l) + P(s) \rightarrow HPO_3^{2-}(aq)$

$3 H_2O(l) + P(s) \rightarrow HPO_3^{2-}(aq) + 5 H^+(aq)$

$[3 H_2O(l) + P(s) \rightarrow HPO_3^{2-}(aq) + 5 H^+(aq) + 3 e^-] \times 2$

                          (oxidation half reaction)

$PO_4^{3-}(aq) \rightarrow HPO_3^{2-}(aq)$

$PO_4^{3-}(aq) \rightarrow HPO_3^{2-}(aq) + H_2O(l)$

$3 H^+(aq) + PO_4^{3-}(aq) \rightarrow HPO_3^{2-}(aq) + H_2O(l)$

$[2 e^- + 3 H^+(aq) + PO_4^{3-}(aq) \rightarrow HPO_3^{2-}(aq) + H_2O(l)] \times 3$

                          (reduction half reaction)

Combine the two half reactions and add OH$^-$.

$6 H_2O(l) + 2 P(s) + 9 H^+(aq) + 3 PO_4^{3-}(aq) \rightarrow$
                             $5 HPO_3^{2-}(aq) + 10 H^+(aq) + 3 H_2O(l)$

$3 H_2O(l) + 2 P(s) + 3 PO_4^{3-}(aq) \rightarrow 5 HPO_3^{2-}(aq) + H^+(aq)$

$3 H_2O(l) + 2 P(s) + 3 PO_4^{3-}(aq) + OH^-(aq) \rightarrow$
                             $5 HPO_3^{2-}(aq) + H^+(aq) + OH^-(aq)$

$3 H_2O(l) + 2 P(s) + 3 PO_4^{3-}(aq) + OH^-(aq) \rightarrow 5 HPO_3^{2-}(aq) + H_2O(l)$

$2 H_2O(l) + 2 P(s) + 3 PO_4^{3-}(aq) + OH^-(aq) \rightarrow 5 HPO_3^{2-}(aq)$

4.102 (a) $Cr^{2+}(aq) + Cr_2O_7^{2-}(aq) \rightarrow Cr^{3+}(aq)$
$[Cr^{2+}(aq) \rightarrow Cr^{3+}(aq) + e^-] \times 6$ (oxidation half reaction)

$Cr_2O_7^{2-}(aq) \rightarrow Cr^{3+}(aq)$
$Cr_2O_7^{2-}(aq) \rightarrow 2\,Cr^{3+}(aq)$
$Cr_2O_7^{2-}(aq) \rightarrow 2\,Cr^{3+}(aq) + 7\,H_2O(l)$
$14\,H^+(aq) + Cr_2O_7^{2-}(aq) \rightarrow 2\,Cr^{3+}(aq) + 7\,H_2O(l)$
$6\,e^- + 14\,H^+(aq) + Cr_2O_7^{2-}(aq) \rightarrow 2\,Cr^{3+}(aq) + 7\,H_2O(l)$ (reduction half reaction)

Combine the two half reactions.
$14\,H^+(aq) + Cr_2O_7^{2-}(aq) + 6\,Cr^{2+}(aq) \rightarrow 8\,Cr^{3+}(aq) + 7\,H_2O(l)$

(b) total volume = 100.0 ml + 20.0 mL = 120.0 mL = 0.1200 L
Initial moles:

$0.120\,\dfrac{mol\ Cr(NO_3)_2}{1\ L} \times 0.1000\ L = 0.0120\ mol\ Cr(NO_3)_2$

$0.500\,\dfrac{mol\ HNO_3}{1\ L} \times 0.1000\ L = 0.0500\ mol\ HNO_3$

$0.250\,\dfrac{mol\ K_2Cr_2O_7}{1\ L} \times 0.0200\ L = 0.005\,00\ mol\ K_2Cr_2O_7$

Check for the limiting reactant. 0.0120 mol of Cr^{2+} requires (0.0120)/6 = 0.00200 mol $Cr_2O_7^{2-}$ and (14/6)(0.0120) = 0.0280 mol H^+. Both are in excess of the required amounts, so Cr^{2+} is the limiting reactant.

	$14\,H^+(aq)$	$+\ Cr_2O_7^{2-}(aq)$	$+\ 6\,Cr^{2+}(aq)$	$\rightarrow 8\,Cr^{3+}(aq) + 7\,H_2O(l)$
Initial moles	0.0500	0.00500	0.0120	0
Change	−14x	−x	−6x	+8x

Because Cr^{2+} is the limiting reactant, 6x = 0.0120 and x = 0.00200

Final moles	0.0220	0.00300	0	0.00160

$mol\ K^+ = 0.00500\ mol\ K_2Cr_2O_7 \times \dfrac{2\ mol\ K^+}{1\ mol\ K_2Cr_2O_7} = 0.0100\ mol\ K^+$

$mol\ NO_3^- = 0.0120\ mol\ Cr(NO_3)_2 \times \dfrac{2\ mol\ NO_3^-}{1\ mol\ Cr(NO_3)_2}$

$+\ 0.0500\ mol\ HNO_3 \times \dfrac{1\ mol\ NO_3^-}{1\ mol\ HNO_3} = 0.0740\ mol\ NO_3^-$

$mol\ H^+ = 0.0220\ mol$; $mol\ Cr_2O_7^{2-} = 0.00300\ mol$; $mol\ Cr^{3+} = 0.01600\ mol$
Check for charge neutrality.
Total moles of +charge = 0.0100 + 0.0220 + 3 x (0.01600) = 0.0800 mol +charge
Total moles of −charge = 0.0740 + 2 x (0.00300) = 0.0800 mol −charge
The charges balance and there is electrical neutrality in the solution after the reaction.

$$K^+ \text{ molarity} = \frac{0.0100 \text{ mol } K^+}{0.1200 \text{ L}} = 0.0833 \text{ M}$$

$$NO_3^- \text{ molarity} = \frac{0.0740 \text{ mol } NO_3^-}{0.1200 \text{ L}} = 0.617 \text{ M}$$

$$H^+ \text{ molarity} = \frac{0.0220 \text{ mol } H^+}{0.1200 \text{ L}} = 0.183 \text{ M}$$

$$Cr_2O_7^{2-} \text{ molarity} = \frac{0.00300 \text{ mol } Cr_2O_7^{2-}}{0.1200 \text{ L}} = 0.0250 \text{ M}$$

$$Cr^{3+} \text{ molarity} = \frac{0.0160 \text{ mol } Cr^{3+}}{0.1200 \text{ L}} = 0.133 \text{ M}$$

5 Periodicity and Atomic Structure

5.1   Gamma ray $\nu = \dfrac{c}{\lambda} = \dfrac{3.00 \times 10^8 \text{ m/s}}{3.56 \times 10^{-11} \text{ m}} = 8.43 \times 10^{18} \text{ s}^{-1} = 8.43 \times 10^{18} \text{ Hz}$

          Radar wave $\nu = \dfrac{c}{\lambda} = \dfrac{3.00 \times 10^8 \text{ m/s}}{10.3 \times 10^{-2} \text{ m}} = 2.91 \times 10^9 \text{ s}^{-1} = 2.91 \times 10^9 \text{ Hz}$

5.2 $\nu = 102.5 \text{ MHz} = 102.5 \times 10^6 \text{ Hz} = 102.5 \times 10^6 \text{ s}^{-1}$

        $\lambda = \dfrac{c}{\nu} = \dfrac{3.00 \times 10^8 \text{ m/s}}{102.5 \times 10^6 \text{ s}^{-1}} = 2.93 \text{ m}$

        $\nu = 9.55 \times 10^{17} \text{ Hz} = 9.55 \times 10^{17} \text{ s}^{-1}$

        $\lambda = \dfrac{c}{\nu} = \dfrac{3.00 \times 10^8 \text{ m/s}}{9.55 \times 10^{17} \text{ s}^{-1}} = 3.14 \times 10^{-10} \text{ m}$

5.3   Violet has the highest frequency. Red has the longest wavelength.

5.4   Balmer series: m = 2; R = $1.097 \times 10^{-2} \text{ nm}^{-1}$

        $\dfrac{1}{\lambda} = R\left[\dfrac{1}{m^2} - \dfrac{1}{n^2}\right]$; $\dfrac{1}{\lambda} = R\left[\dfrac{1}{2^2} - \dfrac{1}{7^2}\right]$; $\dfrac{1}{\lambda} = 2.519 \times 10^{-3} \text{ nm}^{-1}$; $\lambda = 397.0 \text{ nm}$

5.5   Paschen series: m = 3; R = $1.097 \times 10^{-2} \text{ nm}^{-1}$

        $\dfrac{1}{\lambda} = R\left[\dfrac{1}{m^2} - \dfrac{1}{n^2}\right]$; $\dfrac{1}{\lambda} = R\left[\dfrac{1}{3^2} - \dfrac{1}{4^2}\right]$; $\dfrac{1}{\lambda} = 5.333 \times 10^{-4} \text{ nm}^{-1}$; $\lambda - 1875 \text{ nm}$

5.6   Paschen series: m = 3; R = $1.097 \times 10^{-2} \text{ nm}^{-1}$

        $\dfrac{1}{\lambda} = R\left[\dfrac{1}{m^2} - \dfrac{1}{n^2}\right]$; $\dfrac{1}{\lambda} = R\left[\dfrac{1}{3^2} - \dfrac{1}{\infty^2}\right]$; $\dfrac{1}{\lambda} = 1.219 \times 10^{-3} \text{ nm}^{-1}$; $\lambda = 820.4 \text{ nm}$

5.7 $\lambda = 91.2 \text{ nm} = 91.2 \times 10^{-9} \text{ m}$

        $\nu = \dfrac{c}{\lambda} = \dfrac{3.00 \times 10^8 \text{ m/s}}{91.2 \times 10^{-9} \text{ m}} = 3.29 \times 10^{15} \text{ s}^{-1}$

        $E = h\nu = (6.626 \times 10^{-34} \text{ J·s})(3.29 \times 10^{15} \text{ s}^{-1}) = 2.18 \times 10^{-18} \text{ J/photon}$

        $E = (2.18 \times 10^{-18} \text{ J/photon})(6.022 \times 10^{23} \text{ photons/mol}) = 1.31 \times 10^6 \text{ J/mol} = 1310 \text{ kJ/mol}$

5.8   IR, $\lambda = 1.55 \times 10^{-6} \text{ m}$

        $E = h\dfrac{c}{\lambda} = (6.626 \times 10^{-34} \text{ J·s})\left(\dfrac{3.00 \times 10^8 \text{ m/s}}{1.55 \times 10^{-6} \text{ m}}\right)(6.022 \times 10^{23} / \text{mol})$

        $E = 7.72 \times 10^4 \text{ J/mol} = 77.2 \text{ kJ/mol}$

UV, $\quad \lambda = 250 \text{ nm} = 250 \times 10^{-9} \text{ m}$

$$E = h\frac{c}{\lambda} = (6.626 \times 10^{-34} \text{ J·s}) \left(\frac{3.00 \times 10^8 \text{ m/s}}{250 \times 10^{-9} \text{ m}} \right) (6.022 \times 10^{23} \text{ /mol})$$

$E = 4.79 \times 10^5 \text{ J/mol} = 479 \text{ kJ/mol}$

X ray, $\quad \lambda = 5.49 \text{ nm} = 5.49 \times 10^{-9} \text{ m}$

$$E = h\frac{c}{\lambda} = (6.626 \times 10^{-34} \text{ J·s}) \left(\frac{3.00 \times 10^8 \text{ m/s}}{5.49 \times 10^{-9} \text{ m}} \right) (6.022 \times 10^{23} \text{ /mol})$$

$E = 2.18 \times 10^7 \text{ J/mol} = 2.18 \times 10^4 \text{ kJ/mol}$

5.9 $\quad \lambda = \dfrac{h}{mv} = \dfrac{6.626 \times 10^{-34} \text{ kg m}^2 \text{ s}^{-1}}{(1150 \text{ kg})(24.6 \text{ m/s})} = 2.34 \times 10^{-38} \text{ m}$

5.10 $\quad (\Delta x)(\Delta mv) \geq \dfrac{h}{4\pi}$; $\quad$ uncertainty in velocity $= (45 \text{ m/s})(0.02) = 0.9 \text{ m/s}$

$$\Delta x \geq \frac{h}{4\pi(\Delta mv)} = \frac{6.626 \times 10^{-34} \text{ kg m}^2 \text{ s}^{-1}}{4\pi(0.120 \text{ kg})(0.9 \text{ m/s})} = 5 \times 10^{-34} \text{ m}$$

5.11

n	l	m_l	Orbital	No. of Orbitals
5	0	0	5s	1
	1	−1, 0, +1	5p	3
	2	−2, −1, 0, +1, +2	5d	5
	3	−3, −2, −1, 0, +1, +2, +3	5f	7
	4	−4, −3, −2, −1, 0, +1, +2, +3, +4	5g	9

There are 25 possible orbitals in the fifth shell.

5.12 (a) For $n = 2$, $l = 0$ or 1 $\qquad$ (b) For $l = 0$, $m_l = 0$
(c) The l quantum number is a positive integer or zero.

5.13 (a) 2p $\qquad$ (b) 4f $\qquad$ (c) 3d

5.14 (a) 3s orbital: $n = 3$, $l = 0$, $m_l = 0$
(b) 2p orbital: $n = 2$, $l = 1$, $m_l = -1, 0, +1$
(c) 4d orbital: $n = 4$, $l = 2$, $m_l = -2, -1, 0, +1, +2$

5.15 $\quad m = 1$, $n = \infty$; $\quad R = 1.097 \times 10^{-2} \text{ nm}^{-1}$

$$\frac{1}{\lambda} = R\left[\frac{1}{m^2} - \frac{1}{n^2} \right]; \quad \frac{1}{\lambda} = R\left[\frac{1}{1^2} - \frac{1}{\infty^2} \right]; \quad \frac{1}{\lambda} = R\left[\frac{1}{1} \right] = 1.097 \times 10^{-2} \text{ nm}^{-1}; \quad \lambda = 91.2 \text{ nm}$$

$$E = (6.626 \times 10^{-34} \text{ J·s}) \left(\frac{3.00 \times 10^8 \text{ m/s}}{91.2 \times 10^{-9} \text{ m}} \right) (6.022 \times 10^{23} \text{/mol})$$

$E = 1.31 \times 10^6 \text{ J/mol} = 1.31 \times 10^3 \text{ kJ/mol}$

5.16   The g orbitals have four nodal planes.

5.17   (a) Ti, $1s^22s^22p^63s^23p^64s^23d^2$ or [Ar] $4s^23d^2$

[Ar] $\underline{\uparrow\downarrow}$ $\underline{\uparrow}$ $\underline{\uparrow}$ $\underline{}$ $\underline{}$ $\underline{}$
     4s   3d

(b) Zn, $1s^22s^22p^63s^23p^64s^23d^{10}$ or [Ar] $4s^23d^{10}$

[Ar] $\underline{\uparrow\downarrow}$ $\underline{\uparrow\downarrow}$ $\underline{\uparrow\downarrow}$ $\underline{\uparrow\downarrow}$ $\underline{\uparrow\downarrow}$ $\underline{\uparrow\downarrow}$
     4s   3d

(c) Sn, $1s^22s^22p^63s^23p^64s^23d^{10}4p^65s^24d^{10}5p^2$ or [Kr] $5s^24d^{10}5p^2$

[Kr] $\underline{\uparrow\downarrow}$ $\underline{\uparrow\downarrow}$ $\underline{\uparrow\downarrow}$ $\underline{\uparrow\downarrow}$ $\underline{\uparrow\downarrow}$ $\underline{\uparrow\downarrow}$ $\underline{\uparrow}$ $\underline{\uparrow}$ $\underline{}$
     5s   4d   5p

(d) Pb,   [Xe] $6s^24f^{14}5d^{10}6p^2$

5.18   For Na^+, $1s^22s^22p^6$

5.19   Cr, Cu, Nb, Mo, Ru, Rh, Pd, Ag, La, Ce, Gd, Pt, Au, Ac, Th, Pa, U, Np, Cm

5.20   (a) Ra^{2+} [Rn]   (b) La^{3+} [Xe]   (c) Ti^{4+} [Ar]   (d) N^{3-} [Ne]
Each ion has the ground–state electron configuration of the noble gas closest to it in the periodic table.

5.21   The neutral atom contains 30 e^- and is Zn. The ion is Zn^{2+}.

5.22   (a) Ba; atoms get larger as you go down a group.
(b) W; atoms get smaller as you go across a period.
(c) Sn; atoms get larger as you go down a group.
(d) Ce; atoms get smaller as you go across a period.

5.23   (a) O^{2-}; decrease in effective nuclear charge and an increase in electron–electron repulsions lead to the larger anion.
(b) S; atoms get larger as you go down a group.
(c) Fe; in Fe^{3+} electrons are removed from a larger valence shell and there is an increase in effective nuclear charge leading to the smaller cation.
(d) H^-; decrease in effective nuclear charge and an increase in electron–electron repulsions lead to the larger anion.

Understanding Key Concepts

5.24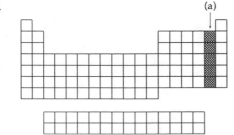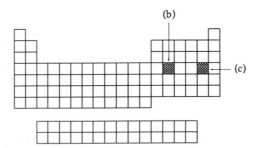

5.25

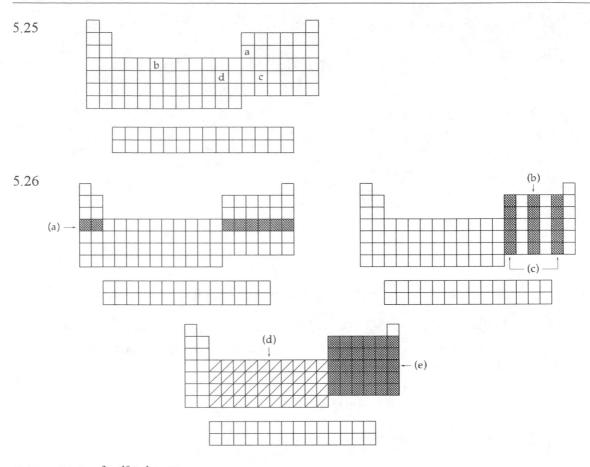

5.26

5.27   [Ar] $4s^23d^{10}4p^1$ is Ga.

5.28   (a) K,   [Ar]   ↑
                              4s

         (b) Mo,   [Kr]   ↑   ↑ ↑ ↑ ↑ ↑
                              5s   4d

         (c) Sn,   [Kr]   ↑↓   ↑↓ ↑↓ ↑↓ ↑↓ ↑↓   ↑ ↑ __
                              5s   4d   5p

         (d) Ir,   [Xe]   ↑↓   ↑↓ ↑↓ ↑↓ ↑↓ ↑↓ ↑↓ ↑↓   ↑↓ ↑↓ ↑ ↑ ↑
                              6s   4f   5d

5.29 K^+, r = 133 pm; Cl^-, r = 184 pm; K, r = 227 pm

5.30   Z = 118   [Rn] $7s^25f^{14}6d^{10}7p^6$

5.31   Order of orbital filling:
        1s→2s→2p→3s→3p→4s→3d→4p→5s→4d→5p→6s→4f→5d→6p→7s→5f→6d→7p→8s→5g
        Z = 121

5.32   (a) $3p_y$ n = 3, l = 1   (b) $4d_{z^2}$ n = 4, l = 2

5.33 The first sphere gets larger on going from reactant to product. This is consistent with it being a nonmetal gaining an electron and becoming an anion. The second sphere gets smaller on going from reactant to product. This is consistent with it being a metal losing an electron and becoming a cation.

Additional Problems
Electromagnetic Radiation

5.34 Violet has the higher frequency and energy. Red has the higher wavelength.

5.36 $\lambda = \dfrac{c}{\nu} = \dfrac{3.00 \times 10^8 \text{ m/s}}{5.5 \times 10^{15} \text{ s}^{-1}} = 5.5 \times 10^{-8} \text{ m}$

5.38 (a) $\nu = 99.5 \text{ MHz} = 99.5 \times 10^6 \text{ s}^{-1}$
$E = h\nu = (6.626 \times 10^{-34} \text{ J·s})(99.5 \times 10^6 \text{ s}^{-1})(6.022 \times 10^{23} \text{ /mol})$
$E = 3.97 \times 10^{-2} \text{ J/mol} = 3.97 \times 10^{-5} \text{ kJ/mol}$
$\nu = 115.0 \text{ kHz} = 115.0 \times 10^3 \text{ s}^{-1}$
$E = h\nu = (6.626 \times 10^{-34} \text{ J·s})(115.0 \times 10^3 \text{ s}^{-1})(6.022 \times 10^{23} \text{ /mol})$
$E = 4.589 \times 10^{-5} \text{ J/mol} = 4.589 \times 10^{-8} \text{ kJ/mol}$
The FM radio wave (99.5 MHz) has the higher energy.

(b) $\lambda = 3.44 \times 10^{-9} \text{ m}$

$E = h\dfrac{c}{\lambda} = (6.626 \times 10^{-34} \text{ J·s})\left(\dfrac{3.00 \times 10^8 \text{ m/s}}{3.44 \times 10^{-9} \text{ m}}\right)(6.022 \times 10^{23}/\text{mol})$

$E = 3.48 \times 10^7 \text{ J/mol} = 3.48 \times 10^4 \text{ kJ/mol}$
$\lambda = 6.71 \times 10^{-2} \text{ m}$

$E = h\dfrac{c}{\lambda} = (6.626 \times 10^{-34} \text{ J·s})\left(\dfrac{3.00 \times 10^8 \text{ m/s}}{6.71 \times 10^{-2} \text{ m}}\right)(6.022 \times 10^{23}/\text{mol})$

$E = 1.78 \text{ J/mol} = 1.78 \times 10^{-3} \text{ kJ/mol}$
The X ray ($\lambda = 3.44 \times 10^{-9}$ m) has the higher energy.

5.40 (a) $E = 90.5 \text{ kJ/mol} \times \dfrac{1000 \text{ J}}{1 \text{ kJ}} \times \dfrac{1 \text{ mol}}{6.02 \times 10^{23}} = 1.50 \times 10^{-19} \text{ J}$

$\nu = \dfrac{E}{h} = \dfrac{1.50 \times 10^{-19} \text{ J}}{6.626 \times 10^{-34} \text{ J·s}} = 2.27 \times 10^{14} \text{ s}^{-1}$

$\lambda = \dfrac{c}{\nu} = \dfrac{3.00 \times 10^8 \text{ m/s}}{2.27 \times 10^{14} \text{ s}^{-1}} = 1.32 \times 10^{-6} \text{ m} = 1320 \times 10^{-9} \text{ m} = 1320 \text{ nm, near IR}$

(b) $E = 8.05 \times 10^{-4} \text{ kJ/mol} \times \dfrac{1000 \text{ J}}{1 \text{ kJ}} \times \dfrac{1 \text{ mol}}{6.02 \times 10^{23}} = 1.34 \times 10^{-24} \text{ J}$

$\nu = \dfrac{E}{h} = \dfrac{1.34 \times 10^{-24} \text{ J}}{6.626 \times 10^{-34} \text{ J·s}} = 2.02 \times 10^9 \text{ s}^{-1}$

$$\lambda = \frac{c}{\nu} = \frac{3.00 \times 10^8 \text{ m/s}}{2.02 \times 10^9 \text{ s}^{-1}} = 0.149 \text{ m, radio wave}$$

(c) $E = 1.83 \times 10^3 \text{ kJ/mol} \times \dfrac{1000 \text{ J}}{1 \text{ kJ}} \times \dfrac{1 \text{ mol}}{6.02 \times 10^{23}} = 3.04 \times 10^{-18} \text{ J}$

$$\nu = \frac{E}{h} = \frac{3.04 \times 10^{-18} \text{ J}}{6.626 \times 10^{-34} \text{ J} \cdot \text{s}} = 4.59 \times 10^{15} \text{ s}^{-1}$$

$$\lambda = \frac{c}{\nu} = \frac{3.00 \times 10^8 \text{ m/s}}{4.59 \times 10^{15} \text{ s}^{-1}} = 6.54 \times 10^{-8} \text{ m} = 65.4 \times 10^{-9} \text{ m} = 65.4 \text{ nm, UV}$$

5.42 $\lambda = \dfrac{h}{mv} = \dfrac{6.626 \times 10^{-34} \text{ kg m}^2 \text{ s}^{-1}}{(9.11 \times 10^{-31} \text{ kg})(0.99 \times 3.00 \times 10^8 \text{ m/s})} = 2.45 \times 10^{-12} \text{ m, } \gamma \text{ ray}$

5.44 $156 \text{ km/h} = 156 \times 10^3 \text{ m}/3600 \text{ s} = 43.3 \text{ m/s}$; $145 \text{ g} = 0.145 \text{ kg}$

$$\lambda = \frac{h}{mv} = \frac{6.626 \times 10^{-34} \text{ kg m}^2 \text{ s}^{-1}}{(0.145 \text{ kg})(43.3 \text{ m/s})} = 1.06 \times 10^{-34} \text{ m}$$

5.46 $145 \text{ g} = 0.145 \text{ kg}$; $0.500 \text{ nm} = 0.500 \times 10^{-9} \text{ m}$

$$\lambda = \frac{h}{mv} = \frac{6.626 \times 10^{-34} \text{ kg m}^2 \text{ s}^{-1}}{(0.145 \text{ kg})(0.500 \times 10^{-9} \text{ m})} = 9.14 \times 10^{-24} \text{ m/s}$$

Atomic Spectra

5.48   For $n = 3$; $\lambda = 656.3 \text{ nm} = 656.3 \times 10^{-9} \text{ m}$

$$E = h\frac{c}{\lambda} = (6.626 \times 10^{-34} \text{ J} \cdot \text{s})\left(\frac{2.998 \times 10^8 \text{ m/s}}{656.3 \times 10^{-9} \text{ m}}\right)\left(\frac{1 \text{ kJ}}{1000 \text{ J}}\right)(6.022 \times 10^{23}/\text{mol})$$

$E = 182.3 \text{ kJ/mol}$

For $n = 4$; $\lambda = 486.1 \text{ nm} = 486.1 \times 10^{-9} \text{ m}$

$$E = h\frac{c}{\lambda} = (6.626 \times 10^{-34} \text{ J} \cdot \text{s})\left(\frac{2.998 \times 10^8 \text{ m/s}}{486.1 \times 10^{-9} \text{ m}}\right)\left(\frac{1 \text{ kJ}}{1000 \text{ J}}\right)(6.022 \times 10^{23}/\text{mol})$$

$E = 246.1 \text{ kJ/mol}$

For $n = 5$; $\lambda = 434.0 \text{ nm} = 434.0 \times 10^{-9} \text{ m}$

$$E = h\frac{c}{\lambda} = (6.626 \times 10^{-34} \text{ J} \cdot \text{s})\left(\frac{2.998 \times 10^8 \text{ m/s}}{434.0 \times 10^{-9} \text{ m}}\right)\left(\frac{1 \text{ kJ}}{1000 \text{ J}}\right)(6.022 \times 10^{23}/\text{mol})$$

$E = 275.6 \text{ kJ/mol}$

5.50 From problem 5.49, for $n = \infty$, $\lambda = 364.6$ nm $= 364.6 \times 10^{-9}$ m

$$E = h\frac{c}{\lambda} = (6.626 \times 10^{-34} \text{ J} \cdot \text{s})\left(\frac{2.998 \times 10^8 \text{ m/s}}{364.6 \times 10^{-9} \text{ m}}\right)\left(\frac{1 \text{ kJ}}{1000 \text{ J}}\right)(6.022 \times 10^{23}/\text{mol})$$

$E = 328.1$ kJ/mol

5.52 $\lambda = 330$ nm $= 330 \times 10^{-9}$ m

$$E = h\frac{c}{\lambda} = (6.626 \times 10^{-34} \text{ J} \cdot \text{s})\left(\frac{3.00 \times 10^8 \text{ m/s}}{330 \times 10^{-9} \text{ m}}\right)\left(\frac{1 \text{ kJ}}{1000 \text{ J}}\right)(6.022 \times 10^{23}/\text{mol})$$

$E = 363$ kJ/mol

Orbitals and Quantum Numbers

5.54 n is the principal quantum number. The size and energy level of an orbital depends on n. l is the angular–momentum quantum number. l defines the three–dimensional shape of an orbital.
m_l is the magnetic quantum number. m_l defines the spatial orientation of an orbital.
m_s is the spin quantum number. m_s indicates the spin of the electron and can have either of two values, $+\frac{1}{2}$ or $-\frac{1}{2}$.

5.56 The probability of finding the electron drops off rapidly as distance from the nucleus increases, although it never drops to zero, even at large distances. As a result, there is no definite boundary or size for an orbital. However, we usually imagine the boundary surface of an orbital enclosing the volume where an electron spends 95% of its time.

5.58 Part of the electron–nucleus attraction is canceled by the electron–electron repulsion, an effect we describe by saying that the electrons are shielded from the nucleus by the other electrons. The net nuclear charge actually felt by an electron is called the effective nuclear charge, Z_{eff}, and is often substantially lower than the actual nuclear charge, Z_{actual}.
$$Z_{eff} = Z_{actual} - \text{electron shielding}$$

5.60 $l = 2$ for d subshells; $l = 3$ for f subshells

5.62 (a) 4s $n = 4$; $l = 0$; $m_l = 0$; $m_s = \pm\frac{1}{2}$
(b) 3p $n = 3$; $l = 1$; $m_l = -1, 0, +1$; $m_s = \pm\frac{1}{2}$
(c) 5f $n = 5$; $l = 3$; $m_l = -3, -2, -1, 0, +1, +2, +3$; $m_s = \pm\frac{1}{2}$
(d) 5d $n = 5$; $l = 2$; $m_l = -2, -1, 0, +1, +2$; $m_s = \pm\frac{1}{2}$

5.64 (a) is not allowed because for $l = 0$, $m_l = 0$ only.
(b) is allowed.
(c) is not allowed because for $n = 4$, $l = 0, 1, 2,$ or 3 only.

5.66 For $n = 5$, the maximum number of electrons will occur when the 5g orbital is filled:
[Rn] $7s^2 5f^{14} 6d^{10} 7p^6 8s^2 5g^{18} = 138$ electrons

5.68 $0.68 \text{ g} = 0.68 \times 10^{-3} \text{ kg}$

$$(\Delta x)(\Delta mv) \geq \frac{h}{4\pi}; \quad \Delta x \geq \frac{h}{4\pi(\Delta mv)} = \frac{6.626 \times 10^{-34} \text{ kg m}^2 \text{ s}^{-1}}{4\pi(0.68 \times 10^{-3} \text{ kg})(0.1 \text{ m/s})} = 8 \times 10^{-31} \text{ m}$$

Electron Configurations

5.70   The number of elements in successive periods of the periodic table increases by the progression 2, 8, 18, 32 because the principal quantum number n increases by 1 from one period to the next. As the principal quantum number increases, the number of orbitals in a shell increases. The progression of elements parallels the number of electrons in a particular shell.

5.72   (a) 5d   (b) 4s   (c) 6s

5.74   (a) 3d after 4s   (b) 4p after 3d   (c) 6d after 5f   (d) 6s after 5p

5.76   (a) Ti, $Z = 22$ $1s^2 2s^2 2p^6 3s^2 3p^6 4s^2 3d^2$
        (b) Ru, $Z = 44$ $1s^2 2s^2 2p^6 3s^2 3p^6 4s^2 3d^{10} 4p^6 5s^2 4d^6$
        (c) Sn, $Z = 50$ $1s^2 2s^2 2p^6 3s^2 3p^6 4s^2 3d^{10} 4p^6 5s^2 4d^{10} 5p^2$
        (d) Sr, $Z = 38$ $1s^2 2s^2 2p^6 3s^2 3p^6 4s^2 3d^{10} 4p^6 5s^2$
        (e) Se, $Z = 34$ $1s^2 2s^2 2p^6 3s^2 3p^6 4s^2 3d^{10} 4p^4$

5.78   (a) Rb, $Z = 37$ [Kr] $\underset{5s}{\uparrow}$

        (b) W, $Z = 74$ [Xe] $\underset{6s}{\uparrow\downarrow}$ $\underset{4f}{\uparrow\downarrow \; \uparrow\downarrow \; \uparrow\downarrow \; \uparrow\downarrow \; \uparrow\downarrow \; \uparrow\downarrow \; \uparrow\downarrow}$ $\underset{5d}{\uparrow \; \uparrow \; \uparrow \; \uparrow \; \_}$

        (c) Ge, $Z = 32$ [Ar] $\underset{4s}{\uparrow\downarrow}$ $\underset{3d}{\uparrow\downarrow \; \uparrow\downarrow \; \uparrow\downarrow \; \uparrow\downarrow \; \uparrow\downarrow}$ $\underset{4p}{\uparrow \; \uparrow \; \_}$

        (d) Zr, $Z = 40$ [Kr] $\underset{5s}{\uparrow\downarrow}$ $\underset{4d}{\uparrow \; \uparrow \; \_ \; \_ \; \_}$

5.80   4s > 4d > 4f

5.82 $Z = 118$ [Rn] $7s^2 5f^{14} 6d^{10} 7p^6$

5.84   (a) O $1s^2 2s^2 2p^4$ $\underset{2p}{\uparrow\downarrow \; \uparrow \; \uparrow}$ 2 unpaired e^-

        (b) Si $1s^2 2s^2 2p^6 3s^2 3p^2$ $\underset{3p}{\uparrow \; \uparrow \; \_}$ 2 unpaired e^-

        (c) K   [Ar] $4s^1$ 1 unpaired e^-

        (d) As   [Ar] $4s^2 3d^{10} 4p^3$ $\underset{4p}{\uparrow \; \uparrow \; \uparrow}$ 3 unpaired e^-

5.86   (a) $Z = 31$, Ga   (b) $Z = 46$, Pd

5.88 (a) La^{3+}, [Xe]   (b) Ag^+, [Kr] $4d^{10}$ (c) Sn^{2+}, [Kr] $5s^2 4d^{10}$

Atomic Radii and Periodic Properties

5.90 Atomic radii increase down a group because the electron shells are farther away from the nucleus.

5.92 F < O < S

5.94 Mg has a higher ionization energy than Na because Mg has a higher Z_{eff} and a smaller size.

5.96 Cu^{2+} has fewer electrons and a larger effective nuclear charge; therefore it has the smaller ionic radius.

General Problems

5.98 Balmer series: m = 2; R = 1.097 x 10^{-2} nm^{-1}

$$\frac{1}{\lambda} = R\left[\frac{1}{m^2} - \frac{1}{n^2}\right]; \quad \frac{1}{\lambda} = R\left[\frac{1}{2^2} - \frac{1}{6^2}\right] = 2.438 \times 10^{-3} \ nm^{-1}$$

λ = 410.2 nm = 410.2 x 10^{-9} m

$$E = h\frac{c}{\lambda} = (6.626 \times 10^{-34} \ J \cdot s)\left(\frac{2.998 \times 10^8 \ m/s}{410.2 \times 10^{-9} \ m}\right)\left(\frac{1 \ kJ}{1000 \ J}\right)(6.022 \times 10^{23}/mol)$$

E = 291.6 kJ/mol

5.100 Pfund series: m = 5; R = 1.097 x 10^{-2} nm^{-1}

$$\frac{1}{\lambda} = R\left[\frac{1}{m^2} - \frac{1}{n^2}\right]$$

n = 6, $\frac{1}{\lambda} = R\left[\frac{1}{5^2} - \frac{1}{6^2}\right]$ = 1.341 x 10^{-4} nm^{-1}; λ = 7458 nm = 7458 x 10^{-9} m

$$E = h\frac{c}{\lambda} = (6.626 \times 10^{-34} \ J \cdot s)\left(\frac{2.998 \times 10^8 \ m/s}{7458 \times 10^{-9} \ m}\right)\left(\frac{1 \ kJ}{1000 \ J}\right)(6.022 \times 10^{23}/mol)$$

E = 16.04 kJ/mol

n = 7, $\frac{1}{\lambda} = R\left[\frac{1}{5^2} - \frac{1}{7^2}\right]$ = 2.149 x 10^{-4} nm^{-1}; λ = 4653 nm = 4653 x 10^{-9} m

$$E = h\frac{c}{\lambda} = (6.626 \times 10^{-34} \ J \cdot s)\left(\frac{2.998 \times 10^8 \ m/s}{4653 \times 10^{-9} \ m}\right)\left(\frac{1 \ kJ}{1000 \ J}\right)(6.022 \times 10^{23}/mol)$$

E = 25.71 kJ/mol

These lines in the Pfund series are in the infrared region of the electromagnetic spectrum.

5.102 (a) $E = \left(142 \dfrac{kJ}{mol}\right)\left(\dfrac{1000\,J}{1\,kJ}\right)\left(\dfrac{1\,mol}{6.02 \times 10^{23}}\right) = 2.36 \times 10^{-19}\,J$

$E = h\dfrac{c}{\lambda}, \quad \lambda = \dfrac{hc}{E} = \dfrac{(6.626 \times 10^{-34}\,J\cdot s)(3.00 \times 10^{8}\,m/s)}{2.36 \times 10^{-19}\,J}$

$\lambda = 8.42 \times 10^{-7}\,m \qquad$ (infrared)

(b) $E = \left(4.55 \times 10^{-2} \dfrac{kJ}{mol}\right)\left(\dfrac{1000\,J}{1\,kJ}\right)\left(\dfrac{1\,mol}{6.02 \times 10^{23}}\right) = 7.56 \times 10^{-23}\,J$

$E = h\dfrac{c}{\lambda}, \quad \lambda = \dfrac{hc}{E} = \dfrac{(6.626 \times 10^{-34}\,J\cdot s)(3.00 \times 10^{8}\,m/s)}{7.56 \times 10^{-23}\,J}$

$\lambda = 2.63 \times 10^{-3}\,m \qquad$ (microwave)

(c) $E = \left(4.81 \times 10^{4} \dfrac{kJ}{mol}\right)\left(\dfrac{1000\,J}{1\,kJ}\right)\left(\dfrac{1\,mol}{6.02 \times 10^{23}}\right) = 7.99 \times 10^{-17}\,J$

$E = h\dfrac{c}{\lambda}, \quad \lambda = \dfrac{hc}{E} = \dfrac{(6.626 \times 10^{-34}\,J\cdot s)(3.00 \times 10^{8}\,m/s)}{7.99 \times 10^{-17}\,J}$

$\lambda = 2.49 \times 10^{-9}\,m \qquad$ (X ray)

5.104 $206.5\,kJ = 206.5 \times 10^{3}\,J; \qquad E = \dfrac{206.5 \times 10^{3}\,J}{1\,mol} \times \dfrac{1\,mol}{6.022 \times 10^{23}} = 3.429 \times 10^{-19}\,J$

$E = h\dfrac{c}{\lambda}, \quad \lambda = \dfrac{hc}{E} = \dfrac{(6.626 \times 10^{-34}\,J\cdot s)(3.00 \times 10^{8}\,m/s)}{3.429 \times 10^{-19}\,J} = 5.797 \times 10^{-7}\,m = 580.\,nm$

5.106 (a) Sr, $Z = 38$ [Kr] $\underline{\uparrow\downarrow}$
5s

(b) Cd, $Z = 48$ [Kr] $\underline{\uparrow\downarrow}$ $\underline{\uparrow\downarrow}\;\underline{\uparrow\downarrow}\;\underline{\uparrow\downarrow}\;\underline{\uparrow\downarrow}\;\underline{\uparrow\downarrow}$
5s   4d

(c) $Z = 22$, Ti   [Ar] $\underline{\uparrow\downarrow}$ $\underline{\uparrow}\;\underline{\uparrow}\;\underline{\;}\;\underline{\;}\;\underline{\;}$
4s   3d

(d) $Z = 34$, Se   [Ar] $\underline{\uparrow\downarrow}$ $\underline{\uparrow\downarrow}\;\underline{\uparrow\downarrow}\;\underline{\uparrow\downarrow}\;\underline{\uparrow\downarrow}\;\underline{\uparrow\downarrow}$ $\underline{\uparrow\downarrow}\;\underline{\uparrow}\;\underline{\uparrow}$
4s   3d   4p

5.108 For K, $Z_{eff} = \sqrt{\dfrac{(418.8\,kJ/mol)(4^{2})}{1312\,kJ/mol}} = 2.26$

For Kr, $Z_{eff} = \sqrt{\dfrac{(1350.7\,kJ/mol)(4^{2})}{1312\,kJ/mol}} = 4.06$

5.110 $q = (350\ \text{g})(4.184\ \text{J/g·°C})(95°C - 20°C) = 109,830\ \text{J}$
$\lambda = 15.0\ \text{cm} = 15.0 \times 10^{-2}\ \text{m}$

$$E = (6.626 \times 10^{-34}\ \text{J·s})\left(\frac{3.00 \times 10^8\ \text{m/s}}{15.0 \times 10^{-2}\ \text{m}}\right) = 1.33 \times 10^{-24}\ \text{J/photon}$$

$$\text{number of photons} = \frac{109,830\ \text{J}}{1.33 \times 10^{-24}\ \text{J/photon}} = 8.3 \times 10^{28}\ \text{photons}$$

5.112 $48.2\ \text{nm} = 48.2 \times 10^{-9}\ \text{m}$

$$E(\text{photon}) = 6.626 \times 10^{-34}\ \text{J·s} \times \frac{3.00 \times 10^8\ \text{m/s}}{48.2 \times 10^{-9}\ \text{m}} \times \frac{1\ \text{kJ}}{1000\ \text{J}} \times \frac{6.022 \times 10^{23}}{\text{mol}} = 2.48 \times 10^3\ \text{kJ/mol}$$

$$E_K = E(\text{electron}) = \tfrac{1}{2}(9.109 \times 10^{-31}\ \text{kg})(2.371 \times 10^6\ \text{m/s})^2\left(\frac{1\ \text{kJ}}{1000\ \text{J}}\right)\left(\frac{6.022 \times 10^{23}}{\text{mol}}\right)$$

$E_K = 1.54 \times 10^3\ \text{kJ/mol}$

$E(\text{photon}) = E_i + E_K;\quad E_i = E(\text{photon}) - E_K = (2.48 \times 10^3) - (1.54 \times 10^3) = 940\ \text{kJ/mol}$

5.114 (a) Ra [Rn] $7s^2$ [Rn] ↑↓
7s

(b) Sc [Ar] $4s^2 3d^1$ [Ar] ↑↓ ↑ _ _ _ _
4s   3d

(c) Lr [Rn] $7s^2 5f^{14} 6d^1$ [Rn] ↑↓ ↑↓ ↑↓ ↑↓ ↑↓ ↑↓ ↑↓ ↑↓ ↑ _ _ _ _
7s   5f   6d

(d) B [He] $2s^2 2p^1$ [He] ↑↓ ↑ _ _
2s   2p

(e) Te [Kr] $5s^2 4d^{10} 5p^4$ [Kr] ↑↓ ↑↓ ↑↓ ↑↓ ↑↓ ↑↓ ↑↓ ↑ ↑
5s   4d   5p

(f) F⁻ [Ne] [He] ↑↓ ↑↓ ↑↓ ↑↓
2s   2p

(g) Co²⁺ [Ar] $3d^7$ [Ar] ↑↓ ↑↓ ↑ ↑ ↑
3d

5.116 Charge on electron $= 1.602 \times 10^{-19}\ \text{C};\quad 1\ \text{V·C} = 1\ \text{J} = 1\ \text{kg m}^2/\text{s}^2$
(a) $E_K = (30,000\ \text{V})(1.602 \times 10^{-19}\ \text{C}) = 4.806 \times 10^{-15}\ \text{J}$

$$E_K = \tfrac{1}{2}mv^2;\quad v = \sqrt{\frac{2\,E_K}{m}} = \sqrt{\frac{2 \times 4.806 \times 10^{-15}\ \text{kg m}^2/\text{s}^2}{9.109 \times 10^{-31}\ \text{kg}}} = 1.03 \times 10^8\ \text{m/s}$$

$$\lambda = \frac{h}{mv} = \frac{6.626 \times 10^{-34}\ \text{kg m}^2/\text{s}}{(9.109 \times 10^{-31}\ \text{kg})(1.03 \times 10^8\ \text{m/s})} = 7.06 \times 10^{-12}\ \text{m}$$

(b) $E = h\dfrac{c}{\lambda} = (6.626 \times 10^{-34}\ \text{J·s})\left(\dfrac{3.00 \times 10^8\ \text{m/s}}{1.54 \times 10^{-10}\ \text{m}}\right) = 1.29 \times 10^{-15}\ \text{J/photon}$

Ionic Bonds and Some Main-Group Chemistry

6.1   (a) Br   (b) S   (c) Se   (d) Ne

6.2   (a) Be $1s^2 2s^2$ N $1s^2 2s^2 2p^3$
Be would have the larger third ionization energy because this electron would come from the 1s orbital.
(b) Ga   [Ar] $4s^2 3d^{10} 4p^1$ Ge   [Ar] $4s^2 3d^{10} 4p^2$
Ga would have the larger fourth ionization energy because this electron would come from the 3d orbitals.

6.3   (b) Cl has the highest E_{i1} and smallest E_{i4}.

6.4   Cr   [Ar] $4s^1 3d^5$ Mn   [Ar] $4s^2 3d^5$ Fe   [Ar] $4s^2 3d^6$
Cr can accept an electron into a 4s orbital. The 4s orbital is lower in energy than a 3d orbital. Both Mn and Fe accept the added electron into a 3d orbital that contains an electron, but Mn has a lower value of Z_{eff}. Therefore, Mn has a less negative E_{ea} than either Cr or Fe.

6.5 $K \rightarrow K^+ + e^-$ + 418.8 kJ/mol
       $F + e^- \rightarrow F^-$ $\underline{-\ 328\quad}$ kJ/mol
                                        + 91   kJ/mol (unfavorable)

6.6 $K(s) \rightarrow K(g)$ +89.2 kJ/mol
       $K(g) \rightarrow K^+(g) + e^-$ +418.8 kJ/mol
       $\frac{1}{2}\ [F_2(g) \rightarrow 2\ F(g)]$ +79 kJ/mol
       $F(g) + e^- \rightarrow F^-(g)$ −328 kJ/mol
       $K^+(g) + F^-(g) \rightarrow KF(s)$ $\underline{−821\quad}$ kJ/mol
                                        Sum = −562 kJ/mol

6.7   (a) KCl has the higher lattice energy because of the smaller K^+.
       (b) CaF_2 has the higher lattice energy because of the smaller Ca^{2+}.
       (c) CaO has the higher lattice energy because of the higher charge on both the cation and anion.

6.8   (a) Li_2O,   O −2   (b) K_2O_2,   O −1   (c) CsO_2,   O −½

6.9   (a) $2\ Cs(s)\ +\ 2\ H_2O(l)\ \rightarrow\ 2\ Cs^+(aq)\ +\ 2\ OH^-(aq)\ +\ H_2(g)$
       (b) $Na(s)\ +\ N_2(g)\ \rightarrow$ N. R.
       (c) $Rb(s)\ +\ O_2(g)\ \rightarrow\ RbO_2(s)$
       (d) $2\ K(s)\ +\ 2\ NH_3(g)\ \rightarrow\ 2\ KNH_2(s)\ +\ H_2(g)$
       (e) $2\ Rb(s)\ +\ H_2(g)\ \rightarrow\ 2\ RbH(s)$

6.10 (a) $Be(s) + Br_2(l) \rightarrow BeBr_2(s)$
(b) $Sr(s) + 2 H_2O(l) \rightarrow Sr(OH)_2(aq) + H_2(g)$
(c) $2 Mg(s) + O_2(g) \rightarrow 2 MgO(s)$

6.11 $BeCl_2 + 2 K \rightarrow Be + 2 KCl$

6.12 $Mg(s) + S(s) \rightarrow MgS(s)$; In MgS, the oxidation number of S is –2.

6.13 $2 Al(s) + 6 H^+(aq) \rightarrow 2 Al^{3+}(aq) + 3 H_2(g)$
H^+ gains electrons and is the oxidizing agent. Al loses electrons and is the reducing agent.

6.14 $2 Al(s) + 3 S(s) \rightarrow Al_2S_3(s)$

6.15 (a) $Br_2(l) + Cl_2(g) \rightarrow 2 BrCl(g)$
(b) $2 Al(s) + 3 F_2(g) \rightarrow 2 AlF_3(s)$
(c) $H_2(g) + I_2(s) \rightarrow 2 HI(g)$

6.16 $Br_2(l) + 2 NaI(s) \rightarrow 2 NaBr(s) + I_2(s)$
Br_2 gains electrons and is the oxidizing agent. I^- (from NaI) loses electrons and is the reducing agent.

6.17 (a) XeF_2 F –1, Xe +2
(b) XeF_4 F –1, Xe +4
(c) $XeOF_4$ F –1, O –2, Xe +6

6.18 (a) Rb would lose one electron and adopt the Kr noble-gas configuration.
(b) Ba would lose two electrons and adopt the Xe noble-gas configuration.
(c) Ga would lose three electrons and adopt an Ar-like noble-gas configuration (note that Ga^{3+} has ten 3d electrons in addition to the two 3s and six 3p electrons).
(d) F would gain one electron and adopt the Ne noble-gas configuration.

6.19 Group 6A elements will gain 2 electrons.

Understanding Key Concepts

6.20 (a) At is in Group 7A. The trend going down the group is gas $\rightarrow$ liquid $\rightarrow$ solid. At, being at the bottom of the group, should be a solid.
(b) At would likely be dark, like I_2, maybe with a metallic sheen.
(c) At is likely to react with Na just like the other halogens, yielding NaAt.

6.21

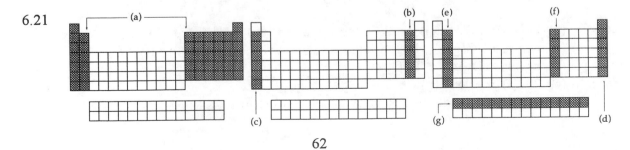

6.22　(a) shows an extended array, which represents an ionic compound.
　　　(b) shows discrete units, which represent a covalent compound.

6.23

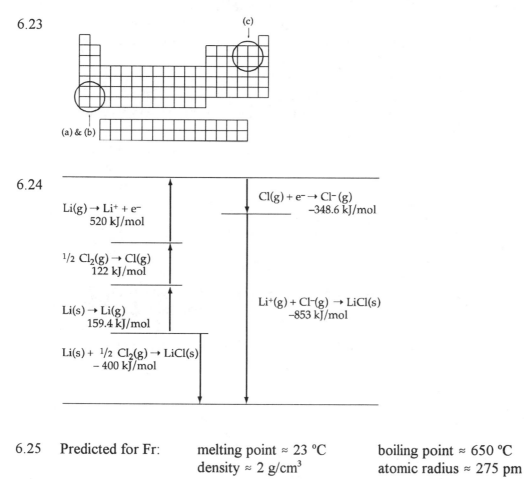

6.24

$Li(g) \rightarrow Li^+ + e^-$
520 kJ/mol

$Cl(g) + e^- \rightarrow Cl^-(g)$
−348.6 kJ/mol

$^{1}/_{2}\ Cl_2(g) \rightarrow Cl(g)$
122 kJ/mol

$Li(s) \rightarrow Li(g)$
159.4 kJ/mol

$Li^+(g) + Cl^-(g) \rightarrow LiCl(s)$
−853 kJ/mol

$Li(s) + {}^{1}/_{2}\ Cl_2(g) \rightarrow LiCl(s)$
− 400 kJ/mol

6.25　Predicted for Fr:　　melting point ≈ 23 °C　　　boiling point ≈ 650 °C
　　　　　　　　　　　density ≈ 2 g/cm³　　　　atomic radius ≈ 275 pm

6.26　(a) I_2　　　(b) Na　　　(c) NaCl　　　(d) Cl_2

6.27　(c) has the largest lattice energy because the charges are closest together.
　　　(a) has the smallest lattice energy because the charges are farthest apart.

Additional Problems
Ionization Energy and Electron Affinity

6.28　Ionization energies have a positive sign because energy is required to remove an electron
　　　from an atom of any element.

6.30　The largest E_{i1} are found in Group 8A because of the largest values of Z_{eff}.
　　　The smallest E_{i1} are found in Group 1A because of the smallest values of Z_{eff}.

6.32　(a) K　　[Ar] $4s^1$　　　　　　　Ca　　[Ar] $4s^2$
　　　Ca has the smaller second ionization energy because it is easier to remove the second 4s

valence electron in Ca than it is to remove the second electron in K from the filled 3p orbitals.
(b) Ca   [Ar] $4s^2$ Ga   [Ar] $4s^2 3d^{10} 4p^1$
Ca has the larger third ionization energy because it is more difficult to remove the third electron in Ca from the filled 3p orbitals than it is to remove the third electron (second 4s valence electron) from Ga.

6.34   (a) $1s^2 2s^2 2p^6 3s^2 3p^3$ is P   (b) $1s^2 2s^2 2p^6 3s^2 3p^6$ is Ar   (c) $1s^2 2s^2 2p^6 3s^2 3p^6 4s^2$ is Ca
Ar has the highest E_{i2}. Ar has a higher Z_{eff} than P. The 4s electrons in Ca are easier to remove than any 3p electrons.
Ar has the lowest E_{i7}. It is difficult to remove 3p electrons from Ca, and it is difficult to remove 2p electrons from P.

6.36   Lowest E_{i1} Highest E_{i1}
        (a)   K   Li
        (b)   B   Cl
        (c)   Ca   Cl

6.38   The relationship between the electron affinity of a univalent cation and the ionization energy of the neutral atom is that they have the same magnitude but opposite sign.

6.40 Na^+ has a more negative electron affinity than either Na or Cl because of its positive charge.

6.42   Energy is usually released when an electron is added to a neutral atom but absorbed when an electron is removed from a neutral atom because of the positive Z_{eff}.

6.44   (a) F; nonmetals have more negative electron affinities than metals.
        (b) Na; Ne (noble gas) has a positive electron affinity.
        (c) Br; nonmetals have more negative electron affinities than metals.

Lattice Energy and Ionic Bonds

6.46 $MgCl_2$ > LiCl > KCl > KBr

6.48   Li → Li^+ + e^- +520 kJ/mol
        Br + e^- → Br^- $\underline{-325 \text{ kJ/mol}}$
                                  +195 kJ/mol

6.50   Li(s) → Li(g)   +159.4 kJ/mol
        Li(s) → Li(g) + e^- +520   kJ/mol
        ½ [Br_2(l) → Br_2(g)]   +15.4 kJ/mol
        ½ [Br_2(g) → 2 Br(g)]   +112   kJ/mol
        Br(g) + e^- → Br^-(g)   −325   kJ/mol
        Li^+ (g) + Br^-(g) → LiBr(s) $\underline{-807 \text{ kJ/mol}}$
                                         Sum = −325 kJ/mol

6.52 $Na(s) \rightarrow Na(g)$ +107.3 kJ/mol
       $Na(g) \rightarrow Na^+(g) + e^-$ +495.8 kJ/mol
       ½ $[H_2(g) \rightarrow 2\ H(g)]$ +218.0 kJ/mol
       $H(g)\ +\ e^- \rightarrow H^-(g)$ −72.8 kJ/mol
       $Na^+(g) + H^-(g) \rightarrow NaH(s)$ <u>−U</u>
                     Sum = −60   kJ/mol

$$-U = -60 - 107.3 - 495.8 - \frac{435.9}{2} + 72.8 = -808 \text{ kJ/mol}; \quad U = 808 \text{ kJ/mol}$$

6.54 $Cs(s) \rightarrow Cs(g)$ +76.1 kJ/mol
       $Cs(g) \rightarrow Cs^+(g)\ +\ e^-$ +375.7 kJ/mol
       ½ $[F_2(g) \rightarrow 2\ F(g)]$ +79 kJ/mol
       $F(g)\ +\ e^- \rightarrow F^-(g)$ −328 kJ/mol
       $Cs^+(g)\ +\ F^-(g)\ \rightarrow CsF(g)$ <u>−740 kJ/mol</u>
                     Sum = −537   kJ/mol

6.56 $Ca(s) \rightarrow Ca(g)$ +178.2 kJ/mol
       $Ca(g) \rightarrow Ca^+(g)\ +\ e^-$ +589.8 kJ/mol
       ½$[Cl_2(g) \rightarrow 2\ Cl(g)]$ +121.5 kJ/mol
       $Cl(g)\ +\ e^- \rightarrow Cl^-(g)$ −348.6 kJ/mol
       $Ca^+(g) + Cl^-(g) \rightarrow CaCl(s)$ <u>−717   kJ/mol</u>
                     Sum = −176   kJ/mol

6.58

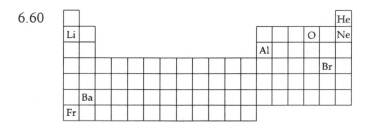

Main-Group Chemistry

6.60

6.62   Solids: I;   Liquids: Br;   Gases: F, Cl, He, Ne, Ar, Kr, Xe

6.64   (a) $2\,NaCl \xrightarrow[580°C]{\text{electrolysis in } CaCl_2} 2\,Na(l) + Cl_2(g)$

     (b) $2\,Al_2O_3 \xrightarrow[980°C]{\text{electrolysis in } Na_3AlF_6} 4\,Al(l) + 3\,O_2(g)$

     (c) Ar is obtained from the distillation of liquid air.

     (d) $2\,Br^-(aq) + Cl_2(g) \rightarrow Br_2(l) + 2\,Cl^-(aq)$

6.66   Main-group elements tend to undergo reactions that leave them with eight valence electrons. That is, main-group elements react so that they attain a noble-gas electron configuration with filled s and p sublevels in their valence electron shell.
The octet rule works for valence-shell electrons because taking electrons away from a filled octet is difficult because they are tightly held by a high Z_{eff}; adding more electrons to a filled octet is difficult because, with s and p sublevels full, there is no low-energy orbital available.

6.68   (a) $2\,K(s) + H_2(g) \rightarrow 2\,KH(s)$
     (b) $2\,K(s) + 2\,H_2O(l) \rightarrow 2\,K^+(aq) + 2\,OH^-(aq) + H_2(g)$
     (c) $2\,K(s) + 2\,NH_3(g) \rightarrow 2\,KNH_2(s) + H_2(g)$
     (d) $2\,K(s) + Br_2(l) \rightarrow 2\,KBr(s)$
     (e) $K(s) + N_2(g) \rightarrow$ N. R.
     (f) $K(s) + O_2(g) \rightarrow KO_2(s)$

6.70   (a) $Cl_2(g) + H_2(g) \rightarrow 2\,HCl(g)$
     (b) $Cl_2(g) + Ar(g) \rightarrow$ N. R.
     (c) $Cl_2(g) + Br_2(l) \rightarrow 2\,BrCl(g)$
     (d) $Cl_2(g) + N_2(g) \rightarrow$ N. R.

6.72 $AlCl_3 + 3\,Na \rightarrow Al + 3\,NaCl$
Al^{3+} (from $AlCl_3$) gains electrons and is reduced. Na loses electrons and is oxidized.

6.74 $CaIO_3$, 215.0 amu; 1.00 kg = 1000 g

     $\%\,I = \dfrac{126.9\ g}{215.0\ g} \times 100\% = 59.02\%$; $(0.5902)(1000\ g) = 590\ g\ I_2$

6.76 $Ca(s) + H_2(g) \rightarrow CaH_2(s)$; H_2, 2.016 amu; CaH_2, 42.09 amu

     $5.65\ g\ Ca \times \dfrac{1\ mol\ Ca}{40.08\ g\ Ca} = 0.141\ mol\ Ca$

     $3.15\ L\ H_2 \times 0.0893\ \dfrac{g\ H_2}{1\ L} \times \dfrac{1\ mol\ H_2}{2.016\ g\ H_2} = 0.140\ mol\ H_2$

Because the reaction stoichiometry between Ca and H_2 is one to one, H_2 is the limiting reactant.

$$0.140 \text{ mol } H_2 \times \frac{1 \text{ mol } CaH_2}{1 \text{ mol } H_2} \times \frac{42.09 \text{ g } CaH_2}{1 \text{ mol } CaH_2} \times 0.943 = 5.56 \text{ g } CaH_2$$

6.78   (a) $Mg(s) + 2 H^+(aq) \rightarrow Mg^{2+}(aq) + H_2(g)$
H^+ gains electrons and is the oxidizing agent. Mg loses electrons and is the reducing agent.
(b) $Kr(g) + F_2(g) \rightarrow KrF_2(s)$
F_2 gains electrons and is the oxidizing agent. Kr loses electrons and is the reducing agent.
(c) $I_2(s) + 3 Cl_2(g) \rightarrow 2 ICl_3(l)$
Cl_2 gains electrons and is the oxidizing agent. I_2 loses electrons and is the reducing agent.

General Problems

6.80
$Mg(s) \rightarrow Mg(g)$	+147.7 kJ/mol
$Mg(g) \rightarrow Mg^+(g) + e^-$	+737.7 kJ/mol
$\frac{1}{2}[F_2(g) \rightarrow 2 F(g)]$	+79 kJ/mol
$F(g) + e^- \rightarrow F^-(g)$	−328 kJ/mol
$Mg^+(g) + F^-(g) \rightarrow MgF(s)$	−930 kJ/mol
Sum =	−294 kJ/mol

$Mg(s) \rightarrow Mg(g)$	+147.7 kJ/mol
$Mg(g) \rightarrow Mg^+(g) + e^-$	+737.7 kJ/mol
$Mg^+(g) \rightarrow Mg^{2+}(g) + e^-$	+1450.7 kJ/mol
$F_2(g) \rightarrow 2 F(g)$	+158 kJ/mol
$2[F(g) + e^- \rightarrow F^-(g)]$	2(−328) kJ/mol
$Mg^{2+}(g) + 2 F^-(g) \rightarrow MgF_2(s)$	−2952 kJ/mol
Sum =	−1114 kJ/mol

6.82   (a) Na is used in table salt (NaCl), glass, rubber, and pharmaceutical agents.
(b) Mg is used as a structural material when alloyed with Al.
(c) F is used in the manufacture of Teflon, $(C_2F_4)_n$, and in toothpaste as SnF_2.

6.84   (a) $2 Li(s) + H_2(g) \rightarrow 2 LiH(s)$
(b) $2 Li(s) + 2 H_2O(l) \rightarrow 2 Li^+(aq) + 2 OH^-(aq) + H_2(g)$
(c) $2 Li(s) + 2 NH_3(g) \rightarrow 2 LiNH_2(s) + H_2(g)$
(d) $2 Li(s) + Br_2(l) \rightarrow 2 LiBr(s)$
(e) $6 Li(s) + N_2(g) \rightarrow 2 Li_3N(s)$
(f) $4 Li(s) + O_2(g) \rightarrow 2 Li_2O(s)$

6.86 $58.4 \text{ nm} = 58.4 \times 10^{-9} \text{ m}$

$$E(\text{photon}) = 6.626 \times 10^{-34} \text{ J·s} \times \frac{3.00 \times 10^8 \text{ m/s}}{58.4 \times 10^{-9} \text{ m}} \times \frac{1 \text{ kJ}}{1000 \text{ J}} \times \frac{6.022 \times 10^{23}}{\text{mol}} = 2049 \text{ kJ/mol}$$

$$E_K = E(\text{electron}) = \tfrac{1}{2}(9.109 \times 10^{-31} \text{ kg})(2.450 \times 10^6 \text{ m/s})^2 \left(\frac{1 \text{ kJ}}{1000 \text{ J}} \right) \left(\frac{6.022 \times 10^{23}}{\text{mol}} \right)$$

$E_K = 1646$ kJ/mol

$E(\text{photon}) = E_i + E_K;$ $\qquad E_i = E(\text{photon}) - E_K = 2049 - 1646 = 403$ kJ/mol

6.88 When moving diagonally down and right on the periodic table, the increase in atomic radius caused by going to a larger shell is offset by a decrease caused by a higher Z_{eff}. Thus, there is little net change.

6.90

$Cl(g) \rightarrow Cl^+(g) + e^-$
1251 kJ/mol

$Na(g) + e^- \rightarrow Na^-(g)$
–52.9 kJ/mol

$Na^-(g) + Cl^+(g) \rightarrow ClNa(s)$
–787 kJ/mol

$\tfrac{1}{2} Cl_2(g) \rightarrow Cl(g)$
122 kJ/mol

$Na(s) + \tfrac{1}{2} Cl_2(g) \rightarrow ClNa(s)$
640 kJ/mol

$Na(s) \rightarrow Na(g)$
107.3 kJ/mol

6.92

$Mg(s) \rightarrow Mg(g)$	+147.7	kJ/mol
$Mg(g) \rightarrow Mg^+(g) + e^-$	+738	kJ/mol
$Mg^+(g) \rightarrow Mg^{2+}(g) + e^-$	+1451	kJ/mol
$\tfrac{1}{2}[O_2(g) \rightarrow 2 O(g)]$	+249.2	kJ/mol
$O(g) + e^- \rightarrow O^-(g)$	–141.0	kJ/mol
$O^-(g) + e^- \rightarrow O^{2-}(g)$	E_{ea2}	
$Mg^{2+}(g) + O^{2-}(g) \rightarrow MgO(s)$	–3791	kJ/mol
$Mg(s) + \tfrac{1}{2}O_2(g) \rightarrow MgO(s)$	–601.7	kJ/mol

$147.7 + 738 + 1451 + 249.2 - 141.0 + E_{ea2} - 3791 = -601.7$

$E_{ea2} = -147.7 - 738 - 1451 - 249.2 + 141.0 + 3791 - 601.7 = +744$ kJ/mol

Because E_{ea2} is positive, O^{2-} is not stable in the gas phase. It is stable in MgO because of the large lattice energy that results from the 2+ and 2– charge of the ions and their small size.

Covalent Bonds and Molecular Structure

7.1   (a) $SiCl_4$ chlorine   EN = 3.0
                     silicon $\underline{EN = 1.8}$
                                ΔEN = 1.2   The Si–Cl bond is polar covalent.

(b) CsBr   bromine   EN = 2.8
           cesium $\underline{EN = 0.7}$
                     ΔEN = 2.1   The Cs^+Br^- bond is ionic.

(c) $FeBr_3$ bromine   EN = 2.8
               iron $\underline{EN = 1.8}$
                         ΔEN = 1.0   The Fe–Br bond is polar covalent.

(d) CH_4 carbon   EN = 2.5
             hydrogen $\underline{EN = 2.1}$
                        ΔEN = 0.4   The C–H bond is polar covalent.

7.2   (a) CCl_4 chlorine   EN = 3.0
                    carbon $\underline{EN = 2.5}$
                               ΔEN = 0.5

(b) $MgCl_2$ chlorine   EN = 3.0
               magnesium $\underline{EN = 1.2}$
                           ΔEN = 1.8

(c) $TiCl_3$ chlorine   EN = 3.0
               titanium $\underline{EN = 1.5}$
                          ΔEN = 1.5

(d) Cl_2O oxygen   EN = 3.5
              chlorine $\underline{EN = 3.0}$
                         ΔEN = 0.5

Increasing ionic character: $CCl_4 \sim ClO_2 < TiCl_3 < MgCl_2$

7.3   (a)   H :S̈: H   (b)   H
                                   :C̈l:C̈:C̈l:
                                      :C̈l:

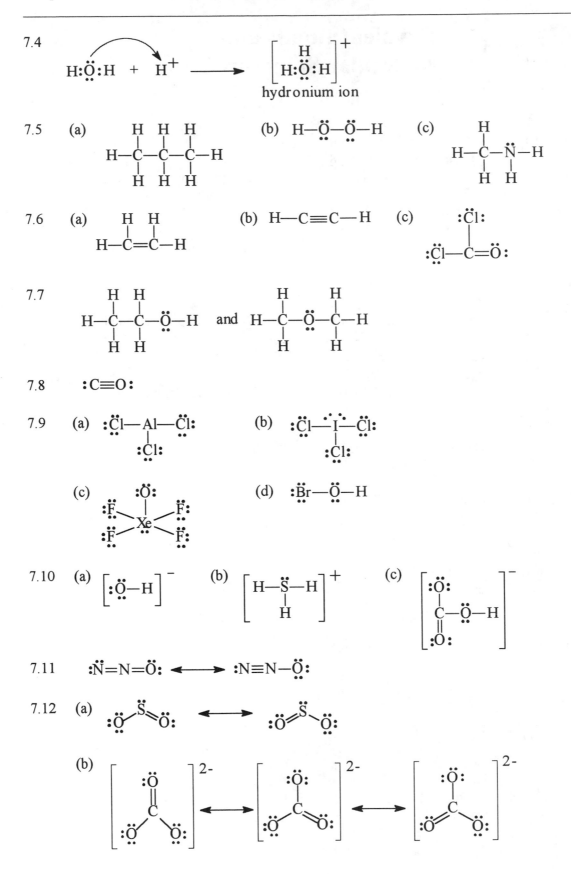

(c)

7.13   For nitrogen: Isolated nitrogen valence electrons   5
                             Bound nitrogen bonding electrons   8
                             Bound nitrogen nonbonding electrons   0
                             Formal charge = 5 − ½(8) − 0 = +1

For singly bound   Isolated oxygen valence electrons   6
oxygen: Bound oxygen bonding electrons   2
                         Bound oxygen nonbonding electrons   6
                         Formal charge = 6 − ½(2) − 6 = −1

For doubly bound   Isolated oxygen valence electrons   6
oxygen: Bound oxygen bonding electrons   4
                         Bound oxygen nonbonding electrons   4
                         Formal charge = 6 − ½(4) − 4 = 0

7.14   (a) $\left[:\ddot{N}=C=\ddot{O}:\right]^{-}$

For nitrogen: Isolated nitrogen valence electrons   5
                        Bound nitrogen bonding electrons   4
                        Bound nitrogen nonbonding electrons   4
                        Formal charge = 5 − ½(4) − 4 = −1

For carbon: Isolated carbon valence electrons   4
                        Bound carbon bonding electrons   8
                        Bound carbon nonbonding electrons   0
                        Formal charge = 4 − ½ (8) − 0 = 0

For oxygen: Isolated oxygen valence electrons   6
                        Bound oxygen bonding electrons   4
                        Bound oxygen nonbonding electrons   4
                        Formal charge = 6 − ½(4) − 4 = 0

(b) $:\ddot{O}-\ddot{O}=\ddot{O}:$

For left oxygen: Isolated oxygen valence electrons   6
                        Bound oxygen bonding electrons   2
                        Bound oxygen nonbonding electrons   6
                        Formal charge = 6 − ½(2) − 6 = −1

For central   Isolated oxygen valence electrons   6
oxygen: Bound oxygen bonding electrons   6

	Bound oxygen nonbonding electrons	2

Formal charge = $6 - \frac{1}{2}(6) - 2 = +1$

For right oxygen:	Isolated oxygen valence electrons	6
	Bound oxygen bonding electrons	4
	Bound oxygen nonbonding electrons	4

Formal charge = $6 - \frac{1}{2}(4) - 4 = 0$

7.15

	Number of Bonded Atoms	Number of Lone Pairs	Shape
(a) O_3	2	1	bent
(b) H_3O^+	3	1	trigonal pyramidal
(c) XeF_2	2	3	linear
(d) PF_6^-	6	0	octahedral
(e) $XeOF_4$	5	1	square pyramidal
(f) AlH_4^-	4	0	tetrahedral
(g) BF_4^-	4	0	tetrahedral
(h) $SiCl_4$	4	0	tetrahedral
(i) ICl_4^-	4	2	square planar
(j) $AlCl_3$	3	0	trigonal planar

7.16

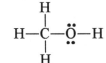

bent about O; tetrahedral about C

7.17

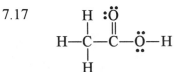 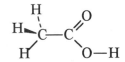

7.18

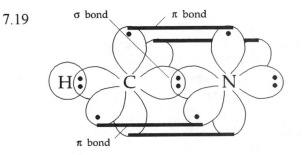

Each C is sp³ hybridized. The C–C bond is formed by the overlap of one singly occupied sp³ hybrid orbital from each C. The C–H bonds are formed by the overlap of one singly occupied sp³ orbital on C with a singly occupied H 1s orbital.

7.19

σ bond   π bond

H   C   N

π bond

In HCN the carbon is sp hybridized.

7.20

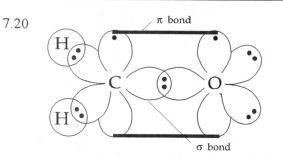

The carbon in formaldehyde is sp^2 hybridized.

7.21   The central I in I_3^- has two single bonds and three lone pairs of electrons. The hybridization of the central I is sp^3d. A sketch of the ion showing the orbitals involved in bonding is shown below.

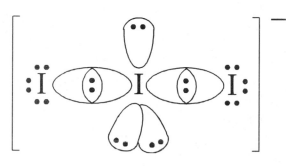

7.22

	Single Bonds	Lone Pairs	S Hybridization
SF_2	2	2	sp^3
SF_4	4	1	sp^3d
SF_6	6	0	sp^3d^2

7.23   For He_2^+ σ^*_{1s} $\uparrow$
                              σ_{1s} $\uparrow\downarrow$

He_2^+ Bond order = $\dfrac{\left(\begin{array}{c}\text{number of}\\\text{bonding electrons}\end{array}\right) - \left(\begin{array}{c}\text{number of}\\\text{antibonding electrons}\end{array}\right)}{2} = \dfrac{2-1}{2} = 1/2$

He_2^+ should be stable with a bond order of 1/2.

7.24   For B_2

σ^*_{2p} ___
π^*_{2p} ___   ___
σ_{2p} ___
π_{2p} $\uparrow$ $\uparrow$
σ^*_{2s} $\uparrow\downarrow$
σ_{2s} $\uparrow\downarrow$

B_2 Bond order = $\dfrac{\left(\begin{array}{c}\text{number of}\\\text{bonding electrons}\end{array}\right) - \left(\begin{array}{c}\text{number of}\\\text{antibonding electrons}\end{array}\right)}{2} = \dfrac{4-2}{2} = 1$

B_2 is paramagnetic because it has two unpaired electrons in the π_{2p} molecular orbitals.

For C_2

σ^*_{2p} ___

π^*_{2p} ___ ___

σ_{2p} ___

π_{2p} ↑↓ ↑↓

σ^*_{2s} ↑↓

σ_{2s} ↑↓

C_2 Bond order $= \dfrac{6-2}{2} = 2$; C_2 is diamagnetic because all electrons are paired.

7.25

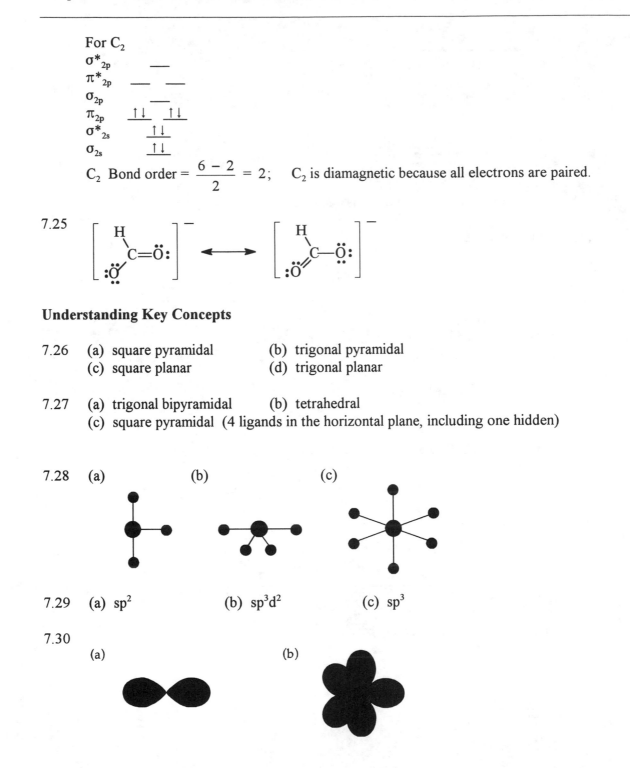

Understanding Key Concepts

7.26 (a) square pyramidal   (b) trigonal pyramidal

(c) square planar   (d) trigonal planar

7.27 (a) trigonal bipyramidal   (b) tetrahedral

(c) square pyramidal (4 ligands in the horizontal plane, including one hidden)

7.28 (a)   (b)   (c)

7.29 (a) sp^2 (b) sp^3d^2 (c) sp^3

7.30

(a)   (b)

7.31 Every carbon is sp^2 hybridized. There are 18 σ bonds and 5 π bonds.

7.32 (a) $C_8H_9NO_2$

(b) & (c)

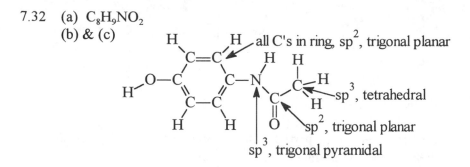

7.33 (a)

H–N–C–C–Ö–H

(b) H–C–H, ~109°;   O–C–O, ~120°;   H–N–H, ~107°

(c) N, sp^3; left C, sp^3; right C, sp^2

Additional Problems
Electronegativity and Polar Covalent Bonds

7.34   Electronegativity increases from left to right across a period and decreases down a group.

7.36   K < Li < Mg < Pb < C < Br

7.38   (a) HF   fluorine   EN = 4.0
                     hydrogen   EN = 2.1
                                     ΔEN = 1.9   HF is polar covalent.

        (b) HI   iodine   EN = 2.5
                     hydrogen   EN = 2.1
                                     ΔEN = 0.4   HI is polar covalent.

        (c) $PdCl_2$ chlorine   EN = 3.0
                     palladium   EN = 2.2
                                     ΔEN = 0.8 $PdCl_2$ is polar covalent.

        (d) BBr_3 bromine   EN = 2.8
                     boron   EN = 2.0
                                     ΔEN = 0.8 BBr_3 is polar covalent.

        (e) NaOH $Na^+ – OH^-$ is ionic
            OH^- oxygen   EN = 3.5
                     hydrogen   EN = 2.1
                                     ΔEN = 1.4 OH^- is polar covalent.

7.40 (a)
$$\overset{\delta-}{C} - \overset{\delta+}{H} \qquad \overset{\delta+}{C} - \overset{\delta-}{Cl}$$
(b)
$$\overset{\delta-}{Si} - \overset{\delta+}{Li} \qquad \overset{\delta+}{Si} - \overset{\delta-}{Cl}$$

(c) N – Cl
$$\overset{\delta-}{N} - \overset{\delta+}{Mg}$$

Electron-Dot Structures and Resonance

7.42 The transition metals are characterized by partially filled d orbitals that can be used to expand their valence shell beyond the normal octet of electrons.

7.44 (a), (b), (c), (d), (e), (f)

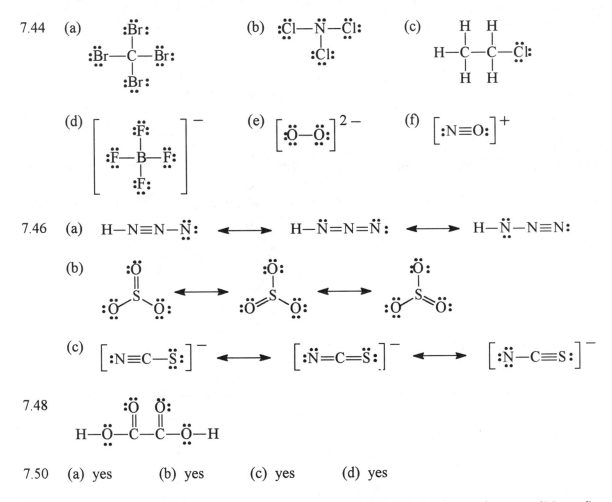

7.46 (a), (b), (c)

7.48

7.50 (a) yes   (b) yes   (c) yes   (d) yes

7.52 (a) The anion has 32 valence electrons. Each Cl has seven valence electrons (28 total). The minus one charge on the anion accounts for one valence electron. This leaves three valence electrons for X. X is Al.
(b) The cation has eight valence electrons. Each H has one valence electron (4 total). X is left with four valence electrons. Since this is a cation, one valence electron was removed from X. X has five valence electrons. X is P.

7.54 (a)

$$\ddot{O}:$$

(image of Lewis structure: Cl=C–O–C with H atoms)

(b)

(image of Lewis structure)

Formal Charges

7.56 :C≡O:

For carbon:	Isolated carbon valence electrons	4
	Bound carbon bonding electrons	6
	Bound carbon nonbonding electrons	2
	Formal charge = 4 – ½(6) – 2 = –1	

For oxygen:	Isolated oxygen valence electrons	6
	Bound oxygen bonding electrons	6
	Bound oxygen nonbonding electrons	2
	Formal charge = 6 – ½(6) – 2 = +1	

7.58

$$\left[:\ddot{O}-\ddot{C}l-\ddot{O}:\right]^{-}$$

For both oxygens:	Isolated oxygen valence electrons	6
	Bound oxygen bonding electrons	2
	Bound oxygen nonbonding electrons	6
	Formal charge = 6 – ½(2) – 6 = –1	

For chlorine:	Isolated chlorine valence electrons	7
	Bound chlorine bonding electrons	4
	Bound chlorine nonbonding electrons	4
	Formal charge = 7 – ½(4) – 4 = +1	

$$\left[:\ddot{O}-\ddot{C}l=\ddot{O}\right]^{-}$$

For left oxygen:	Isolated oxygen valence electrons	6
	Bound oxygen bonding electrons	2
	Bound oxygen nonbonding electrons	6
	Formal charge = 6 – ½(2) – 6 = –1	

For right oxygen:	Isolated oxygen valence electrons	6
	Bound oxygen bonding electrons	4
	Bound oxygen nonbonding electrons	4
	Formal charge = 6 – ½(4) – 4 = 0	

For chlorine:	Isolated chlorine valence electrons	7
	Bound chlorine bonding electrons	6
	Bound chlorine nonbonding electrons	4
	Formal charge = 7 – ½(6) – 4 = 0	

7.60　(a)　$\underset{H}{\overset{H}{\diagdown}}C{=}N{=}\ddot{N}:$

For hydrogen:	Isolated hydrogen valence electrons	1
	Bound hydrogen bonding electrons	2
	Bound hydrogen nonbonding electrons	0
	Formal charge = $1 - \frac{1}{2}(2) - 0 = 0$	

For nitrogen: (central)	Isolated nitrogen valence electrons	5
	Bound nitrogen bonding electrons	8
	Bound nitrogen nonbonding electrons	0
	Formal charge = $5 - \frac{1}{2}(8) - 0 = +1$	

For nitrogen: (terminal)	Isolated nitrogen valence electrons	5
	Bound nitrogen bonding electrons	4
	Bound nitrogen nonbonding electrons	4
	Formal charge = $5 - \frac{1}{2}(4) - 4 = -1$	

For carbon:	Isolated carbon valence electrons	4
	Bound carbon bonding electrons	8
	Bound carbon nonbonding electrons	0
	Formal charge = $4 - \frac{1}{2}(8) - 0 = 0$	

(b)　$\underset{H}{\overset{H}{\diagdown}}C{-}\ddot{N}{=}\ddot{N}:$

For hydrogen:	Isolated hydrogen valence electrons	1
	Bound hydrogen bonding electrons	2
	Bound hydrogen nonbonding electrons	0
	Formal charge = $1 - \frac{1}{2}(2) - 0 = 0$	

For nitrogen: (central)	Isolated nitrogen valence electrons	5
	Bound nitrogen bonding electrons	6
	Bound nitrogen nonbonding electrons	2
	Formal charge = $5 - \frac{1}{2}(6) - 2 = 0$	

For nitrogen: (terminal)	Isolated nitrogen valence electrons	5
	Bound nitrogen bonding electrons	4
	Bound nitrogen nonbonding electrons	4
	Formal charge = $5 - \frac{1}{2}(4) - 4 = -1$	

For carbon:	Isolated carbon valence electrons	4
	Bound carbon bonding electrons	6
	Bound carbon nonbonding electrons	0
	Formal charge = $4 - \frac{1}{2}(6) - 0 = +1$	

Structure (a) is more important because of the octet of electrons around carbon.

The VSEPR Model

7.62   From data in Table 7.4:
        (a) trigonal planar   (b) trigonal bipyramidal   (c) linear   (d) octahedral

7.64   From data in Table 7.4:
        (a) tetrahedral, 4   (b) octahedral, 6   (c) bent, 3 or 4
        (d) linear, 2 or 5   (e) square pyramidal, 6   (f) trigonal pyramidal, 4

7.66

		Number of Bonded Atoms	Number of Lone Pairs	Shape
(a)	H_2Se	2	2	bent
(b)	$SiCl_4$	4	0	tetrahedral
(c)	O_3	2	1	bent
(d)	GaH_3	3	0	trigonal planar

7.68

		Number of Bonded Atoms	Number of Lone Pairs	Shape
(a)	SbF_5	5	0	trigonal bipyramidal
(b)	IF_4^+	4	1	see saw
(c)	SeO_3^{2-}	3	1	trigonal pyramidal
(d)	CrO_4^{2-}	4	0	tetrahederal

7.70

		Number of Bonded Atoms	Number of Lone Pairs	Shape
(a)	PO_4^{3-}	4	0	tetrahedral
(b)	MnO_4^-	4	0	tetrahedral
(c)	SO_4^{2-}	4	0	tetrahedral
(d)	SO_3^{2-}	3	1	trigonal pyramidal
(e)	ClO_4^-	4	0	tetrahedral

7.72   (a) In SF_2 the sulfur is bound to two fluorines and contains two lone pairs of electrons.
        SF_2 is bent and the F–S–F bond angle is approximately $109°$.
        (b) In N_2H_2 each nitrogen is bound to the other nitrogen and one hydrogen. Each
        nitrogen has one lone pair of electrons. The H–N–N bond angle is approximately $120°$
        (c) In KrF_4 the krypton is bound to four fluorines and contains two lone pairs of
        electrons. KrF_4 is square planar, and the F–Kr–F bond angle is $90°$.
        (d) In NOCl the nitrogen is bound to one oxygen and one chlorine and contains one lone
        pair of electrons. NOCl is bent, and the Cl–N–O bond angle is approximately $120°$.

7.74

$$\begin{array}{c} H \\ \diagdown \\ {}^{H}\diagup C = \underset{a}{C} - \underset{b}{C} \equiv \underset{c}{N} \end{array}$$

H – C_a – H	~ $120°$	C_b – C_c – N	$180°$
H – C_a – C_b	~ $120°$	C_a – C_b – H	~ $120°$
C_a – C_b – C_c	~ $120°$	H – C_b – C_c	~ $120°$

7.76   All six carbons in cyclohexane are bonded to two other carbons and two hydrogens (i.e. four charge clouds). The geometry about each carbon is tetrahedral with a C–C–C bond angle of approximately 109°. Because the geometry about each carbon is tetrahedral, the cyclohexane ring cannot be flat.

Hybrid Orbitals and Molecular Orbital Theory

7.78   In a π bond, the shared electrons occupy a region above and below a line connecting the two nuclei. A σ bond has its shared electrons located along the axis between the two nuclei.

7.80   See Table 7.5.
    (a) sp   (b) sp^3d (c) sp^3d^2 (d) sp^3

7.82   See Table 7.5.
    (a) sp^3 (b) sp^3d^2 (c) sp^2 or sp^3 (d) sp or sp^3d (e) sp^3d^2

7.84   (a) sp^2 (b) sp^3 (c) sp^3d^2 (d) sp^2

7.86

         O
         ‖ ~120°   The C is sp^2 hybridized and the N atoms are sp^3 hybridized.
         C
  H_2N N—H
   ~109°   ~109°
        H

7.88

	O_2^+	O_2	O_2^-
σ^*_{2p}	—	—	—
π^*_{2p}	↑	↑ ↑	↑↓ ↑
π_{2p}	↑↓ ↑↓	↑↓ ↑↓	↑↓ ↑↓
σ_{2p}	↑↓	↑↓	↑↓
σ^*_{2s}	↑↓	↑↓	↑↓
σ_{2s}	↑↓	↑↓	↑↓

$$\text{Bond order} = \frac{\left(\begin{array}{c}\text{number of}\\ \text{bonding electrons}\end{array}\right) - \left(\begin{array}{c}\text{number of}\\ \text{antibonding electrons}\end{array}\right)}{2}$$

$$O_2^+ \text{ bond order} = \frac{8-3}{2} = 2.5; \qquad O_2 \text{ bond order} = \frac{8-4}{2} = 2$$

$$O_2^- \text{ bond order} = \frac{8-5}{2} = 1.5$$

All are stable with bond orders between 1.5 and 2.5. All have unpaired electrons.

7.90   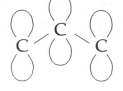   p orbitals in allyl cation

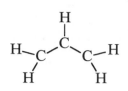

allyl cation showing only the σ bonds (each C is sp² hybridized)

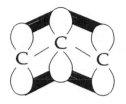

delocalized MO model for π bonding in the allyl cation

General Problems

7.92

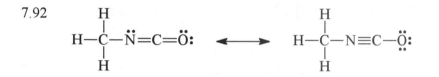

7.94 (a)

$$F{-}B{-}\overset{F}{\underset{F}{\vert}}\overset{CH_3}{\underset{CH_3}{O}}$$

For boron: Isolated boron valence electrons   3
               Bound boron bonding electrons   8
               Bound boron nonbonding electrons   0
               Formal charge $= 3 - \frac{1}{2}(8) - 0 = -1$

For oxygen: Isolated oxygen valence electrons   6
               Bound oxygen bonding electrons   6
               Bound oxygen nonbonding electrons   2
               Formal charge $= 6 - \frac{1}{2}(6) - 2 - +1$

(b) In BF₃ the B has three bonding pairs of electrons and no lone pairs. The B is sp² hybridized and BF₃ is trigonal planar.

 $H_3C{-}\overset{..}{\underset{..}{O}}{-}CH_3$ is bent about the oxygen because of two bonding pairs and two lone pairs of electrons. The O is sp³ hybridized.

In the product, B is sp³ hybridized (with four bonding pairs of electrons), and the geometry about it is tetrahedral. The O is also sp³ hybridized (with three bonding pairs and one lone pair of electrons), and the geometry about it is trigonal pyramidal.

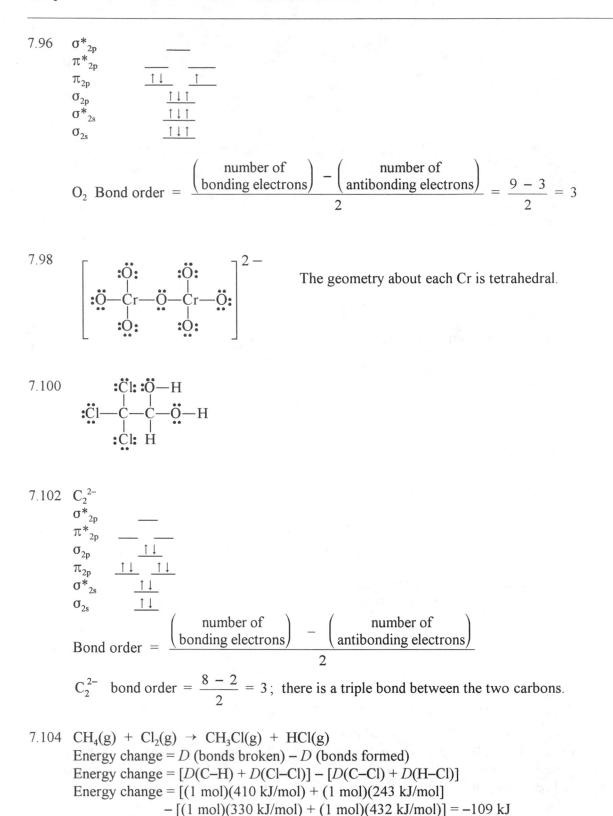

7.96

$$\text{O}_2 \text{ Bond order} = \frac{\left(\begin{array}{c}\text{number of}\\\text{bonding electrons}\end{array}\right) - \left(\begin{array}{c}\text{number of}\\\text{antibonding electrons}\end{array}\right)}{2} = \frac{9-3}{2} = 3$$

7.98

The geometry about each Cr is tetrahedral.

7.100

7.102 C_2^{2-}

$$\text{Bond order} = \frac{\left(\begin{array}{c}\text{number of}\\\text{bonding electrons}\end{array}\right) - \left(\begin{array}{c}\text{number of}\\\text{antibonding electrons}\end{array}\right)}{2}$$

C_2^{2-} bond order $= \dfrac{8-2}{2} = 3$; there is a triple bond between the two carbons.

7.104 $CH_4(g) + Cl_2(g) \rightarrow CH_3Cl(g) + HCl(g)$
Energy change $= D$ (bonds broken) $- D$ (bonds formed)
Energy change $= [D(C\text{–}H) + D(Cl\text{–}Cl)] - [D(C\text{–}Cl) + D(H\text{–}Cl)]$
Energy change $= [(1 \text{ mol})(410 \text{ kJ/mol}) + (1 \text{ mol})(243 \text{ kJ/mol}]$
$\qquad\qquad - [(1 \text{ mol})(330 \text{ kJ/mol}) + (1 \text{ mol})(432 \text{ kJ/mol})] = -109 \text{ kJ}$

7.106 (a)   (b)   (c)

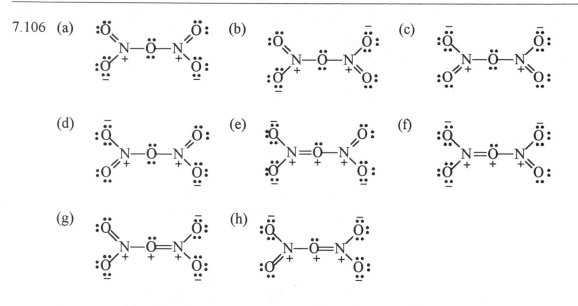

(d)   (e)   (f)

(g)   (h)

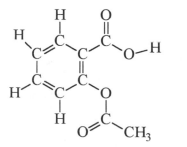

Structures (a) - (d) make more important contributions to the resonance hybrid because of only −1 and 0 formal charges on the oxygens.

7.108

21 σ bonds
5 π bonds

Thermochemistry: Chemical Energy

8.1 $E = \frac{1}{2}mv^2 = \frac{1}{2}(2 \text{ kg})(4 \text{ m/s})^2 = 16 \text{ kg} \cdot \text{m}^2/\text{s}^2 = 16 \text{ J}$

8.2 Convert lb to kg. $2300 \text{ lb} \times \dfrac{453.59 \text{ g}}{1 \text{ lb}} \times \dfrac{1 \text{ kg}}{1000 \text{ g}} = 1043 \text{ kg}$

Convert mi/h to m/s. $55 \dfrac{\text{mi}}{\text{h}} \times \dfrac{1 \text{ km}}{0.62137 \text{ mi}} \times \dfrac{1000 \text{ m}}{1 \text{km}} \times \dfrac{1 \text{ h}}{3600 \text{ s}} = 24.6 \text{ m/s}$

$1 \text{ kg} \cdot \text{m}^2/\text{s}^2 = 1 \text{ J};$ $E = \frac{1}{2}mv^2 = \frac{1}{2}(1043 \text{ kg})(24.6 \text{ m/s})^2 = 3.2 \times 10^5 \text{ J}$

$E = 3.2 \times 10^5 \text{ J} \times \dfrac{1 \text{ kJ}}{1000 \text{ J}} = 3.2 \times 10^2 \text{ kJ}$

8.3 (a) and (b) are state functions; (c) is not.

8.4 $\Delta V = (4.3 \text{ L} - 8.6 \text{ L}) = -4.3 \text{ L}$
$w = -P\Delta V = -(44 \text{ atm})(-4.3 \text{ L}) = +189.2 \text{ L} \cdot \text{atm}$

$w = (189.2 \text{ L} \cdot \text{atm})(101 \dfrac{\text{J}}{\text{L} \cdot \text{atm}}) = +1.9 \times 10^4 \text{ J}$

The positive sign for the work indicates that the surroundings does work on the system. Energy flows into the system.

8.5 $\Delta H° = -484 \dfrac{\text{kJ}}{2 \text{ mol H}_2}$

$P\Delta V = (1.00 \text{ atm})(-5.6 \text{ L}) = -5.6 \text{ L} \cdot \text{atm}$

$P\Delta V = (-5.6 \text{ L} \cdot \text{atm})(101 \dfrac{\text{J}}{\text{L} \cdot \text{atm}}) = -565.6 \text{ J} = -570 \text{ J} = -0.57 \text{ kJ}$

$w = -P\Delta V = 570 \text{ J} = 0.57 \text{ kJ}$

$\Delta H = -121 \dfrac{\text{kJ}}{0.50 \text{ mol H}_2}$

$\Delta E = \Delta H - P\Delta V = -121 \text{ kJ} - (-0.57 \text{ kJ}) = -120.43 \text{ kJ} = -120 \text{ kJ}$

8.6 $\Delta V = 448 \text{ L}$ and assume $P = 1.00 \text{ atm}$
$w = -P\Delta V = -(1.00 \text{ atm})(448 \text{ L}) = -448 \text{ L} \cdot \text{atm}$

$w = -(448 \text{ L} \cdot \text{atm})(101 \dfrac{\text{J}}{\text{L} \cdot \text{atm}}) = -4.52 \times 10^4 \text{ J}$

$w = -4.52 \times 10^4 \text{ J} \times \dfrac{1 \text{ kJ}}{1000 \text{ J}} = -45.2 \text{ kJ}$

8.7   (a) C_3H_8, 44.10 amu; $\Delta H° = -2219$ kJ/mol C_3H_8

15.5 g x $\dfrac{1 \text{ mol } C_3H_8}{44.10 \text{ g } C_3H_8}$ x $\dfrac{-2219 \text{ kJ}}{1 \text{ mol } C_3H_8}$ $= -780.$ kJ

780. kJ of heat is evolved.

(b) $Ba(OH)_2 \cdot 8\,H_2O$, 315.5 amu; $\Delta H° = +80.3$ kJ/mol $Ba(OH)_2 \cdot 8\,H_2O$

4.88 g x $\dfrac{1 \text{ mol } Ba(OH)_2 \cdot 8\,H_2O}{315.5 \text{ g } Ba(OH)_2 \cdot 8\,H_2O}$ x $\dfrac{80.3 \text{ kJ}}{1 \text{ mol } Ba(OH)_2 \cdot 8\,H_2O}$ $= +1.24$ kJ

1.24 kJ of heat is absorbed.

8.8 $q = (\text{specific heat}) \times m \times \Delta T = (4.18 \; \dfrac{J}{g \cdot °C})(350 \text{ g})(3°C - 25°C) = -3.2 \times 10^4$ J

$q = -3.2 \times 10^4 \times \dfrac{1 \text{ kJ}}{1000 \text{ J}} = -32$ kJ

8.9 $q = (\text{specific heat}) \times m \times \Delta T$;   specific heat $= \dfrac{q}{m \times \Delta T} = \dfrac{96 \text{ J}}{(75 \text{ g})(10°C)} = 0.13$ J/(g · °C)

8.10 $q = (\text{specific heat}) \times m \times \Delta T$

$m = (25.0 \text{ mL} + 50.0 \text{ mL})(1.00 \; \dfrac{g}{mL}) = 75.0$ g

$q = (4.18 \; \dfrac{J}{g \cdot °C})(75.0 \text{ g})(33.9°C - 25.0°C) = 2790$ J

mol $H_2SO_4 = 0.0250$ L x $1.00 \; \dfrac{\text{mol}}{L} \; H_2SO_4 = 0.0250$ mol H_2SO_4

Heat evolved per mole of $H_2SO_4 = \dfrac{2.79 \times 10^3 \text{ J}}{0.0250 \text{ mol } H_2SO_4} = 1.1 \times 10^5$ J/mol H_2SO_4

Since the reaction evolves heat, the sign for ΔH is negative.

$\Delta H = -1.1 \times 10^5$ J x $\dfrac{1 \text{ kJ}}{1000 \text{ J}} = -1.1 \times 10^2$ kJ

8.11 $CH_4(g) + Cl_2(g) \rightarrow CH_3Cl(g) + HCl(g)$ $\Delta H°_1 = -98.3$ kJ
             $\underline{CH_3Cl(g) + Cl_2(g) \rightarrow CH_2Cl_2(g) + HCl(g)}$ $\Delta H°_2 = -104$ kJ
     Sum $CH_4(g) + 2\,Cl_2(g) \rightarrow CH_2Cl_2(g) + 2\,HCl(g)$
             $\Delta H° = \Delta H°_1 + \Delta H°_2 = -202$ kJ

8.12 $CH_4(g) + Cl_2(g) \rightarrow CH_3Cl(g) + HCl(g)$ $\Delta H°_1 = -98.3$ kJ
             $CH_3Cl(g) + Cl_2(g) \rightarrow CH_2Cl_2(g) + HCl(g)$ $\Delta H°_2 = -104$ kJ
             $\underline{CH_2Cl_2(g) \rightarrow CH_2Cl_2(l)}$ $\Delta H°_3 = -\Delta H_{vap} = -29.0$ kJ
     Sum $CH_4(g) + 2\,Cl_2(g) \rightarrow CH_2Cl_2(l) + 2\,HCl(g)$
             $\Delta H° = \Delta H°_1 + \Delta H°_2 + \Delta H°_3 = -231$ kJ

8.13   Reactants CH_4 + 2 Cl_2

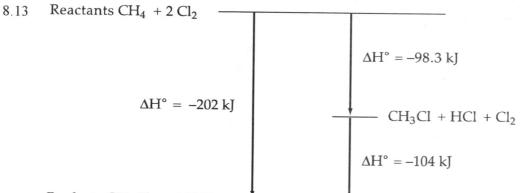

$\Delta H° = -98.3$ kJ

$\Delta H° = -202$ kJ

CH_3Cl + HCl + Cl_2

$\Delta H° = -104$ kJ

Products CH_2Cl_2 + 2 HCl

8.14 $4\,NH_3(g)\ +\ 5\,O_2(g)\ \rightarrow\ 4\,NO(g)\ +\ 6\,H_2O(g)$
$\Delta H°_{rxn} = [4\ \Delta H°_f\,(NO)\ +\ 6\ \Delta H°_f\,(H_2O)]\ -\ [4\ \Delta H°_f\,(NH_3)]$
$\Delta H°_{rxn} = [(4\ mol)(90.2\ kJ/mol)\ +\ (6\ mol)(-\,241.8\ kJ/mol)]\ -\ [(4\ mol)(-\,46.1\ kJ/mol)]$
$\Delta H°_{rxn} = -905.6$ kJ

8.15 $6\,CO_2(g)\ +\ 6\,H_2O(l)\ \rightarrow\ C_6H_{12}O_6(s)\ +\ 6\,O_2(g)$
$\Delta H°_{rxn} = \Delta H°_f(C_6H_{12}O_6)\ -\ [6\ \Delta H°_f(CO_2)\ +\ 6\ \Delta H°_f(H_2O(l)]$
$\Delta H°_{rxn} = [(1\ mol)(-1260\ kJ/mol)]\ -\ [(6\ mol)(-393.5\ kJ/mol)\ +\ (6\ mol)(-285.8\ kJ/mol)]$
$\Delta H°_{rxn} = +2815.8$ kJ $= +2816$ kJ

8.16 $\Delta H°_{rxn} = D$ (bonds broken) $-\ D$ (bonds formed)
$H_2C{=}CH_2(g)\ +\ H_2O(g)\ \rightarrow\ C_2H_5OH(g)$
$\Delta H°_{rxn} = [D(C{=}C)\ +\ D(O{-}H)]\ -\ [D(C{-}C)\ +\ D(C{-}O)\ +\ D(C{-}H)]$
$\Delta H°_{rxn} = [(1\ mol)(611\ kJ/mol)\ +\ (1\ mol)(460\ kJ/mol)]$
$\quad -\ [(1\ mol)(350\ kJ/mol)\ +\ (1\ mol)(\ 350\ kJ/mol)\ +\ (1\ mol)(410\ kJ/mol)] = -39$ kJ

8.17 $2\,NH_3(g)\ +\ Cl_2(g)\ \rightarrow\ N_2H_4(g)\ +\ 2\,HCl(g)$
$\Delta H°_{rxn} = D$ (bonds broken) $-\ D$ (bonds formed)
$\Delta H°_{rxn} = [2\ x\ D(N{-}H)\ +\ D(Cl{-}Cl)]\ -\ [D(N{-}N)\ +\ 2\ x\ D(H{-}Cl)]$
$\Delta H°_{rxn} = [(2\ mol)(390\ kJ/mol)\ +\ (1\ mol)(243\ kJ/mol)]$
$\quad -\ [(1\ mol)(240\ kJ/mol)\ +\ (2\ mol)(432\ kJ/mol)] = -81$ kJ

8.18 $C_4H_{10}(l)\ +\ \dfrac{13}{2}\ O_2(g)\ \rightarrow\ 4\,CO_2(g)\ +\ 5\,H_2O(g)$

$\Delta H°_{rxn} = [4\ \Delta H°_f\,(CO_2)\ +\ 5\ \Delta H°_f\,(H_2O)]\ -\ \Delta H°_f\,(C_4H_{10})$
$\Delta H°_{rxn} = [(4\ mol)(-395.5\ kJ/mol)\ +\ (5\ mol)(-241.8\ kJ/mol)]\ -\ [(1\ mol)(-147.5\ kJ/mol)]$
$\Delta H°_{rxn} = -\,2643.5$ kJ
$\Delta H°_C\ = -\,2643.5$ kJ/mol

C_4H_{10}, 58.12 amu; $\Delta H°_C\ = \left(-2643.5\ \dfrac{kJ}{mol}\right)\!\left(\dfrac{1\ mol}{58.12\ g}\right) = -45.48$ kJ/g

$\Delta H°_C\ = \left(-45.48\ \dfrac{kJ}{g}\right)\!\left(0.579\ \dfrac{g}{mL}\right) = -26.3$ kJ/mL

Chapter 8 – Thermochemistry: Chemical Energy

8.19 $\Delta S° < 0$ because the reaction decreases the number of moles of gaseous molecules.

8.20   (a) Since $\Delta G°$ is negative, the reaction is spontaneous.
        (b) Since $\Delta G°$ is positive, the reaction is nonspontaneous.

8.21 $\Delta G° = \Delta H° - T\Delta S° = (-92.2 \text{ kJ}) - (298 \text{ K})(-0.199 \text{ kJ/K}) = -32.9 \text{ kJ}$
        Because $\Delta G°$ is negative, the reaction is spontaneous.
        Set $\Delta G° = 0$ and solve for T.
        $\Delta G° = 0 = \Delta H° - T\Delta S°$
        $$T = \frac{\Delta H°}{\Delta S°} = \frac{-92.2 \text{ kJ}}{-0.199 \text{ kJ/K}} = 463 \text{ K} = 190°C$$

Understanding Key Concepts

8.22.   (a) $w = -P\Delta V$, $\Delta V > 0$; therefore $w < 0$ and the system is doing work on the surroundings.
        (b) Since the temperature has increased there has been an enthalpy change. The system evolved heat, the reaction is exothermic, and $\Delta H < 0$.

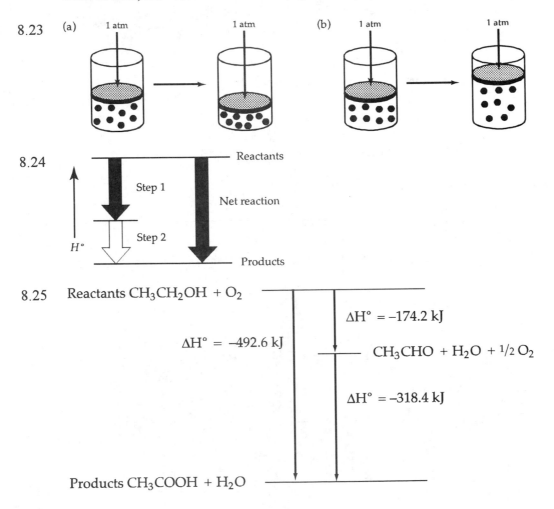

8.25   Reactants $CH_3CH_2OH + O_2$

$\Delta H° = -492.6 \text{ kJ}$

$\Delta H° = -174.2 \text{ kJ}$

$CH_3CHO + H_2O + \frac{1}{2} O_2$

$\Delta H° = -318.4 \text{ kJ}$

Products $CH_3COOH + H_2O$

8.26 (a) The entropy change is positive. (b) The entropy change is negative.

8.27 Since the process is spontaneous, $\Delta G < 0$. The system is more disordered, $\Delta S > 0$.
$\Delta H \approx 0$ for the mixing of gaseous molecules.

8.28 An air conditioner works by compressing a gas to a liquid (this produces heat) and then allowing the liquid to evaporate to a gas (this absorbs heat). The amount of heat absorbed from the room and given back to the room should be the same because of the conservation of energy. If we assume that the compressor does not generate any heat (it does) then there should be no net temperature change in the room. Under normal operating conditions, the heat that an air conditioner produces is pumped outside.

8.29 $\Delta H = \Delta E + P\Delta V$
$\Delta H - \Delta E = P\Delta V$
$$\Delta V = \frac{\Delta H - \Delta E}{P} = \frac{[-35.0\ kJ - (-34.8\ kJ)]}{1\ atm} \times \frac{1\ L \cdot atm}{101 \times 10^{-3}\ kJ} = -2\ L$$
$\Delta V = -2\ L = V_{final} - V_{initial} = V_{final} - 5\ L;$ $V_{final} = -2\ L - (-5\ L) = 3\ L$
The volume decreases from 5 L to 3 L.

Additional Problems
Heat, Work, and Energy

8.30 Heat is the energy transferred from one object to another as the result of a temperature difference between them. Temperature is a measure of the kinetic energy of molecular motion.
Energy is the capacity to do work or supply heat. Work is defined as the distance moved times the force that opposes the motion (w = d x f).
Kinetic energy is the energy of motion. Potential energy is stored energy.

8.32 Car: $E_K = \frac{1}{2}(1400\ kg)\left(\frac{115 \times 10^3\ m}{3600\ s}\right)^2 = 7.1 \times 10^5\ J$

Truck: $E_K = \frac{1}{2}(12{,}000\ kg)\left(\frac{38 \times 10^3\ m}{3600\ s}\right)^2 = 6.7 \times 10^5\ J$

The car has more kinetic energy.

8.34 $w = -P\Delta V = -(3.6 \text{ atm})(3.4 \text{ L} - 3.2 \text{ L}) = -0.72 \text{ L} \cdot \text{atm}$

   $w = (-0.72 \text{ L} \cdot \text{atm})\left(\dfrac{101 \text{ J}}{1 \text{ L} \cdot \text{atm}}\right) = -72.7 \text{ J} = -70 \text{ J}$;   The energy change is negative.

Energy and Enthalpy

8.36 $\Delta E = q_v$ is the heat change associated with a reaction at constant volume. Since $\Delta V = 0$, no PV work is done.
   $\Delta H = q_p$ is the heat change associated with a reaction at constant pressure. Since $\Delta V \neq 0$, PV work can also be done.

8.38 $\Delta H = \Delta E + P\Delta V$; ΔH and ΔE are nearly equal when there are no gases involved in a chemical reaction, or, if gases are involved, $\Delta V = 0$ (that is, there are the same number of reactant and product gas molecules).

8.40 $P\Delta V = -7.6 \text{ J}$ (from Problem 8.35)
   $\Delta H = \Delta E + P\Delta V$
   $\Delta E = \Delta H - P\Delta V = -0.31 \text{ kJ} - (-7.6 \times 10^{-3} \text{ kJ}) = -0.30 \text{ kJ}$

8.42 $\Delta H = -1255.5 \text{ kJ/mol } C_2H_2$; C_2H_2, 26.04 amu
   $w = -P\Delta V = -(1.00 \text{ atm})(-2.80 \text{ L}) = 2.80 \text{ L} \cdot \text{atm}$

   $w = (2.80 \text{ L} \cdot \text{atm})\left(\dfrac{101 \text{ J}}{1 \text{ L} \cdot \text{atm}}\right) = 283 \text{ J} = 0.283 \text{ kJ}$

   $6.50 \text{ g} \times \dfrac{1 \text{ mol } C_2H_2}{26.04 \text{ g } C_2H_2} = 0.250 \text{ mol } C_2H_2$

   $q = (-1255.5 \text{ kJ/mol})(0.250 \text{ mol}) = -314 \text{ kJ}$
   $\Delta E = \Delta H - P\Delta V = -314 \text{ kJ} - (-0.283 \text{ kJ}) = -314 \text{ kJ}$

8.44   mass of $C_4H_{10}O = \left(0.7138 \dfrac{\text{g}}{\text{mL}}\right)(100 \text{ mL}) = 71.38 \text{ g}$

   $C_4H_{10}O$, 74.12 amu

   mol $C_4H_{10}O = 71.38 \text{ g} \times \dfrac{1 \text{ mol}}{74.12 \text{ g}} = 0.9626 \text{ mol}$

   $q = n \times \Delta H_{vap} = 0.9626 \text{ mol} \times 26.5 \text{ kJ/mol} = 25.5 \text{ kJ}$

8.46   Al, 26.98 amu
   mol Al $= 5.00 \text{ g} \times \dfrac{1 \text{ mol}}{26.98 \text{ g}} = 0.1853 \text{ mol}$

   $q = n \times \Delta H° = 0.1853 \text{ mol Al} \times \dfrac{-1408.4 \text{ kJ}}{2 \text{ mol Al}} = -131 \text{ kJ}$

8.48 Fe_2O_3, 159.7 amu

$$mol\ Fe_2O_3 = 2.50\ g \times \frac{1\ mol}{159.7\ g} = 0.015\ 65\ mol$$

$$q = n \times \Delta H° = 0.015\ 65\ mol\ Fe_2O_3 \times \frac{-24.8\ kJ}{1\ mol\ Fe_2O_3} = -0.388\ kJ$$

Because ΔH is negative, the reaction is exothermic.

Calorimetry and Heat Capacity

8.50   Heat capacity is the amount of heat required to raise the temperature of a substance a given amount. Specific heat is the amount of heat necessary to raise the temperature of exactly 1 g of a substance by exactly 1°C.

8.52   Na, 22.99 amu

$$specific\ heat = 28.2\ \frac{J}{mol \cdot °C} \times \frac{1\ mol}{22.99\ g} = 1.23\ J/(g \cdot °C)$$

8.54   Mass of solution = 50.0 g + 1.045 g = 51.0 g
q = (specific heat) x m x ΔT

$$q = \left(4.18\ \frac{J}{g \cdot °C}\right)(51.0\ g)(32.3°C - 25.0°C) = 1.56 \times 10^3\ J = 1.56\ kJ$$

CaO, 56.08 amu; $mol\ CaO = 1.045\ g \times \frac{1\ mol}{56.08\ g} = 0.018\ 63\ mol$

Heat evolved per mole of CaO $= \frac{1.56\ kJ}{0.018\ 63\ mol} = 83.7\ kJ/mol\ CaO$

Because the reaction evolves heat, the sign for ΔH is negative. $\Delta H = -83.7\ kJ$

8.56   NaOH, 40.00 amu;   HCl, 36.46 amu

$$8.00\ g\ NaOH \times \frac{1\ mol\ NaOH}{40.00\ g\ NaOH} = 0.200\ mol\ NaOH$$

$$8.00\ g\ HCl \times \frac{1\ mol\ HCl}{36.46\ g\ HCl} = 0.219\ mol\ HCl$$

Because the reaction stoichiometry between NaOH and HCl is one to one, the NaOH is the limiting reactant.

$$q_P = -q_{soln} = -(specific\ heat) \times m \times \Delta T = -\left(4.18\ \frac{J}{g \cdot °C}\right)(316\ g)(33.5°C - 25.0°C) = -11.2\ kJ$$

$\Delta H = q_P/n = (-11.2\ kJ)/(0.200\ mol) = -56.0\ kJ/mol$
When 10.00 g of HCl in 248.0 g of water are added the same temperature increase is observed because the mass of NaOH is the same and it is still the limiting reactant. The mass of the solution is also the same.

Hess's Law and Heats of Formation

8.58   The standard state of an element is its most stable form at 1 atm and 25°C.

8.60   Hess's Law — the overall enthalpy change for a reaction is equal to the sum of the
enthalpy changes for the individual steps in the reaction.
Hess's Law works because of the law of conservation of energy.

8.62
$$S(s) + O_2(g) \rightarrow SO_2(g) \qquad \Delta H°_1 = -296.8 \text{ kJ}$$
$$\underline{SO_2 + \tfrac{1}{2}O_2(g) \rightarrow SO_3(g)} \qquad \Delta H°_2 = -98.9 \text{ kJ}$$
Sum $S(s) + 3/2\,O_2(g) \rightarrow SO_3(g) \qquad \Delta H°_3 = \Delta H°_1 + \Delta H°_2$
$$\Delta H°_f = \Delta H°_3 = -296.8 \text{ kJ} + (-98.9 \text{ kJ}) = -395.7 \text{ kJ/mol}$$

8.64
$$SO_3(g) + H_2O(l) \rightarrow H_2SO_4(aq) \qquad \Delta H°_1 = -227.8 \text{ kJ}$$
$$H_2(g) + \tfrac{1}{2}O_2(g) \rightarrow H_2O(l) \qquad \Delta H°_2 = \Delta H°_f = -285.8 \text{ kJ}$$
$$\underline{S(s) + 3/2\,O_2(g) \rightarrow SO_3(g)} \qquad \Delta H°_3 = \Delta H°_f = -395.7$$
Sum $S(s) + H_2(g) + 2\,O_2(g) \rightarrow H_2SO_4(aq) \quad \Delta H°_f\,(H_2SO_4) = ?$
$$\Delta H°_f\,(H_2SO_4) = \Delta H°_1 + \Delta H°_2 + \Delta H°_3 = -909.3 \text{ kJ}$$

8.66 $C_8H_8(l) + 10\,O_2(g) \rightarrow 8\,CO_2(g) + 4\,H_2O(l)$
$\Delta H°_{rxn} = \Delta H°_c = -4395.2 \text{ kJ}$
$\Delta H°_{rxn} = [8\,\Delta H°_f(CO_2) + 4\,\Delta H°_f(H_2O)] - \Delta H°_f(C_8H_8)$
$-4395.2 \text{ kJ} = [(8 \text{ mol})(-393.5 \text{ kJ/mol}) + (4 \text{ mol})(-285.8 \text{ kJ/mol})] - [(1 \text{ mol})(\Delta H°_f(C_8H_8))]$
Solve for $\Delta H°_f(C_8H_8)$
$-4395.2 \text{ kJ} = -4291.2 \text{ kJ} - (1 \text{ mol})(\Delta H°_f(C_8H_8))$
$-104.0 \text{ kJ} = -(1 \text{ mol})(\Delta H°_f(C_8H_8))$
$$\Delta H°_f(C_8H_8) = \frac{-104.0 \text{ kJ}}{-1 \text{ mol}} = +104.0 \text{ kJ/mol}$$

8.68 $\Delta H°_{rxn} = \Delta H°_f(MTBE) - [\Delta H°_f(\text{2-Methylpropene}) + \Delta H°_f(CH_3OH)]$
$-57.8 \text{ kJ} = -313.6 \text{ kJ} - [(1 \text{ mol})(\Delta H°_f(\text{2-Methylpropene})) + (-238.7 \text{ kJ})]$
Solve for $\Delta H°_f(\text{2-Methylpropene})$.
$-17.1 \text{ kJ} = (1 \text{ mol})(\Delta H°_f(\text{2-Methylpropene}))$
$\Delta H°_f(\text{2-Methylpropene}) = -17.1 \text{ kJ/mol}$

Bond Dissociation Energies

8.70 $H_2C{=}CH_2(g) + H_2(g) \rightarrow CH_3CH_3(g)$
$\Delta H°_{rxn} = D \text{ (bonds broken)} - D \text{ (bonds formed)}$
$\Delta H°_{rxn} = [D(C{=}C) + D(H_2)] - [2 \times D(C{-}H) + D(C{-}C)]$
$\Delta H°_{rxn} = [(1 \text{ mol})(611 \text{ kJ/mol}) + (1 \text{ mol})(436 \text{ kJ/mol})]$
$\qquad\qquad - [(2 \text{ mol})(410 \text{ kJ/mol}) + (1 \text{ mol})(350 \text{ kJ/mol})] = -123 \text{ kJ}$

8.72 $C_4H_{10} + 13/2\ O_2 \rightarrow 4\ CO_2 + 5\ H_2O$
       $\Delta H^\circ_{rxn} = D\ \text{(bonds broken)} - D\ \text{(bonds formed)}$
       $\Delta H^\circ_{rxn} = [3 \times D(C–C) + 10 \times D(C–H) + 13/2 \times D(O_2)] - [8 \times D(C=O) + 10 \times D(O–H)]$
       $\Delta H^\circ_{rxn} = [(3\ mol)(350\ kJ/mol) + (10\ mol)(410\ kJ/mol) + (13/2\ mol)(498\ kJ/mol)]$
              $- [(8\ mol)(804\ kJ/mol) + (10\ mol)(460\ kJ/mol)] = -2645\ kJ$

Free Energy and Entropy

8.74   Entropy is a measure of molecular disorder.

8.76   A reaction can be spontaneous yet endothermic if ΔS is positive (more disorder) and the $T\Delta S$ term is larger than ΔH.

8.78   (a) positive   (more disorder)   (b) negative   (more order)
       (c) positive   (more disorder)   (d) positive   (more disorder)

8.80 $\Delta S > 0$. The reaction increases the total number of molecules.

8.82 $\Delta G = \Delta H - T\Delta S$
       (a) $\Delta G = -48\ kJ - (400\ K)(135 \times 10^{-3}\ kJ/K) = -102\ kJ$
       $\Delta G < 0$, spontaneous; $\Delta H < 0$, exothermic.
       (b) $\Delta G = -48\ kJ - (400\ K)(-135 \times 10^{-3}\ kJ/K) = +6\ kJ$
       $\Delta G > 0$, nonspontaneous; $\Delta H < 0$, exothermic.
       (c) $\Delta G = +48\ kJ - (400\ K)(135 \times 10^{-3}\ kJ/K) = -6\ kJ$
       $\Delta G < 0$, spontaneous; $\Delta H > 0$, endothermic.
       (d) $\Delta G = +48\ kJ - (400\ K)(-135 \times 10^{-3}\ kJ/K) = +102\ kJ$
       $\Delta G > 0$, nonspontaneous; $\Delta H > 0$, endothermic.

8.84 $\Delta G = \Delta H - T\Delta S$;   Set $\Delta G = 0$ and solve for T (the crossover temperature).
       $T = \dfrac{\Delta H}{\Delta S} = \dfrac{-33\ kJ}{-0.058\ kJ/K} = 570\ K$

8.86   (a) $\Delta H < 0$ and $\Delta S > 0$; reaction is spontaneous at all temperatures.
       (b) $\Delta H < 0$ and $\Delta S < 0$; reaction has a crossover temperature.
       (c) $\Delta H > 0$ and $\Delta S > 0$; reaction has a crossover temperature.
       (d) $\Delta H > 0$ and $\Delta S < 0$; reaction is nonspontaneous at all temperatures.

General Problems

8.88 $Mg(s) + 2\ HCl(aq) \rightarrow MgCl_2(aq) + H_2(g)$
       $mol\ Mg = 1.50\ g \times \dfrac{1\ mol}{24.3\ g} = 0.0617\ mol\ Mg$
       $mol\ HCl = 0.200\ L \times 6.00\ \dfrac{mol}{L} = 1.20\ mol\ HCl$
       There is an excess of HCl. Mg is the limiting reactant.

$$q = \left(4.18\ \frac{J}{g \cdot {}^{\circ}C}\right)(200\ g)(42.9\ {}^{\circ}C - 25.0\ {}^{\circ}C) + \left(776\ \frac{J}{{}^{\circ}C}\right)(42.9\ {}^{\circ}C - 25.0\ {}^{\circ}C) = 2.89 \times 10^4\ J$$

$$q = 2.89 \times 10^4\ J \times \frac{1\ kJ}{1000\ J} = 28.9\ kJ$$

Heat evolved per mole of Mg $= \dfrac{28.9\ kJ}{0.0617\ mol} = 468\ kJ/mol$

Because the reaction evolves heat, the sign for ΔH is negative. $\Delta H = -468\ kJ$

8.90
$$2\ NO(g) + O_2(g) \rightarrow 2\ NO_2\ (g) \qquad \Delta H^{\circ}_1 = 2(-57.0\ kJ)$$
$$\underline{2\ NO_2\ (g) \rightarrow N_2O_4(g)} \qquad\qquad \Delta H^{\circ}_2 = -57.2\ kJ$$
Sum $\quad 2\ NO(g) + O_2(g) \rightarrow N_2O_4(g)$
$$\Delta H^{\circ} = \Delta H^{\circ}_1 + \Delta H^{\circ}_2 = -171.2\ kJ$$

8.92 $\Delta G_{fus} = \Delta H_{fus} - T\Delta S_{fus}$; at the melting point $\Delta G = 0$. Set $\Delta G = 0$ and solve for T (the melting point).
$$\Delta G = 0 = \Delta H_{fus} - T\Delta S_{fus}$$
$$T = \frac{\Delta H_{fus}}{\Delta S_{fus}} = \frac{9.95\ kJ}{0.0357\ kJ/K} = 279\ K$$

8.94 $\Delta H^{\circ}_{rxn} = D$ (bonds broken) $- D$ (bonds formed)
(a) $2\ CH_4(g) \rightarrow C_2H_6(g) + H_2(g)$
$\Delta H^{\circ}_{rxn} = [2 \times D(C–H)] - [D(C–C) + D(H–H)]$
$\Delta H^{\circ}_{rxn} = [(2\ mol)(410\ kJ/mol)] - [(1\ mol)(350\ kJ/mol) + (1\ mol)(436\ kJ/mol)] = +34\ kJ$
(b) $C_2H_6(g) + F_2(g) \rightarrow C_2H_5F(g) + HF(g)$
$\Delta H^{\circ}_{rxn} = [D(C–H) + D(F–F)] - [D(C–F) + D(H–F)]$
$\Delta H^{\circ}_{rxn} = [(1\ mol)(410\ kJ/mol) + (1\ mol)(159\ kJ/mol)]$
$\qquad - [(1\ mol)(450\ kJ/mol) + (1\ mol)(570\ kJ/mol)] = -451\ kJ$
(c) $N_2(g) + 3\ H_2(g) \rightarrow 2\ NH_3(g)$
The bond dissociation energy for N_2 is 945 kJ/mol.
$\Delta H^{\circ}_{rxn} = [D(N{\equiv}N) + 3 \times D(H–H)] - [6 \times D(N–H)]$
$\Delta H^{\circ}_{rxn} = [(1\ mol)(945\ kJ/mol) + (3\ mol)(436\ kJ/mol)] - [(6\ mol)(390\ kJ/mol)] = -87\ kJ$

8.96 (a) $2\ C_8H_{18}(l) + 25\ O_2(g) \rightarrow 16\ CO_2(g) + 18\ H_2O(g)$
(b) $C_8H_{18}(l) + 25/2\ O_2(g) \rightarrow 8\ CO_2(g) + 9\ H_2O(g)$
$\Delta H^{\circ}_{rxn} = \Delta H^{\circ}_c = -5456.6\ kJ$
$\Delta H^{\circ}_{rxn} = [8\ \Delta H^{\circ}_f(CO_2) + 9\ \Delta H^{\circ}_f(H_2O)] - \Delta H^{\circ}_f(C_8H_{18})$
$-5456.6\ kJ = [(8\ mol)(-393.5\ kJ/mol) + (9\ mol)(-241.8\ kJ/mol)] - [(1\ mol)(\Delta H^{\circ}_f(C_8H_{18}))]$
Solve for $\Delta H^{\circ}_f(C_8H_{18})$.
$-5456.6\ kJ = -5324\ kJ - [(1\ mol)(\Delta H^{\circ}_f(C_8H_{18}))]$
$-132.4\ kJ = -(1\ mol)(\Delta H^{\circ}_f(C_8H_{18}))$
$\Delta H^{\circ}_f(C_8H_{18}) = +132.4\ kJ/mol$

8.98   (a) $\Delta S_{total} = \Delta S_{system} + \Delta S_{surr}$ and $\Delta S_{surr} = -\Delta H/T$

$\Delta S_{total} = \Delta S_{system} + (-\Delta H/T) = \Delta S_{system} - \Delta H/T$

$\Delta S_{system} = \Delta S_{total} + \Delta H/T$

$\Delta G = \Delta H - T\Delta S$ (substitute ΔS_{system} for ΔS in this equation)

$\Delta G = \Delta H - T(\Delta S_{total} + \Delta H/T) = -T\Delta S_{total}$

$\Delta G = -T\Delta S_{total}$ For a spontaneous reaction, if $\Delta S_{total} > 0$ then $\Delta G < 0$.

(b) $\Delta G° = \Delta H° - T\Delta S°$

$\Delta H° = \Delta G° + T\Delta S°$

$$\Delta S_{surr} = -\frac{\Delta H°}{T} = -\frac{[\Delta G° + T\Delta S°]}{T} = -\frac{[2879 \times 10^3 \text{ J/mol} + (298 \text{ K})(-210 \text{J/(K}\cdot\text{mol))}]}{298 \text{ K}}$$

$\Delta S_{surr} = -9451$ J/(K$\cdot$ mol)

Gases: Their Properties and Behavior

9.1 1.00 atm = 14.7 psi

$$1.00 \text{ mm Hg} \times \frac{1 \text{ atm}}{760 \text{ mm Hg}} \times \frac{14.7 \text{ psi}}{1 \text{ atm}} = 1.93 \times 10^{-2} \text{ psi}$$

9.2 1.00 atmosphere pressure can support a column of Hg 0.760 m high. Because the density of H_2O is 1.00 g/mL and that of Hg is 13.6 g/mL, 1.00 atmosphere pressure can support a column of H_2O 13.6 times higher than that of Hg. The column of H_2O supported by 1.00 atmosphere will be (0.760 m)(13.6) = 10.3 m.

9.3 The pressure in the flask is less than 0.975 atm. The 24.7 cm of Hg is the difference between the two pressures.

$$\text{Pressure difference} = 24.7 \text{ cm Hg} \times \frac{1.00 \text{ atm}}{76.0 \text{ cm Hg}} = 0.325 \text{ atm}$$

Pressure in flask = 0.975 atm − 0.325 atm = 0.650 atm

9.4 $$n = \frac{PV}{RT} = \frac{(1.000 \text{ atm})(1.000 \times 10^5 \text{ L})}{\left(0.082\ 06 \dfrac{\text{L} \cdot \text{atm}}{\text{mol} \cdot \text{K}}\right)(273.15 \text{ K})} = 4.461 \times 10^3 \text{ mol CH}_4$$

$$CH_4, 16.04 \text{ amu}; \quad \text{mass CH}_4 = (4.461 \times 10^3 \text{ mol})\left(\frac{16.04 \text{ g}}{1 \text{ mol}}\right) = 7.155 \times 10^4 \text{ g CH}_4$$

9.5 C_3H_8, 44.10 amu; $\quad$ V = 350 mL = 0.350 L; $\qquad$ T = 20°C = 293 K

$$n = 3.2 \text{ g} \times \frac{1 \text{ mol C}_3\text{H}_8}{44.10 \text{ g C}_3\text{H}_8} = 0.073 \text{ mol C}_3\text{H}_8$$

$$P = \frac{nRT}{V} = \frac{(0.073 \text{ mol})\left(0.082\ 06 \dfrac{\text{L} \cdot \text{atm}}{\text{mol} \cdot \text{K}}\right)(293 \text{ K})}{0.350 \text{ L}} = 5.0 \text{ atm}$$

9.6 $$P = 1.51 \times 10^4 \text{ kPa} \times \frac{1 \text{ atm}}{101.325 \text{ k Pa}} = 149 \text{ atm}; \qquad T = 25.0°C = 298 \text{ K}$$

$$n = \frac{PV}{RT} = \frac{(149 \text{ atm})(43.8 \text{ L})}{\left(0.082\ 06 \dfrac{\text{L} \cdot \text{atm}}{\text{mol} \cdot \text{K}}\right)(298 \text{ K})} = 267 \text{ mol He}$$

9.7 The volume and number of moles of gas remain constant.

$$\frac{nR}{V} = \frac{P_i}{T_i} = \frac{P_f}{T_f}; \qquad T_f = \frac{P_f T_i}{P_i} = \frac{(2.37\ \text{atm})(273\ \text{K})}{2.15\ \text{atm}} = 301\ \text{K} = 28°\text{C}$$

9.8 $CaCO_3(s) + 2\ HCl(aq) \rightarrow CaCl_2(aq) + CO_2(g) + H_2O(l)$
$CaCO_3$, 100.1 amu; CO_2, 44.01 amu

$$\text{mole } CO_2 = 33.7\ \text{g } CaCO_3 \times \frac{1\ \text{mol } CaCO_3}{100.1\ \text{g } CaCO_3} \times \frac{1\ \text{mol } CO_2}{1\ \text{mol } CaCO_3} = 0.337\ \text{mol } CO_2$$

$$\text{mass } CO_2 = 0.337\ \text{mol } CO_2 \times \frac{44.01\ \text{g } CO_2}{1\ \text{mol } CO_2} = 14.8\ \text{g } CO_2$$

$$V = \frac{nRT}{P} = \frac{(0.337\ \text{mol})\left(0.082\ 06\ \dfrac{\text{L} \cdot \text{atm}}{\text{mol} \cdot \text{K}}\right)(273\ \text{K})}{1.00\ \text{atm}} = 7.55\ \text{L}$$

9.9 $C_3H_8(g) + 5\ O_2(g) \rightarrow 3\ CO_2(g) + 4\ H_2O(l)$

$$n_{propane} = \frac{PV}{RT} = \frac{(4.5\ \text{atm})(15.0\ \text{L})}{\left(0.082\ 06\ \dfrac{\text{L} \cdot \text{atm}}{\text{mol} \cdot \text{K}}\right)(298\ \text{K})} = 2.76\ \text{mol } C_3H_8$$

$$2.76\ \text{mol } C_3H_8 \times \frac{3\ \text{mol } CO_2}{1\ \text{mol } C_3H_8} = 8.28\ \text{mol } CO_2$$

$$V = \frac{nRT}{P} = \frac{(8.28\ \text{mol})\left(0.082\ 06\ \dfrac{\text{L} \cdot \text{atm}}{\text{mol} \cdot \text{K}}\right)(273\ \text{K})}{1.00\ \text{atm}} = 186\ \text{L} = 190\ \text{L}$$

9.10 $$n = \frac{PV}{RT} = \frac{(1.00\ \text{atm})(1.00\ \text{L})}{\left(0.082\ 06\ \dfrac{\text{L} \cdot \text{atm}}{\text{mol} \cdot \text{K}}\right)(273\ \text{K})} = 0.0446\ \text{mol}$$

$$\text{molar mass} = \frac{1.52\ \text{g}}{0.0446\ \text{mol}} = 34.1\ \text{g/mol}; \quad \text{molecular weight} = 34.1\ \text{amu}$$

$Na_2S(aq) + 2\ HCl(aq) \rightarrow H_2S(g) + 2\ NaCl(aq)$
The foul-smelling gas is H_2S, hydrogen sulfide.

9.11 $$12.45\ \text{g } H_2 \times \frac{1\ \text{mol } H_2}{2.016\ \text{g } H_2} = 6.176\ \text{mol } H_2$$

$$60.67\ \text{g } N_2 \times \frac{1\ \text{mol } N_2}{28.01\ \text{g } N_2} = 2.166\ \text{mol } N_2$$

$$2.38\ \text{g } NH_3 \times \frac{1\ \text{mol } NH_3}{17.03\ \text{g } NH_3} = 0.140\ \text{mol } NH_3$$

$$n_{total} = n_{H_2} + n_{N_2} + n_{NH_3} = 6.176 \text{ mol} + 2.166 \text{ mol} + 0.140 \text{ mol} = 8.482 \text{ mol}$$

$$X_{H_2} = \frac{6.176 \text{ mol}}{8.482 \text{ mol}} = 0.7281; \quad X_{N_2} = \frac{2.166 \text{ mol}}{8.482 \text{ mol}} = 0.2554; \quad X_{NH_3} = \frac{0.140 \text{ mol}}{8.482 \text{ mol}} = 0.0165$$

9.12 $n_{total} = 8.482 \text{ mol}$ (from Problem 9.11). $T = 90°C = 363 \text{ K}$

$$P_{total} = \frac{n_{total}RT}{V} = \frac{(8.482 \text{ mol})\left(0.082\ 06 \ \dfrac{L \cdot atm}{mol \cdot K}\right)(363 \text{ K})}{10.00 \text{ L}} = 25.27 \text{ atm}$$

$$P_{H_2} = X_{H_2} \cdot P_{total} = (0.7281)(25.27 \text{ atm}) = 18.4 \text{ atm}$$

$$P_{N_2} = X_{N_2} \cdot P_{total} = (0.2554)(25.27 \text{ atm}) = 6.45 \text{ atm}$$

$$P_{NH_3} = X_{NH_3} \cdot P_{total} = (0.0165)(25.27 \text{ atm}) = 0.417 \text{ atm}$$

9.13 $P_{H_2O} = X_{H_2O} \cdot P_{Total} = (0.0287)(0.977 \text{ atm}) = 0.0280 \text{ atm}$

9.14 $u = \sqrt{\dfrac{3RT}{M}}$, M = molar mass, $R = 8.314 \text{ J/(K} \cdot \text{mol)}$, $1 \text{ J} = 1 \text{ kg} \cdot \text{m}^2/\text{s}^2$

at 37°C = 310 K, $u = \sqrt{\dfrac{3 \times 8.314 \text{ kg m}^2/(s^2 \text{ K mol}) \times 310 \text{ K}}{28.01 \times 10^{-3} \text{ kg/mol}}} = 525 \text{ m/s}$

at −25°C = 248 K, $u = \sqrt{\dfrac{3 \times 8.314 \text{ kg m}^2/(s^2 \text{ K mol}) \times 248 \text{ K}}{28.01 \times 10^{-3} \text{ kg/mol}}} = 470 \text{ m/s}$

9.15   (a) $\dfrac{\text{rate } O_2}{\text{rate Kr}} = \dfrac{\sqrt{M_{Kr}}}{\sqrt{M_{O_2}}} = \dfrac{\sqrt{83.8}}{\sqrt{32.0}}; \quad \dfrac{\text{rate } O_2}{\text{rate Kr}} = 1.62$

O_2 diffuses 1.62 times faster than Kr.

(b) $\dfrac{\text{rate } C_2H_2}{\text{rate } N_2} = \dfrac{\sqrt{M_{N_2}}}{\sqrt{M_{C_2H_2}}} - \dfrac{\sqrt{28.0}}{\sqrt{26.0}}; \quad \dfrac{\text{rate } C_2H_2}{\text{rate } N_2} = 1.04$

C_2H_2 diffuses 1.04 times faster than N_2.

9.16 $\dfrac{\text{rate } ^{20}Ne}{\text{rate } ^{22}Ne} = \dfrac{\sqrt{M\ ^{22}Ne}}{\sqrt{M\ ^{20}Ne}} = \dfrac{\sqrt{22}}{\sqrt{20}} = 1.05; \quad \dfrac{\text{rate } ^{21}Ne}{\text{rate } ^{22}Ne} = \dfrac{\sqrt{M\ ^{22}Ne}}{\sqrt{M\ ^{21}Ne}} = \dfrac{\sqrt{22}}{\sqrt{21}} = 1.02$

Thus, the relative rates of diffusion are $^{20}Ne(1.05) > {}^{21}Ne(1.02) > {}^{22}Ne(1.00)$.

9.17   The amount of ozone is assumed to be constant.

Therefore $nR = \dfrac{P_iV_i}{T_i} = \dfrac{P_fV_f}{T_f}$

Because $V \propto h$, then $\dfrac{P_i h_i}{T_i} = \dfrac{P_f h_f}{T_f}$ where h is the thickness of the O_3 layer.

$$h_f = \frac{P_i}{P_f} \times \frac{T_f}{T_i} \times h_i = \left(\frac{1.6 \times 10^{-9} \text{ atm}}{1 \text{ atm}} \right) \left(\frac{273 \text{ K}}{230 \text{ K}} \right) (20 \times 10^3 \text{ m}) = 3.8 \times 10^{-5} \text{ m}$$

(Actually, $V = 4\pi r^2 h$, where r = the radius of the earth. When you go out ~30 km to get to the ozone layer, the change in r^2 is less than 1%. Therefore you can neglect the change in r^2 and assume that V is proportional to h.)

Understanding Key Concepts

9.18 (a)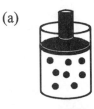

The volume of a gas is proportional to the kelvin temperature at constant pressure. As the temperature increases from 300 K to 450 K, the volume will increase by a factor of 1.5.

(b)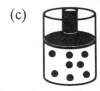

The volume of a gas is inversely proportional to pressure at constant temperature. As the pressure increases from 1 atm to 2 atm, the volume will decrease by a factor of 2.

(c)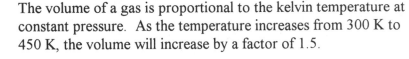

$PV = nRT$; The amount of gas (n) is constant.

Therefore $nR = \dfrac{P_i V_i}{T_i} = \dfrac{P_f V_f}{T_f}$.

Assume $V_i = 1$ L and solve for V_f.

$$\frac{P_i V_i T_f}{T_i P_f} = \frac{(3 \text{ atm})(1 \text{ L})(200 \text{ K})}{(300 \text{ K})(2 \text{ atm})} = V_f = 1 \text{ L}$$

There is no change in volume.

9.19 If the sample remains a gas at 150 K, then drawing (c) represents the gas at this temperature. The gas molecules still fill the container.

9.20 The two gases should mix randomly and homogeneously [by drawing (c)].

9.21 The two gases, on average, will be equally distributed among the three flasks.

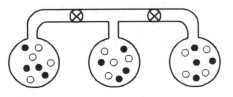

9.22   The gas pressure in the bulb in mm Hg is equal to the difference in the height of the Hg in the two arms of the manometer.

9.23   A

When stopcock A is opened, the pressure in the flask will equal the external pressure, and the level of mercury will be the same in both arms of the manometer.

Additional Problems
Gases and Gas Pressure

9.24   Collisions of gas molecules with the walls of a container exert a force per unit area that is measured as gas pressure.

9.26   Gases are much more compressible than solids or liquids because there is a large amount of empty space between individual gas molecules.

9.28 $P = 480 \text{ mm Hg } \times \dfrac{1.00 \text{ atm}}{760 \text{ mm Hg}} = 0.632 \text{ atm}$

$P = 480 \text{ mm Hg } \times \dfrac{101,325 \text{ Pa}}{760 \text{ mm Hg}} = 6.40 \times 10^4 \text{ Pa}$

9.30 $P_{flask} > 754.3 \text{ mm Hg}; \ P_{flask} = 754.3 \text{ mm Hg} + 176 \text{ mm Hg} = 930 \text{ mm Hg}$

9.32 $P_{flask} > 752.3 \text{ mm Hg}$
If the pressure in the flask can support a column of ethyl alcohol (d = 0.7893 g/mL) 55.1 cm high, then it can only support a column of Hg that is much shorter because of the higher density of Hg.

$55.1 \text{ cm } \times \dfrac{0.7893 \text{ g/mL}}{13.546 \text{ g/mL}} = 3.21 \text{ cm Hg} = 32.1 \text{ mm Hg}$

$P_{flask} = 752.3 \text{ mm Hg} + 32.1 \text{ mm } \text{ Hg} = 784.4 \text{ mm Hg}$

$P_{flask} = 784.4 \text{ mm Hg} \times \dfrac{101,325 \text{ Pa}}{760 \text{ mm Hg}} = 1.046 \times 10^5 \text{ Pa}$

9.34

	% Volume
N_2	78.08
O_2	20.95
Ar	0.93
CO_2	0.034

The % volume for a particular gas is proportional to the number of molecules of that gas in a mixture of gases.

Average molecular weight of air
$$= (0.7808)(\text{mol. wt. } N_2) + (0.2095)(\text{mol. wt. } O_2)$$
$$+ (0.0093)(\text{at. wt. Ar}) + (0.000\ 34)(\text{mol. wt. } CO_2)$$
$$= (0.7808)(28.01\ \text{amu}) + (0.2095)(32.00\ \text{amu})$$
$$+ (0.0093)(39.95\ \text{amu}) + (0.000\ 34)(44.01\ \text{amu}) = 28.96\ \text{amu}$$

The Gas Laws

9.36 (a) $\dfrac{nR}{V} = \dfrac{P_i}{T_i} = \dfrac{P_f}{T_f};$ $\qquad \dfrac{P_i T_f}{T_i} = P_f$

Let $P_i = 1$ atm, $T_i = 100$ K, $T_f = 300$ K

$$P_f = \frac{P_i T_f}{T_i} = \frac{(1\ \text{atm})(300\ \text{K})}{(100\ \text{K})} = 3\ \text{atm}$$

The pressure would triple.

(b) $\dfrac{RT}{V} = \dfrac{P_i}{n_i} = \dfrac{P_f}{n_f};$ $\qquad \dfrac{P_i n_f}{n_i} = P_f$

Let $P_i = 1$ atm, $n_i = 3$ mol, $n_f = 1$ mol

$$P_f = \frac{P_i n_f}{n_i} = \frac{(1\ \text{atm})(1\ \text{mol})}{(3\ \text{mol})} = \frac{1}{3}\ \text{atm}$$

The pressure would be $\dfrac{1}{3}$ the initial pressure.

(c) $nRT = P_i V_i = P_f V_f;$ $\qquad \dfrac{P_i V_i}{V_f} = P_f$

Let $P_i = 1$ atm, $V_i = 1$ L, $V_f = 1 - 0.45$ L $= 0.55$ L

$$P_f = \frac{P_i V_i}{V_f} = \frac{(1\ \text{atm})(1\ \text{L})}{(0.55\ \text{L})} = 1.8\ \text{atm}$$

The pressure would increase by 1.8 times.

(d) $nR = \dfrac{P_i V_i}{T_i} = \dfrac{P_f V_f}{T_f};$ $\qquad \dfrac{P_i V_i T_f}{T_i V_f} = P_f$

Let $P_i = 1$ atm, $V_i = 1$ L, $T_i = 200$ K, $V_f = 3$ L, $T_i = 100$ K

$$P_f = \frac{P_i V_i T_f}{T_i V_f} = \frac{(1\ \text{atm})(1\ \text{L})(100\ \text{K})}{(200\ \text{K})(3\ \text{L})} = 0.17\ \text{atm}$$

The pressure would be 0.17 times the initial pressure.

9.38 They all contain the same number of gas molecules.

9.40   n and T are constant; therefore $nRT = P_iV_i = P_fV_f$

$$V_f = \frac{P_iV_i}{P_f} = \frac{(150 \text{ atm})(49.0 \text{ L})}{(1.02 \text{ atm})} = 7210 \text{ L}$$

n and P are constant; therefore $\dfrac{nR}{P} = \dfrac{V_i}{T_i} = \dfrac{V_f}{T_f}$

$$V_f = \frac{V_iT_f}{T_i} = \frac{(49.0 \text{ L})(308 \text{ K})}{(293 \text{ K})} = 51.5 \text{ L}$$

9.42 $15.0 \text{ g CO}_2 \times \dfrac{1 \text{ mol CO}_2}{44.0 \text{ g CO}_2} = 0.341 \text{ mol CO}_2$

$$P = \frac{nRT}{V} = \frac{(0.341 \text{ mol})\left(0.082\ 06\ \dfrac{\text{L} \cdot \text{atm}}{\text{mol} \cdot \text{K}}\right)(300 \text{ K})}{(0.30 \text{ L})} = 27.98 \text{ atm}$$

$27.98 \text{ atm} \times \dfrac{760 \text{ mm Hg}}{1 \text{ atm}} = 2.1 \times 10^4 \text{ mm Hg}$

9.44 $\dfrac{1 \text{ H atom}}{\text{cm}^3} \times \dfrac{1 \text{ mol H}}{6.02 \times 10^{23} \text{ atoms}} \times \dfrac{1000 \text{ cm}^3}{1 \text{ L}} = 1.7 \times 10^{-21} \text{ mol H/L}$

$$P = \frac{nRT}{V} = \frac{(1.7 \times 10^{-21} \text{ mol})\left(0.082\ 06\ \dfrac{\text{L} \cdot \text{atm}}{\text{mol} \cdot \text{K}}\right)(100 \text{ K})}{(1 \text{ L})} = 1.4 \times 10^{-20} \text{ atm}$$

$P = 1.4 \times 10^{-20} \text{ atm} \times \dfrac{760 \text{ mm Hg}}{1.0 \text{ atm}} = 1 \times 10^{-17} \text{ mm Hg}$

9.46 $n = \dfrac{PV}{RT} = \dfrac{\left(17{,}180 \text{ kPa} \times \dfrac{1000 \text{ Pa}}{1 \text{ kPa}} \times \dfrac{1 \text{ atm}}{101{,}325 \text{ Pa}}\right)(43.8 \text{ L})}{\left(0.082\ 06\ \dfrac{\text{L} \cdot \text{atm}}{\text{mol} \cdot \text{K}}\right)(293 \text{K})} - 308.9 \text{ mol}$

$\text{mass Ar} = 308.9 \text{ mol} \times \dfrac{39.948 \text{ g}}{1 \text{ mol}} = 12340 \text{ g} = 1.23 \times 10^4 \text{ g}$

Gas Stoichiometry

9.48   For steam, $T = 123.0°C = 396 \text{ K}$

$n = \dfrac{PV}{RT} = \dfrac{(0.93 \text{ atm})(15.0 \text{ L})}{\left(0.082\ 06\ \dfrac{\text{L} \cdot \text{atm}}{\text{mol} \cdot \text{K}}\right)(396 \text{ K})} = 0.43 \text{ mol steam}$

For ice, H_2O, 18.02 amu; $\quad n = 10.5 \text{ g} \times \dfrac{1 \text{ mol}}{18.02 \text{ g}} = 0.583 \text{ mol ice}$

Because the number of moles of ice is larger than the number of moles of steam, the ice contains more H_2O molecules.

9.50   The containers are identical. Both containers contain the same number of gas molecules. Weigh the containers. Since the molecular weight for O_2 is greater than the molecular weight for H_2, the heavier container contains O_2.

9.52   room volume = 4.0 m x 5.0 m x 2.5 m = 50 m³

room volume = 50 m³ $\times \dfrac{1 \text{ L}}{10^{-3} \text{ m}^3} = 5.0 \times 10^4 \text{ L}$

$$n_{total} = \frac{PV}{RT} = \frac{(1.0 \text{ atm})(5.0 \times 10^4 \text{ L})}{\left(0.082\ 06 \dfrac{\text{L} \cdot \text{atm}}{\text{mol} \cdot \text{K}}\right)(273 \text{ K})} = 2.23 \times 10^3 \text{ mol}$$

$n_{O_2} = (0.2095)n_{total} = (0.2095)(2.23 \times 10^3 \text{ mol}) = 467 \text{ mol } O_2$

mass $O_2 = 467 \text{ mol} \times \dfrac{32.0 \text{ g}}{1 \text{ mol}} = 1.5 \times 10^4 \text{ g } O_2$

9.54   (a) CH_4, 16.04 amu; $\qquad d = \dfrac{16.04 \text{ g}}{22.4 \text{ L}} = 0.716 \text{ g/L}$

(b) CO_2, 44.01 amu; $\qquad d = \dfrac{44.01 \text{ g}}{22.4 \text{ L}} = 1.96 \text{ g/L}$

(c) O_2, 32.00 amu; $\qquad d = \dfrac{32.00 \text{ g}}{22.4 \text{ L}} = 1.43 \text{ g/L}$

(d) UF_6, 352.0 amu; $\qquad d = \dfrac{352.0 \text{ g}}{22.4 \text{ L}} = 15.7 \text{ g/L}$

9.56 $n = \dfrac{PV}{RT} = \dfrac{\left(356 \text{ mm Hg} \times \dfrac{1.00 \text{ atm}}{760 \text{ mm Hg}}\right)(1.500 \text{ L})}{\left(0.082\ 06 \dfrac{\text{L} \cdot \text{atm}}{\text{mol} \cdot \text{K}}\right)(295.5 \text{ K})} = 0.0290 \text{ mol}$

molar mass $= \dfrac{0.9847 \text{ g}}{0.0290 \text{ mol}} = 34.0 \text{ g/mol};$ molecular weight = 34.0 amu

9.58 $2 \text{ HgO(s)} \rightarrow 2 \text{ Hg(l)} + O_2(g);$ $\qquad$ HgO, 216.59 amu

10.57 g HgO $\times \dfrac{1 \text{ mol HgO}}{216.59 \text{ g HgO}} \times \dfrac{1 \text{ mol } O_2}{2 \text{ mol HgO}} = 0.024\ 40 \text{ mol } O_2$

$$V = \frac{nRT}{P} = \frac{(0.024\ 40 \text{ mol})\left(0.082\ 06 \dfrac{\text{L} \cdot \text{atm}}{\text{mol} \cdot \text{K}}\right)(273.15 \text{ K})}{1.000 \text{ atm}} = 0.5469 \text{ L}$$

9.60 $Zn(s) + 2 HCl(aq) \rightarrow ZnCl_2(aq) + H_2(g)$

(a) $25.5 \text{ g Zn} \times \dfrac{1 \text{ mol Zn}}{65.39 \text{ g Zn}} \times \dfrac{1 \text{ mol H}_2}{1 \text{ mol Zn}} = 0.390 \text{ mol H}_2$

$$V = \frac{nRT}{P} = \frac{(0.390 \text{ mol})\left(0.082\ 06 \dfrac{\text{L} \cdot \text{atm}}{\text{mol} \cdot \text{K}}\right)(288 \text{ K})}{\left(742 \text{ mm Hg} \times \dfrac{1.00 \text{ atm}}{760 \text{ mm Hg}}\right)} = 9.44 \text{ L}$$

(b) $n = \dfrac{PV}{RT} = \dfrac{\left(350 \text{ mm Hg} \times \dfrac{1.00 \text{ atm}}{760 \text{ mm Hg}}\right)(5.00 \text{ L})}{\left(0.082\ 06 \dfrac{\text{L} \cdot \text{atm}}{\text{mol} \cdot \text{K}}\right)(303.15 \text{ K})} = 0.092\ 56 \text{ mol H}_2$

$0.092\ 56 \text{ mol H}_2 \times \dfrac{1 \text{ mol Zn}}{1 \text{ mol H}_2} \times \dfrac{65.39 \text{ g Zn}}{1 \text{ mol Zn}} = 6.05 \text{ g Zn}$

9.62 (a) $V_{24h} = (4.50 \text{ L/min})(60 \text{ min/h})(24 \text{ h/day}) = 6480 \text{ L}$
  $V_{CO_2} = (0.034)V_{24h} = (0.034)(6480 \text{ L}) = 220 \text{ L}$

$$n = \frac{PV}{RT} = \frac{\left(735 \text{ mm Hg} \times \dfrac{1.00 \text{ atm}}{760 \text{ mm Hg}}\right)(220 \text{ L})}{\left(0.082\ 06 \dfrac{\text{L} \cdot \text{atm}}{\text{mol} \cdot \text{K}}\right)(298 \text{ K})} = 8.70 \text{ mol CO}_2$$

$8.70 \text{ mol CO}_2 \times \dfrac{44.01 \text{ g CO}_2}{1 \text{ mol CO}_2} = 383 \text{ g} = 380 \text{ g CO}_2$

(b) $2 Na_2O_2(s) + 2 CO_2(g) \rightarrow 2 Na_2CO_3(s) + O_2(g);$ $Na_2O_2, 77.98 \text{ amu}$

$3650 \text{ g Na}_2O_2 \times \dfrac{1 \text{ mol Na}_2O_2}{77.98 \text{ g Na}_2O_2} \times \dfrac{2 \text{ mol CO}_2}{2 \text{ mol Na}_2O_2} \times \dfrac{1 \text{ day}}{8.70 \text{ mol CO}_2} = 5.4 \text{ days}$

Dalton's Law and Mole Fraction

9.64 Because of Avogadro's Law ($V \propto n$), the % volumes are also % moles.

	% mole
N$_2$	78.08
O$_2$	20.95
Ar	0.93
CO$_2$	0.034

In decimal form, % mole = mole fraction.

$P_{N_2} = X_{N_2} \cdot P_{total} = (0.7808)(1.000 \text{ atm}) = 0.7808 \text{ atm}$

$P_{O_2} = X_{O_2} \cdot P_{total} = (0.2095)(1.000 \text{ atm}) = 0.2095 \text{ atm}$

$P_{Ar} = X_{Ar} \cdot P_{total} = (0.0093)(1.000 \text{ atm}) = 0.0093 \text{ atm}$

$P_{CO_2} = X_{CO_2} \cdot P_{total} = (0.000\ 34)(1.000 \text{ atm}) = 0.000\ 34 \text{ atm}$

Pressures of the rest are negligible.

9.66   Assume a 100.0 g sample.   g CO_2 = 1.00 g and g O_2 = 99.0 g

mol CO_2 = 1.00 g x $\dfrac{1 \text{ mol}}{44.01 \text{ g}}$ = 0.0227 mol CO_2

mol O_2 = 99.0 g x $\dfrac{1 \text{ mol}}{32.00 \text{ g}}$ = 3.094 mol O_2

n_{total} = 3.094 mol + 0.0227 mol = 3.117 mol

$X_{O_2} = \dfrac{3.094 \text{ mol}}{3.117 \text{ mol}} = 0.993$ $X_{CO_2} = \dfrac{0.0227 \text{ mol}}{3.117 \text{ mol}} = 0.007\ 28$

$P_{O_2} = X_{O_2} \cdot P_{total} = (0.993)(0.977 \text{ atm}) = 0.970 \text{ atm}$

$P_{CO_2} = X_{CO_2} \cdot P_{total} = (0.007\ 28)(0.977 \text{ atm}) = 0.007\ 11 \text{ atm}$

9.68   Assume a 100.0 g sample.

g HCl = (0.0500)(100.0 g) = 5.00 g;   5.00 g HCl x $\dfrac{1 \text{ mol HCl}}{36.5 \text{ g HCl}}$ = 0.137 mol HCl

g H_2 = (0.0100)(100.0 g) = 1.00 g;   1.00 g H_2 x $\dfrac{1 \text{ mol } H_2}{2.016 \text{ g } H_2}$ = 0.496 mol H_2

g Ne = (0.94)(100.0 g) = 94 g;   94 g Ne x $\dfrac{1 \text{ mol Ne}}{20.18 \text{ g Ne}}$ = 4.66 mol Ne

n_{total} = 0.137 + 0.496 + 4.66 = 5.3 mol

$X_{HCl} = \dfrac{0.137 \text{ mol}}{5.3 \text{ mol}} = 0.026$ $X_{H_2} = \dfrac{0.496 \text{ mol}}{5.3 \text{ mol}} = 0.094$ $X_{Ne} = \dfrac{4.66 \text{ mol}}{5.3 \text{ mol}} = 0.88$

9.70 $P_{total} = P_{H_2} + P_{H_2O}$; $P_{H_2} = P_{total} - P_{H_2O} = 747 \text{ mm Hg} - 23.8 \text{ mm Hg} = 723 \text{ mm Hg}$

$$n = \dfrac{PV}{RT} = \dfrac{\left(723 \text{ mm Hg} \times \dfrac{1.00 \text{ atm}}{760 \text{ mm Hg}}\right)(3.557 \text{ L})}{\left(0.082\ 06 \dfrac{L \cdot atm}{mol \cdot K}\right)(298 \text{ K})} = 0.1384 \text{ mol } H_2$$

0.1384 mol H_2 x $\dfrac{1 \text{ mol Mg}}{1 \text{ mol } H_2}$ x $\dfrac{24.3 \text{ g Mg}}{1 \text{ mol Mg}}$ = 3.36 g Mg

Kinetic-Molecular Theory and Graham's Law

9.72   The kinetic-molecular theory is based on the following assumptions:
1. A gas consists of tiny particles, either atoms or molecules, moving about at random.
2. The volume of the particles themselves is negligible compared with the total volume of the gas; most of the volume of a gas is empty space.

3. The gas particles act independently; there are no attractive or repulsive forces between particles.

4. Collisions of the gas particles, either with other particles or with the walls of the container, are elastic; that is, the total kinetic energy of the gas particles is constant at constant T.

5. The average kinetic energy of the gas particles is proportional to the Kelvin temperature of the sample.

9.74 Heat is the energy transferred from one object to another as the result of a temperature difference between them.

Temperature is a measure of the kinetic energy of molecular motion.

9.76 $$u = \sqrt{\frac{3\,RT}{M}} = \sqrt{\frac{3 \times 8.314\ kg\ m^2/(s^2\ K\ mol) \times 220\ K}{28.0 \times 10^{-3}\ kg/mol}} = 443\ m/s$$

9.78 For H_2, $$u = \sqrt{\frac{3\,RT}{M}} = \sqrt{\frac{3 \times 8.314\ kg\ m^2/(s^2\ K\ mol) \times 150\ K}{2.02 \times 10^{-3}\ kg/mol}} = 1360\ m/s$$

For He, $$u = \sqrt{\frac{3 \times 8.314\ kg\ m^2/(s^2\ K\ mol) \times 648\ K}{4.00 \times 10^{-3}\ kg/mol}} = 2010\ m/s$$

He at 375°C has the higher average speed.

9.80 $$\frac{rate_{H_2}}{rate_X} = \frac{\sqrt{M_X}}{\sqrt{M_{H_2}}}; \qquad \frac{2.92}{1} = \frac{\sqrt{M_X}}{\sqrt{2.02}}; \qquad 2.92\sqrt{2.02} = \sqrt{M_X}$$

Solve for M_X: $M_X = 17.2$ g/mol; molecular weight = 17.2 amu

9.82 HCl, 36.5 amu; F_2, 38.0 amu; Ar, 39.9 amu

$$\frac{rate\ HCl}{rate\ Ar} = \frac{\sqrt{M_{Ar}}}{\sqrt{M_{HCl}}} = \frac{\sqrt{39.9}}{\sqrt{36.5}} = 1.05 \qquad \frac{rate\ F_2}{rate\ Ar} = \frac{\sqrt{M_{Ar}}}{\sqrt{M_{F_2}}} = \frac{\sqrt{39.9}}{\sqrt{38.0}} = 1.02$$

The relative rates of diffusion are HCl(1.05) > F_2(1.02) > Ar(1.00).

9.84 $$u = 45\ m/s = \sqrt{\frac{3 \times 8.314\ kg\ m^2/(s^2\ K\ mol) \times T}{4.00 \times 10^{-3}\ kg/mol}}$$

Square both sides of the equation and solve for T.

$$2025\ m^2/s^2 = \frac{3 \times 8.314\ kg\ m^2/(s^2\ K\ mol) \times T}{4.00 \times 10^{-3}\ kg/mol}$$

T = 0.325 K = −272.83°C   (near absolute zero)

General Problems

9.86
$$\frac{\text{rate } ^{35}\text{Cl}_2}{\text{rate } ^{37}\text{Cl}_2} = \frac{\sqrt{M\ ^{37}\text{Cl}_2}}{\sqrt{M\ ^{35}\text{Cl}_2}} = \frac{\sqrt{74.0}}{\sqrt{70.0}} = 1.03$$

$$\frac{\text{rate } ^{35}\text{Cl}\ ^{37}\text{Cl}}{\text{rate } ^{37}\text{Cl}_2} = \frac{\sqrt{M\ ^{37}\text{Cl}_2}}{\sqrt{M\ ^{35}\text{Cl}^{37}\text{Cl}}} = \frac{\sqrt{74.0}}{\sqrt{72.0}} = 1.01$$

The relative rates of diffusion are $^{35}\text{Cl}_2(1.03) > {}^{35}\text{Cl}^{37}\text{Cl}(1.01) > {}^{37}\text{Cl}_2(1.00)$.

9.88
$$V = \frac{nRT}{P} = \frac{(1.00\ \text{mol})\left(0.082\ 06\ \dfrac{\text{L}\cdot\text{atm}}{\text{mol}\cdot\text{K}}\right)(1050\ \text{K})}{(75\ \text{atm})} = 1.1\ \text{L}$$

9.90
$$n = \frac{PV}{RT} = \frac{(2.15\ \text{atm})(7.35\ \text{L})}{\left(0.082\ 06\ \dfrac{\text{L}\cdot\text{atm}}{\text{mol}\cdot\text{K}}\right)(293\ \text{K})} = 0.657\ \text{mol Ar}$$

$$0.657\ \text{mol Ar} \times \frac{39.948\ \text{g Ar}}{1\ \text{mol Ar}} = 26.2\ \text{g Ar}$$

$$m_{\text{total}} = 478.1\ \text{g} + 26.2\ \text{g} = 504.3\ \text{g}$$

9.92 (a) Bulb A contains $CO_2(g)$ and $N_2(g)$;   Bulb B contains $CO_2(g)$, $N_2(g)$, and $H_2O(s)$.

(b) Initial moles of gas $= n = \dfrac{PV}{RT} = \dfrac{\left(564\ \text{mm Hg} \times \dfrac{1.00\ \text{atm}}{760\ \text{mm Hg}}\right)(1.000\ \text{L})}{\left(0.082\ 06\ \dfrac{\text{L}\cdot\text{atm}}{\text{mol}\cdot\text{K}}\right)(298\ \text{K})}$

Initial moles of gas $= 0.030\ 35$ mol

mol gas in Bulb A $= n = \dfrac{PV}{RT} = \dfrac{\left(219\ \text{mm Hg} \times \dfrac{1.00\ \text{atm}}{760\ \text{mm Hg}}\right)(1.000\ \text{L})}{\left(0.082\ 06\ \dfrac{\text{L}\cdot\text{atm}}{\text{mol}\cdot\text{K}}\right)(298\ \text{K})} = 0.011\ 78$ mol

mol gas in Bulb B $= n = \dfrac{PV}{RT} = \dfrac{\left(219\ \text{mm Hg} \times \dfrac{1.00\ \text{atm}}{760\ \text{mm Hg}}\right)(1.000\ \text{L})}{\left(0.082\ 06\ \dfrac{\text{L}\cdot\text{atm}}{\text{mol}\cdot\text{K}}\right)(203\ \text{K})} = 0.017\ 29$ mol

$n_{H_2O} = n_{\text{initial}} - n_A - n_B = 0.030\ 35 - 0.011\ 78 - 0.017\ 29 = 0.001\ 28\ \text{mol} = 0.0013\ \text{mol } H_2O$

(c) Bulb A contains $N_2(g)$.
    Bulb B contains $N_2(g)$ and $H_2O(s)$.
    Bulb C contains $N_2(g)$ and $CO_2(s)$.

(d) $n_A = \dfrac{PV}{RT} = \dfrac{\left(33.5 \text{ mm Hg} \times \dfrac{1.00 \text{ atm}}{760 \text{ mm Hg}}\right)(1.000 \text{ L})}{\left(0.082\ 06 \dfrac{\text{L} \cdot \text{atm}}{\text{mol} \cdot \text{K}}\right)(298 \text{ K})} = 0.001\ 803 \text{ mol}$

$n_B = \dfrac{PV}{RT} = \dfrac{\left(33.5 \text{ mm Hg} \times \dfrac{1.00 \text{ atm}}{760 \text{ mm Hg}}\right)(1.000 \text{ L})}{\left(0.082\ 06 \dfrac{\text{L} \cdot \text{atm}}{\text{mol} \cdot \text{K}}\right)(203 \text{ K})} = 0.002\ 646 \text{ mol}$

$n_C = \dfrac{PV}{RT} = \dfrac{\left(33.5 \text{ mm Hg} \times \dfrac{1.00 \text{ atm}}{760 \text{ mm Hg}}\right)(1.000 \text{ L})}{\left(0.082\ 06 \dfrac{\text{L} \cdot \text{atm}}{\text{mol} \cdot \text{K}}\right)(83 \text{ K})} = 0.006\ 472 \text{ mol}$

$n_{N_2} = n_A + n_B + n_C = 0.001\ 803 + 0.002\ 646 + 0.006\ 472 = 0.010\ 92 \text{ mol } N_2$

(e) $n_{CO_2} = n_{initial} - n_{H_2O} - n_{N_2} = 0.030\ 35 - 0.0013 - 0.010\ 92 = 0.0181 \text{ mol } CO_2$

9.94 $P = \dfrac{nRT}{V} = \dfrac{(0.60 \text{ mol})\left(0.082\ 06 \dfrac{\text{L} \cdot \text{atm}}{\text{mol} \cdot \text{K}}\right)(293 \text{ K})}{(0.200 \text{ L})} = 72 \text{ atm}$

$P = \dfrac{nRT}{(V - nb)} - \dfrac{an^2}{V^2}$

$P = \dfrac{(0.60 \text{ mol})\left(0.082\ 06 \dfrac{\text{L} \cdot \text{atm}}{\text{mol} \cdot \text{K}}\right)(293 \text{ K})}{[(0.200 \text{ L}) - (0.60 \text{ mol})(0.0391 \text{ L/mol})]} - \dfrac{\left(1.39 \dfrac{\text{L}^2 \cdot \text{atm}}{\text{mol}^2}\right)(0.60 \text{ mol})^2}{(0.200 \text{ L})^2}$

$P = 81.716 \text{ atm} - 12.51 \text{ atm} = 69 \text{ atm}$

9.96 (a) $2 \text{ C}_8\text{H}_{18}(l) + 25 \text{ O}_2(g) \rightarrow 16 \text{ CO}_2(g) + 18 \text{ H}_2\text{O}(g)$

(b) $4.6 \times 10^{10} \text{ L C}_8\text{H}_{18} \times \dfrac{1000 \text{ mL}}{1 \text{ L}} \times \dfrac{0.792 \text{ g}}{1 \text{ mL}} = 3.64 \times 10^{13} \text{ g C}_8\text{H}_{18}$

$3.64 \times 10^{13} \text{ g C}_8\text{H}_{18} \times \dfrac{1 \text{ mol C}_8\text{H}_{18}}{114.2 \text{ g C}_8\text{H}_{18}} \times \dfrac{16 \text{ mol CO}_2}{2 \text{ mol C}_8\text{H}_{18}} = 2.55 \times 10^{12} \text{ mol CO}_2$

$2.55 \times 10^{12} \text{ mol CO}_2 \times \dfrac{44.0 \text{ g CO}_2}{1 \text{ mol CO}_2} \times \dfrac{1 \text{ kg}}{1000 \text{ g}} = 1.1 \times 10^{11} \text{ kg CO}_2$

(c) $V = \dfrac{nRT}{P} = \dfrac{(2.55 \times 10^{12} \text{ mol})\left(0.082\ 06 \dfrac{\text{L} \cdot \text{atm}}{\text{mol} \cdot \text{K}}\right)(273 \text{ K})}{(1.00 \text{ atm})} = 5.7 \times 10^{13} \text{ L of CO}_2$

9.98   Freezing point of H_2O on the Rankine scale is $(9/5)(273.15) = 492°R$.

$$R = \frac{PV}{nT} = \frac{(1.00 \text{ atm})(22.414 \text{ L})}{(1.00 \text{ mol})(492°R)} = 0.0456 \frac{L \cdot atm}{mol \cdot °R}$$

9.100 $n = \dfrac{PV}{RT} = \dfrac{(1 \text{ atm})(1323 \text{ L})}{\left(0.082 \ 06 \dfrac{L \cdot atm}{mol \cdot K}\right)(2223 \text{ K})} = 7.25$ mol of all gases

(a) 0.004 00 mol "nitro" x $\dfrac{7.25 \text{ mol gases}}{1 \text{ mol }''nitro''} = 0.0290$ mol hot gases

(b) $n = \dfrac{PV}{RT} = \dfrac{\left(623 \text{ mm Hg } x \ \dfrac{1.00 \text{ atm}}{760 \text{ mm Hg}}\right)(0.500 \text{ L})}{\left(0.082 \ 06 \dfrac{L \cdot atm}{mol \cdot K}\right)(263 \text{ K})} = 0.0190$ mol B + C + D

$n_A = n_{total} - n_{(B+C+D)} = 0.0290 - 0.0190 = 0.0100$ mol A; A = H_2O

(c) $n = \dfrac{PV}{RT} = \dfrac{\left(260 \text{ mm Hg } x \ \dfrac{1.00 \text{ atm}}{760 \text{ mm Hg}}\right)(0.500 \text{ L})}{\left(0.082 \ 06 \dfrac{L \cdot atm}{mol \cdot K}\right)(298 \text{ K})} = 0.007 \ 00$ mol C + D

$n_B = n_{(B+C+D)} - n_{(C+D)} = 0.0190 - 0.007 \ 00 = 0.0120$ mol B; B = CO_2

(d) $n = \dfrac{PV}{RT} = \dfrac{\left(223 \text{ mm Hg } x \ \dfrac{1.00 \text{ atm}}{760 \text{ mm Hg}}\right)(0.500 \text{ L})}{\left(0.082 \ 06 \dfrac{L \cdot atm}{mol \cdot K}\right)(298 \text{ K})} = 0.006 \ 00$ mol D

$n_C = n_{(C+D)} - n_D = 0.007 \ 00 - 0.006 \ 00 = 0.001 \ 00$ mol C; C = O_2

molar mass D = $\dfrac{0.168 \text{ g}}{0.006 \ 00 \text{ mol}} = 28.0$ g/mol; D = N_2

(e) $0.004 \ C_3H_5N_3O_9(l) \rightarrow 0.0100 \ H_2O(g) + 0.012 \ CO_2(g) + 0.001 \ O_2(g) + 0.006 \ N_2(g)$
Multiply each coefficient by 1000 to obtain integers.
   $4 \ C_3H_5N_3O_9(l) \rightarrow 10 \ H_2O(g) + 12 \ CO_2(g) + O_2(g) + 6 \ N_2(g)$

9.102   (a) average molecular weight for natural gas
                $= (0.915)(16.04 \text{ amu}) + (0.085)(30.07 \text{ amu}) = 17.2$ amu

         total moles of gas = 15.50 g x $\dfrac{1 \text{ mol gas}}{17.2 \text{ g gas}} = 0.901$ mol gas

(b) $P = \dfrac{(0.901 \text{ mol})\left(0.082\ 06\ \dfrac{\text{L} \cdot \text{atm}}{\text{mol} \cdot \text{K}}\right)(293\ \text{K})}{(15.00\ \text{L})} = 1.44\ \text{atm}$

(c) $P_{CH_4} = X_{CH_4} \cdot P_{total} = (1.44\ \text{atm})(0.915) = 1.32\ \text{atm}$

$\quad P_{C_2H_6} = X_{C_2H_6} \cdot P_{total} = (1.44\ \text{atm})(0.085) = 0.12\ \text{atm}$

(d) $\Delta H_{comb}(CH_4) = -802.3\ \text{kJ/mol}$ and $\Delta H_{comb}(C_2H_6) = -1427.7\ \text{kJ/mol}$

Heat liberated $= (0.915)(0.901\ \text{mol})(-802.3\ \text{kJ/mol})$

$\qquad\qquad\qquad\quad + (0.085)(0.901)(-1427.7\ \text{kJ/mol}) = -771\ \text{kJ}$

10 Liquids, Solids, and Changes of State

10.1 $\mu = Q \times r = (1.60 \times 10^{-19}\ \text{C})(92 \times 10^{-12}\ \text{m})\left(\dfrac{1\ \text{D}}{3.336 \times 10^{-30}\ \text{C}\cdot\text{m}}\right) = 4.41\ \text{D}$

% ionic character for HF $= \dfrac{1.82\ \text{D}}{4.41\ \text{D}} \times 100\% = 41\%$

HF has more ionic character than HCl. HCl has only 17% ionic character.

10.2 net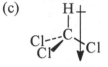

10.3 (a) SF_6 has polar covalent bonds but the molecule is symmetrical (octahedral). The individual bond polarities cancel, and the molecule has no dipole moment.
(b) $H_2C=CH_2$ can be assumed to have nonpolar C–H bonds. In addition, the molecule is symmetrical. The molecule has no dipole moment.
(c)   The C–Cl bonds in $CHCl_3$ are polar covalent bonds, and the molecule is polar.

(d)   The C–Cl bonds in CH_2Cl_2 are polar covalent bonds, and the molecule is polar.

10.4 (a) Of the four substances, only HNO_3 has a net dipole moment.
(b) Only HNO_3 can hydrogen bond.
(c) Ar has fewer electrons than Cl_2 and CCl_4, and has the smallest dispersion forces.

10.5 H_2S dipole–dipole, dispersion
CH_3OH hydrogen bonding, dipole–dipole, dispersion
CBr_4 dispersion
Ne   dispersion
$Ne < H_2S < CH_3OH < CBr_4$
Even though CBr_4 experiences only dispersion forces, it has the highest boiling point of the group because of its large number of electrons.

10.6 (a) $CO_2(s) \rightarrow CO_2(g)$, ΔS is positive.
(b) $H_2O(g) \rightarrow H_2O(l)$, ΔS is negative.
(c) ΔS is positive (more disorder).

10.7 $\Delta G = \Delta H - T\Delta S$; at the boiling point (phase change), $\Delta G = 0$.

$$\Delta H = T\Delta S; \quad T = \frac{\Delta H_{vap}}{\Delta S_{vap}} = \frac{29.2 \text{ kJ/mol}}{87.5 \times 10^{-3} \text{ kJ/(K·mol)}} = 334 \text{ K}$$

10.8 The boiling point is the temperature where the vapor pressure of a liquid equals the external pressure.

$P_1 = 760 \text{ mm Hg}; \quad P_2 = 260 \text{ mm Hg}; \qquad T_1 = 80.1°C$

$\Delta H_{vap} = 30.8 \text{ kJ/mol}$

$$\log P_2 = \log P_1 + \frac{\Delta H_{vap}}{2.303 \text{ R}}\left(\frac{1}{T_1} - \frac{1}{T_2}\right)$$

$$(\log P_2 - \log P_1)\left(\frac{2.303 \text{ R}}{\Delta H_{vap}}\right) = \frac{1}{T_1} - \frac{1}{T_2}$$

Solve for T_2 (the boiling point for benzene at 260 mm Hg).

$$\frac{1}{T_1} - (\log P_2 - \log P_1)\left(\frac{2.303 \text{ R}}{\Delta H_{vap}}\right) = \frac{1}{T_2}$$

$$\frac{1}{353.2 \text{ K}} - [\log(260) - \log(760)]\left(\frac{2.303\left(8.3145 \frac{J}{K·mol}\right)}{30,800 \text{ J/mol}}\right) = \frac{1}{T_2}$$

$$\frac{1}{T_2} = 0.003\ 121 \text{ K}^{-1}; \ T_2 = 320 \text{ K} = 47°C \text{ (boiling point is lower at lower pressure)}$$

10.9 $$\Delta H_{vap} = \frac{(\log P_2 - \log P_1)(2.303 \text{ R})}{\left(\dfrac{1}{T_1} - \dfrac{1}{T_2}\right)}$$

$P_1 = 400 \text{ mm Hg}; \qquad T_1 = 41.0 °C = 314.2 \text{ K}$

$P_2 = 760 \text{ mm Hg}; \qquad T_2 = 331.9 \text{ K}$

$$\Delta H_{vap} = \frac{[\log(760) - \log(400)][2.303\,(8.3145 \frac{J}{K·mol})]}{\left(\dfrac{1}{314.2 \text{ K}} - \dfrac{1}{331.9 \text{ K}}\right)} = 31,448 \text{ J/mol} = 31.4 \text{ kJ/mol}$$

10.10 Two cubes share a common face, and four cubes have a common edge.

10.11 (a) 1/8 atom at 8 corners and 1 atom at body center = 2 atoms
       (b) 1/8 atom at 8 corners and 1/2 atom at 6 faces = 4 atoms

10.12 For a simple cube, d = 2r; $\quad r = \dfrac{d}{2} = \dfrac{334 \text{ pm}}{2} = 167 \text{ pm}$

10.13   For a simple cube, there is one atom per unit cell.

mass of one Po atom = 209 g/mol x $\dfrac{1 \text{ mol}}{6.022 \times 10^{23} \text{ atoms}}$ = 3.4706 x 10^{-22} g/atom

unit cell edge = d = 334 pm = 334 x 10^{-12} m = 3.34 x 10^{-8} cm

unit cell volume = d^3 = (3.34 x 10^{-8} cm)^3 = 3.7260 x 10^{-23} cm^3

density = $\dfrac{\text{mass}}{\text{volume}}$ = $\dfrac{3.4706 \times 10^{-22} \text{ g}}{3.7260 \times 10^{-23} \text{ cm}^3}$ = 9.31 g/cm^3

10.14   Because Ar crystallizes in a face–centered cubic unit cell, there are four Ar atoms in the unit cell.

mass of one Ar atom = 39.95 g/mol x $\dfrac{1 \text{ mol}}{6.022 \times 10^{23} \text{ atom}}$ = 6.634 x 10^{-23} g/atom

unit cell mass = 4 atoms x mass of one Ar atom
        = 4 atoms x 6.634 x 10^{-23} g/atom = 2.654 x 10^{-22} g

density = $\dfrac{\text{mass}}{\text{volume}}$

unit cell volume = $\dfrac{\text{unit cell mass}}{\text{density}}$ = $\dfrac{2.654 \times 10^{-22} \text{ g}}{1.623 \text{ g/cm}^3}$ = 1.635 x 10^{-22} cm^3

unit cell edge = d = $\sqrt[3]{1.635 \times 10^{-22} \text{ cm}^3}$ = 5.468 x 10^{-8} cm

d = 5.468 x 10^{-8} cm x $\dfrac{1 \text{m}}{100 \text{ cm}}$ = 5.468 x 10^{-10} m = 546.8 x 10^{-12} m = 546.8 pm

r = $\sqrt{\dfrac{d^2}{8}}$ = $\sqrt{\dfrac{(546.8 \text{ pm})^2}{8}}$ = 193.3 pm

10.15   For CuCl:
1/8 Cl$^-$ at 8 corners and 1/2 Cl$^-$ at 6 faces = 4 Cl$^-$ (4 minuses)
4 Cu$^+$ inside (4 pluses)
For BaCl$_2$:
1/8 Ba^{2+} at 8 corners and 1/2 Ba^{2+} at 6 faces = 4 Ba^{2+} (8 pluses)
8 Cl$^-$ inside (8 minuses)

10.16   The minimum pressure at which liquid CO$_2$ can exist is its triple point pressure of 5.11 atm.

10.17   (a) CO$_2$(s) → CO$_2$(g)
(b) CO$_2$(l) → CO$_2$(g)
(c) CO$_2$(g) → CO$_2$(l) → supercritical CO$_2$

Understanding Key Concepts

10.18.   (a) toluene, C$_7$H$_8$(l); dispersion forces   (b) mercury, Hg(l); dispersion forces
(c) NH$_3$(l); hydrogen bonding

10.19

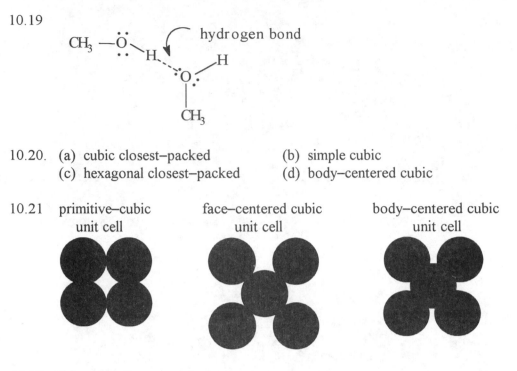

10.20. (a) cubic closest–packed   (b) simple cubic
(c) hexagonal closest–packed   (d) body–centered cubic

10.21   primitive–cubic   face–centered cubic   body–centered cubic
unit cell   unit cell   unit cell

10.22   (a) cubic closest–packed
(b) 1/8 S^{2-} at 8 corners and 1/2 S^{2-} at 6 faces = 4 S^{2-};   4 Zn^{2+} inside

10.23   1/8 Ca^{2+} at 8 corners = 1 Ca^{2+}; 1/2 O^{2-} at 6 faces = 3 O^{2-}; 1 Ti^{4+} inside
The oxidation number of Ti is +4 to maintain charge neutrality in the unit cell.
The formula for perovskite is $CaTiO_3$.

10.24   (a) normal boiling point ≈ 300 K; normal melting point ≈ 180 K
(b) (i) solid   (ii) gas   (iii) supercritical fluid

10.25

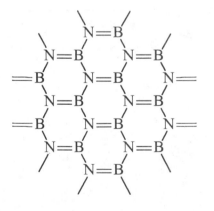

Additional Problems
Dipole Moments and Intermolecular Forces

10.26   The measure of net molecular polarity is called the dipole moment, μ.
μ = Q x r, where Q is the magnitude of the charge at either end of the dipole, and r is the distance between the charges.

10.28   ion–dipole forces, dipole–dipole forces, dispersion forces, and hydrogen bonding

10.30   (a) $CHCl_3$ has a permanent dipole moment. Dipole–dipole intermolecular forces are important. London dispersion forces are also present.
(b) O_2 has no dipole moment. London dispersion intermolecular forces are important.
(c) polyethylene, C_nH_{2n+2}. London dispersion intermolecular forces are important.
(d) CH_3OH has a permanent dipole moment. Dipole–dipole intermolecular forces and hydrogen bonding are important. London dispersion forces are also present.

10.32   For CH_3OH and CH_4, dispersion forces are small. CH_3OH can hydrogen bond; CH_4 cannot. This accounts for the large difference in boiling points.
For 1-decanol and decane, dispersion forces are comparable and relatively large along the C–H chain. 1-decanol can hydrogen bond; decane cannot. This accounts for the 55°C higher boiling point for 1-decanol.

10.34   (a)

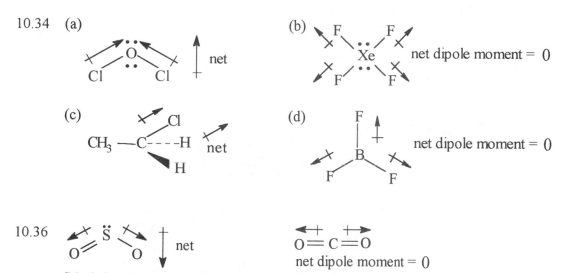

(b)   net dipole moment = 0

(c)   net

(d)   net dipole moment = 0

10.36   net   net dipole moment = 0
SO_2 is bent and the individual bond dipole moments add to give the molecule a net dipole moment.
CO_2 is linear and the individual bond dipole moments point in opposite directions to cancel each other out. CO_2 has no net dipole moment.

10.38   hydrogen bond

Vapor Pressure and Changes of State

10.40 ΔH_{vap} is the heat necessary to convert 1 mole of a liquid to 1 mole of its vapor (gas) at the normal boiling point.
ΔH_{fusion} is the heat necessary to convert 1 mole of a solid to 1 mol of the liquid at the normal melting point.
Both ΔH_{vap} and ΔH_{fusion} are positive (endothermic).

10.42   It takes more energy to convert liquid water to vapor at 100°C than to heat liquid water from 0°C to 100°C because vaporization involves the breaking of many more hydrogen bonds.

10.44   (a) $Hg(l) \rightarrow Hg(g)$ (b) no change of state, Hg remains a liquid
         (c) $Hg(g) \rightarrow Hg(l) \rightarrow Hg(s)$

10.46   As the pressure over the liquid H_2O is lowered, H_2O vapor is removed by the pump. As H_2O vapor is removed, more of the liquid H_2O is converted to H_2O vapor. This conversion is an endothermic process and the temperature decreases. The combination of both a decrease in pressure and temperature takes the system across the liquid/solid boundary in the phase diagram so the H_2O that remains turns to ice.

10.48 H_2O, 18.02 amu; $5.00 \text{ g } H_2O \times \dfrac{1 \text{ mol } H_2O}{18.02 \text{ g } H_2O} = 0.2775 \text{ mol } H_2O$

$q_1 = (0.2775 \text{ mol})[36.6 \times 10^{-3} \text{ kJ/(K} \cdot \text{mol)}](273 \text{ K} - 263 \text{ K}) = 0.1016 \text{ kJ}$
$q_2 = (0.2775 \text{ mol})(6.01 \text{ kJ/mol}) = 1.668 \text{ kJ}$
$q_3 = (0.2775 \text{ mol})(75.3 \times 10^{-3} \text{ kJ/(K} \cdot \text{mol)}](303 \text{ K} - 273 \text{ K}) = 0.6269 \text{ kJ}$
$q_{total} = q_1 + q_2 + q_3 = 2.40 \text{ kJ};$ 2.40 kJ of heat is required.

10.50 H_2O, 18.02 amu; $7.55 \text{ g } H_2O \times \dfrac{1 \text{ mol } H_2O}{18.02 \text{ g } H_2O} = 0.4190 \text{ mol } H_2O$

$q_1 = (0.4190 \text{ mol})[75.3 \times 10^{-3} \text{ kJ/(K} \cdot \text{mol)}](273.15 \text{ K} - 306.65 \text{ K}) = -1.057 \text{ kJ}$
$q_2 = -(0.4190 \text{ mol})(6.01 \text{ kJ/mol}) = -2.518 \text{ kJ}$
$q_3 = (0.4190 \text{ mol})[36.6 \times 10^{-3} \text{ kJ/(K} \cdot \text{mol)}](263.15 \text{ K} - 273.15 \text{ K}) = -0.1534 \text{ kJ}$
$q_{total} = q_1 + q_2 + q_3 = -3.73 \text{ kJ};$ 3.73 kJ of heat is released.

10.52

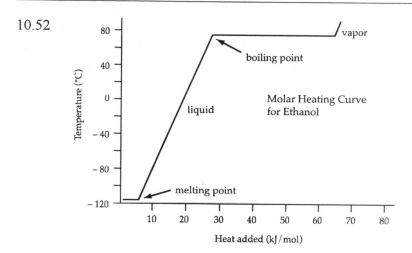

10.54   boiling point = 218°C = 491 K

$\Delta G = \Delta H_{vap} - T\Delta S_{vap};$ At the boiling point (phase change), $\Delta G = 0$

$\Delta H_{vap} = T\Delta S_{vap};$ $\Delta S_{vap} = \dfrac{\Delta H_{vap}}{T} = \dfrac{43.3 \text{ kJ/mol}}{491 \text{ K}} = 0.0882 \text{ kJ/(K} \cdot \text{mol)} = 88.2 \text{ J/(K} \cdot \text{mol)}$

10.56 $\Delta H_{vap} = \dfrac{(\log P_2 - \log P_1)(2.303)(R)}{\left(\dfrac{1}{T_1} - \dfrac{1}{T_2}\right)}$

$T_1 = -5.1°C = 268.0 \text{ K};$ $P_1 = 100 \text{ mm Hg}$
$T_2 = 46.5°C = 319.6 \text{ K};$ $P_2 = 760 \text{ mm Hg}$

$\Delta H_{vap} = \dfrac{[\log(760) - \log(100)](2.303)[8.3145 \times 10^{-3} \text{ kJ/(K} \cdot \text{mol)}]}{\left(\dfrac{1}{268.0 \text{ K}} - \dfrac{1}{319.6 \text{ K}}\right)} = 28.0 \text{ kJ/mol}$

10.58 $\log P_2 = \log P_1 + \dfrac{\Delta H_{vap}}{2.303 \text{ R}}\left(\dfrac{1}{T_1} - \dfrac{1}{T_2}\right)$

$\Delta H_{vap} = 28.0 \text{ kJ/mol}$
$P_1 = 100 \text{ mm Hg};$ $T_1 = -5.1°C = 268.0 \text{ K};$ $T_2 = 20.0°C = 293.2 \text{ K}$
Solve for P_2.

$\log P_2 = \log (100) + \dfrac{28.0 \text{ kJ/mol}}{(2.303)[8.3145 \times 10^{-3} \text{ kJ/(K} \cdot \text{mol)}]}\left(\dfrac{1}{268.0 \text{ K}} - \dfrac{1}{293.2 \text{ K}}\right)$

$\log P_2 = 2.46895;$ $P_2 = 294.4 \text{ mm Hg} = 294 \text{ mm Hg}$

10.60	T(K)	P_{vap}(mm Hg)	log P_{vap}	1/T
	263	80.1	1.904	0.003 802
	273	133.6	2.1258	0.003 663
	283	213.3	2.3290	0.003 534
	293	329.6	2.5180	0.003 413
	303	495.4	2.6950	0.003 300
	313	724.4	2.8600	0.003 195

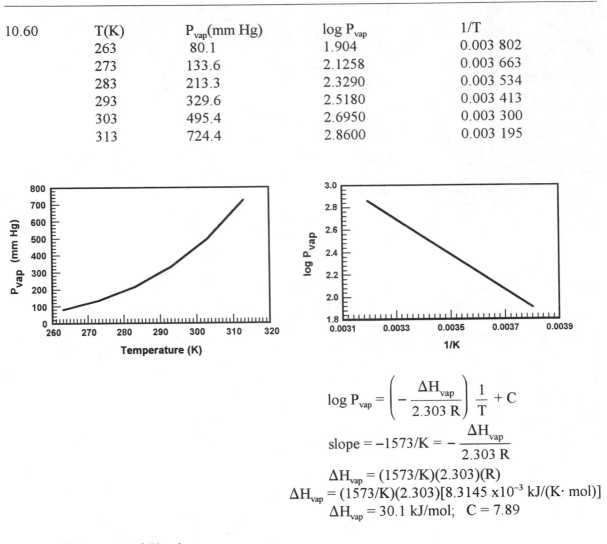

$$\log P_{vap} = \left(-\frac{\Delta H_{vap}}{2.303\ R}\right)\frac{1}{T} + C$$

$$\text{slope} = -1573/\text{K} = -\frac{\Delta H_{vap}}{2.303\ R}$$

$$\Delta H_{vap} = (1573/\text{K})(2.303)(R)$$
$$\Delta H_{vap} = (1573/\text{K})(2.303)[8.3145 \times 10^{-3}\ \text{kJ}/(\text{K} \cdot \text{mol})]$$
$$\Delta H_{vap} = 30.1\ \text{kJ/mol}; \quad C = 7.89$$

10.62 $\Delta H_{vap} = 30.1$ kJ/mol

10.64 $\Delta H_{vap} = \dfrac{(\log P_2 - \log P_1)(2.303)(R)}{\left(\dfrac{1}{T_1} - \dfrac{1}{T_2}\right)}$

$P_1 = 80.1$ mm Hg; $T_1 = 263$ K
$P_2 = 724.4$ mm Hg; $T_2 = 313$ K

$\Delta H_{vap} = \dfrac{[\log(724.4) - \log(80.1)](2.303)(8.3145 \times 10^{-3}\ \text{kJ/K}\cdot \text{mol})}{\left(\dfrac{1}{263\ \text{K}} - \dfrac{1}{313\ \text{K}}\right)} = 30.1$ kJ/mol

The calculated ΔH_{vap} and that obtained from the plot in Problem 10.60 are the same.

Structures of Solids

10.66   molecular solid, CO_2, I_2
        metallic solid, any metallic element
        covalent network solid, diamond
        ionic solid, NaCl

10.68   The unit cell is the smallest repeating unit in a crystal.

10.70   Cu is face–centered cubic. d = 362 pm; $r = \sqrt{\dfrac{d^2}{8}} = \sqrt{\dfrac{(362 \text{ pm})^2}{8}} = 128$ pm

   362 pm = 362 x 10^{-12} m = 3.62 x 10^{-8} cm
   unit cell volume = (3.62 x 10^{-8} cm)^3 = 4.74 x 10^{-23} cm^3

   mass of one Cu atom = 63.55 g/mol x $\dfrac{1 \text{ mol}}{6.022 \times 10^{23} \text{ atom}}$ = 1.055 x 10^{-22} g/atom

   Cu is face–centered cubic; there are therefore four Cu atoms in the unit cell.
   unit cell mass = (4 atoms)(1.055 x 10^{-22} g/atom) = 4.22 x 10^{-22} g

   density = $\dfrac{\text{mass}}{\text{volume}}$ = $\dfrac{4.22 \times 10^{-22} \text{g}}{4.74 \times 10^{-23} \text{ cm}^3}$ = 8.90 g/cm^3

10.72   mass of one Al atom = 26.98 g/mol x $\dfrac{1 \text{ mol}}{6.022 \times 10^{23} \text{ atom}}$ = 4.480 x 10^{-23} g/atom

   Al is face–centered cubic; there are therefore four Al atoms in the unit cell.
   unit cell mass = (4 atoms)(4.480 x 10^{-23} g/atom) = 1.792 x 10^{-22} g

   density = $\dfrac{\text{mass}}{\text{volume}}$

   unit cell volume = $\dfrac{\text{unit cell mass}}{\text{density}}$ = $\dfrac{1.792 \times 10^{-22} \text{ g}}{2.699 \text{ g/cm}^3}$ = 6.640 x 10^{-23} cm^3

   unit cell edge = d = $\sqrt[3]{6.640 \times 10^{-23} \text{ cm}^3}$ = 4.049 x 10^{-8} cm

   d = 4.049 x 10^{-8} cm x $\dfrac{1 \text{m}}{100 \text{ cm}}$ = 4.049 x 10^{-10} m = 404.9 x 10^{-12} m = 404.9 pm

10.74   unit cell body diagonal = 4r = 549 pm
        For W, r = $\dfrac{549 \text{ pm}}{4}$ = 137 pm

10.76   mass of one Ti atom = 47.88 g/mol x $\dfrac{1 \text{ mol}}{6.022 \times 10^{23} \text{ atoms}}$ = 7.951 x 10^{-23} g/atom

   r = 144.8 pm = 144.8 x 10^{-12} m

   r = 144.8 x 10^{-12} m x $\dfrac{100 \text{ cm}}{1 \text{ m}}$ = 1.448 x 10^{-8} cm

   Calculate the volume and then the density for Ti assuming it is primitive cubic, body-centered cubic, and face-centered cubic. Compare the calculated density with the actual density to identify the unit cell.

For primitive cubic:

$$d = 2r; \text{ volume} = d^3 = [2(1.448 \times 10^{-8} \text{ cm})]^3 = 2.429 \times 10^{-23} \text{ cm}^3$$

$$\text{density} = \frac{\text{unit cell mass}}{\text{volume}} = \frac{7.951 \times 10^{-23} \text{ g}}{2.429 \times 10^{-23} \text{ cm}^3} = 3.273 \text{ g/cm}^3$$

For face-centered cubic:

$$d = 2\sqrt{2}r; \text{ volume} = d^3 = [2\sqrt{2}(1.448 \times 10^{-8} \text{ cm})]^3 = 6.870 \times 10^{-23} \text{ cm}^3$$

$$\text{density} = \frac{4(7.951 \times 10^{-23} \text{ g})}{6.870 \times 10^{-23} \text{ cm}^3} = 4.630 \text{ g/cm}^3$$

For body-centered cubic:

From Problems 10.73 and 10.74,

$$d = \frac{4r}{\sqrt{3}}; \text{ volume} = d^3 = \left[\frac{4(1.448 \times 10^{-8} \text{ cm})}{\sqrt{3}}\right]^3 = 3.739 \times 10^{-23} \text{ cm}^3$$

$$\text{density} = \frac{2(7.951 \times 10^{-23} \text{ g})}{3.739 \times 10^{-23} \text{ cm}^3} = 4.253 \text{ g/cm}^3$$

The calculated density for a face–centered cube (4.630 g/cm^3) is closest to the actual density of 4.54 g/cm^3. Ti crystallizes in the face–centered cubic unit cell.

10.78 Six Na$^+$ ions touch each H$^-$ ion and six H$^-$ ions touch each Na$^+$ ion.

10.80   Na$^+$ H$^-$ Na$^+$
          $\leftarrow$ 488 pm $\rightarrow$ unit cell edge = d = 488 pm; Na–H bond = d/2 = 244 pm

Phase Diagrams

10.82 (a) gas   (b) liquid   (c) solid

10.84

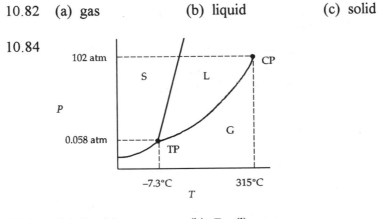

10.86 (a) Br$_2$(s)   (b) Br$_2$(l)

10.88 Solid O$_2$ does not melt when pressure is applied because the solid is denser than the liquid and the solid/liquid boundary in the phase diagram slopes to the right.

10.90

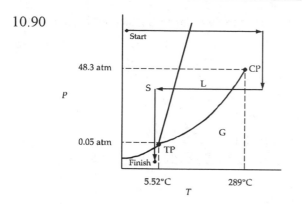

The starting phase is benzene as a solid, and the final phase is benzene as a gas.

10.92   solid → liquid → supercritical fluid → liquid → solid → gas

General Problems

10.94   Because chlorine is larger than fluorine, the charge separation is larger in CH_3Cl compared to CH_3F resulting in CH_3Cl having a slightly larger dipole moment.

10.96 $7.50 \text{ g} \times \dfrac{1 \text{ mol}}{200.6 \text{ g}} = 0.037\ 39 \text{ mol Hg}$

$q_1 = (0.037\ 39 \text{ mol})[28.2 \times 10^{-3} \text{ kJ/(K} \cdot \text{mol)}](234.2 \text{ K} - 223.2 \text{ K}) = 0.011\ 60 \text{ kJ}$
$q_2 = (0.037\ 39 \text{ mol})(2.33 \text{ kJ/mol}) = 0.087\ 12 \text{ kJ}$
$q_3 = (0.037\ 39 \text{ mol})[27.9 \times 10^{-3} \text{ kJ/(K} \cdot \text{mol)}](323.2 \text{ K} - 234.2 \text{ K}) = 0.092\ 84 \text{ kJ}$
$q_{total} = q_1 + q_2 + q_3 = 0.192 \text{ kJ};$ $0.192 \text{ kJ of heat is required.}$

10.98 $\log P_2 = \log P_1 + \dfrac{\Delta H_{vap}}{2.303 \text{ R}}\left(\dfrac{1}{T_1} - \dfrac{1}{T_2}\right)$

$\Delta H_{vap} = 40.67 \text{ kJ/mol}$
At 1 atm, H_2O boils at 100°C; therefore set
$T_1 = 100°C = 373 \text{ K}$, and $P_1 = 1.00 \text{ atm}.$
Let $T_2 = 95°C = 368 \text{ K}$, and solve for $P_2.$ (P_2 is the atmospheric pressure in Denver.)

$\log P_2 = \log(1) + \dfrac{40.67 \text{ kJ/mol}}{(2.303)[8.3145 \times 10^{-3} \text{ kJ/(K} \cdot \text{mol)}]}\left(\dfrac{1}{373 \text{ K}} - \dfrac{1}{368 \text{ K}}\right)$

$\log P_2 = -0.077\ 37;$ $P_2 = 0.837 \text{ atm}$

10.100 $\Delta G = \Delta H - T\Delta S$; at the melting point (phase change), $\Delta G = 0.$

$\Delta H = T\Delta S;$ $T = \dfrac{\Delta H_{vap}}{\Delta S_{vap}} = \dfrac{9.037 \text{ kJ/mol}}{9.79 \times 10^{-3} \text{ kJ/(K} \cdot \text{mol)}} = 923 \text{ K} = 650°C$

10.102 $\Delta H_{vap} = \dfrac{(\log P_2 - \log P_1)(2.303\,R)}{\left(\dfrac{1}{T_1} - \dfrac{1}{T_2}\right)}$

$P_1 = 40.0$ mm Hg; $\qquad T_1 = 191.6$ K
$P_2 = 400$ mm Hg; $\qquad T_2 = 229.2$ K

$\Delta H_{vap} = \dfrac{[\log(400) - \log(40.0)]\left[(2.303)\left(8.3145 \times 10^{-3}\dfrac{kJ}{K\cdot mol}\right)\right]}{\left(\dfrac{1}{191.6\,K} - \dfrac{1}{229.2\,K}\right)} = 22.36$ kJ/mol

Using $\Delta H_{vap} = 22.36$ kJ/mol

$\log P_2 = \log P_1 + \dfrac{\Delta H_{vap}}{2.303\,R}\left(\dfrac{1}{T_1} - \dfrac{1}{T_2}\right)$

$(\log P_2 - \log P_1)\left(\dfrac{2.303\,R}{\Delta H_{vap}}\right) = \dfrac{1}{T_1} - \dfrac{1}{T_2}$

$\dfrac{1}{T_1} - (\log P_2 - \log P_1)\left(\dfrac{2.303\,R}{\Delta H_{vap}}\right) = \dfrac{1}{T_2}$

$P_1 = 40.0$ mm Hg; $\qquad T_1 = 191.6$ K
$P_2 = 760$ mm Hg
Solve for T_2 (the normal boiling point).

$\dfrac{1}{191.6\,K} - [\log(760) - \log(40.0)]\left(\dfrac{2.303\left(8.3145 \times 10^{-3}\dfrac{kJ}{K\cdot mol}\right)}{22.36\ kJ/mol}\right) = \dfrac{1}{T_2}$

$\dfrac{1}{T_2} = 0.004\,124\,19;\ \ T_2 = 242.47$ K $= -30.7°C$

10.104 $\Delta H_{vap} = \dfrac{(\log P_2 - \log P_1)(2.303\,R)}{\left(\dfrac{1}{T_1} - \dfrac{1}{T_2}\right)}$

$P_1 = 100$ mm Hg; $\qquad T_1 = 162.85$ K
$P_2 = 760$ mm Hg; $\qquad T_2 = 184.65$ K

$\Delta H_{vap} = \dfrac{[\log(760) - \log(100)]\left[(2.303)\left(8.3145 \times 10^{-3}\dfrac{kJ}{K\cdot mol}\right)\right]}{\left(\dfrac{1}{162.85\,K} - \dfrac{1}{184.65\,K}\right)} = 23.3$ kJ/mol

10.106

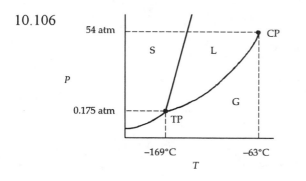

Kr cannot be liquified at room temperature because room temperature is above T_c (−63°C).

10.108 For a body–centered cube

$$4r = \sqrt{3} \text{ edge}; \qquad \text{edge} = \frac{4r}{\sqrt{3}}$$

$$\text{volume of sphere} = \frac{4}{3}\pi r^3$$

$$\text{volume of unit cell} = \left(\frac{4r}{\sqrt{3}}\right)^3 = \frac{64\,r^3}{3\sqrt{3}}$$

$$\text{volume of 2 spheres} = 2\left(\frac{4}{3}\pi r^3\right) = \frac{8}{3}\pi r^3$$

$$\% \text{ volume occupied} = \frac{\left(\frac{8}{3}\pi r^3\right)}{\left(\frac{64r^3}{3\sqrt{3}}\right)} \times 100\% = 68\%$$

10.110 unit cell edge = d = 287 pm = 287 x 10^{-12} m = 2.87 x 10^{-8} cm
unit cell volume = d^3 = (2.87 x 10^{-8} cm)^3 = 2.364 x 10^{-23} cm^3
unit cell mass = (2.364 x 10^{-23} cm^3)(7.86 g/cm^3) = 1.858 x 10^{-22} g
Fe is body-centered cubic; therefore there are two Fe atoms per unit cell.

$$\text{mass of one Fe atom} = \frac{1.858 \times 10^{-22} \text{ g}}{2 \text{ Fe atoms}} = 9.290 \times 10^{-23} \text{ g/atom}$$

$$\text{Avogadro's number} = 55.85 \text{ g/mol} \times \frac{1 \text{ atom}}{9.290 \times 10^{-23} \text{ g}} = 6.01 \times 10^{23} \text{ atoms/mol}$$

10.112 unit cell edge = $2r_{Cl^-}$ + $2r_{Na^+}$ = 2(181 pm) + 2(97 pm) = 556 pm

10.114 (a) $n_{H_2} = \dfrac{PV}{RT} = \dfrac{\left(740 \text{ mm Hg} \times \dfrac{1.00 \text{ atm}}{760 \text{ mm Hg}}\right)(4.00 \text{ L})}{\left(0.08206 \dfrac{L \cdot atm}{mol \cdot K}\right)(296 \text{ K})} = 0.160 \text{ mol } H_2$

M = Group 3A metal; $2\,M(s) + 6\,H^+(aq) \rightarrow 2\,M^{3+}(aq) + 3\,H_2(g)$

$n_M = 0.160\;\text{mol}\;H_2 \times \dfrac{2\;\text{mol M}}{3\;\text{mol}\;H_2} = 0.107\;\text{mol M}$

mass M = $1.07\;\text{cm}^3 \times 2.70\;\text{g/cm}^3 = 2.89\;\text{g M}$

molar mass M = $\dfrac{2.89\;\text{g M}}{0.107\;\text{mol M}} = 27.0\;\text{g/mol};$ The Group 3A metal is Al

(b) mass of one Al atom = $26.98\;\text{g/mol} \times \dfrac{1\;\text{mol}}{6.022 \times 10^{23}\;\text{atoms}} = 4.48 \times 10^{-23}\;\text{g/atom}$

unit cell edge = d = 404 pm = 404×10^{-12} m

d = $404 \times 10^{-12}\;\text{m} \times \dfrac{100\;\text{cm}}{1\;\text{m}} = 4.04 \times 10^{-8}\;\text{cm}$

unit cell volume = $d^3 = (4.04 \times 10^{-8}\;\text{cm})^3 = 6.59 \times 10^{-23}\;\text{cm}^3$

Calculate the density of Al assuming it is primitive cubic, body-centered cubic, and face-centered cubic. Compare the calculated density with the actual density to identify the unit cell.

For primitive cubic:

$\text{density} = \dfrac{\text{unit cell mass}}{\text{unit cell volume}} = \dfrac{(1\;\text{Al})(4.48 \times 10^{-23}\;\text{g/Al atom})}{6.59 \times 10^{-23}\;\text{cm}^3} = 0.680\;\text{g/cm}^3$

For body-centered cubic:

$\text{density} = \dfrac{\text{unit cell mass}}{\text{unit cell volume}} = \dfrac{(2\;\text{Al})(4.48 \times 10^{-23}\;\text{g/Al atom})}{6.59 \times 10^{-23}\;\text{cm}^3} = 1.36\;\text{g/cm}^3$

For face-centered cubic:

$\text{density} = \dfrac{\text{unit cell mass}}{\text{unit cell volume}} = \dfrac{(4\;\text{Al})(4.48 \times 10^{-23}\;\text{g/Al atom})}{6.59 \times 10^{-23}\;\text{cm}^3} = 2.72\;\text{g/cm}^3$

The calculated density for a face-centered cube ($2.72\;\text{g/cm}^3$) is closest to the actual density of $2.70\;\text{g/cm}^3$. Al crystallizes in the face-centered cubic unit cell.

(c) $r = \sqrt{\dfrac{d^2}{8}} = \sqrt{\dfrac{(404\;\text{pm})^2}{8}} = 143\;\text{pm}$

10.116 unit cell edge = d = 839 pm = 839×10^{-12} m

d = $839 \times 10^{-12}\;\text{m} \times \dfrac{100\;\text{cm}}{1\;\text{m}} = 8.39 \times 10^{-8}\;\text{cm}$

unit cell volume = $d^3 = (8.39 \times 10^{-8}\;\text{cm})^3 = 5.91 \times 10^{-22}\;\text{cm}^3$

unit cell mass = $(5.91 \times 10^{-22}\;\text{cm}^3)(5.20\;\text{g/cm}^3) = 3.07 \times 10^{-21}\;\text{g}$

mass of Fe in unit cell = $(0.7236)(3.07 \times 10^{-21}\;\text{g}) = 2.22 \times 10^{-21}\;\text{g Fe}$

mass of O in unit cell = $(1 - 0.7236)(3.07 \times 10^{-21}\;\text{g}) = 8.49 \times 10^{-22}\;\text{g O}$

Fe atoms in unit cell = $2.22 \times 10^{-21}\;\text{g} \times \dfrac{6.022 \times 10^{23}\;\text{atoms/mol}}{55.847\;\text{g/mol}} = 24\;\text{Fe atoms}$

O atoms in unit cell = $8.49 \times 10^{-22}\;\text{g} \times \dfrac{6.022 \times 10^{23}\;\text{atoms/mol}}{16.00\;\text{g/mol}} = 32\;\text{O atoms}$

Solutions and Their Properties

11.1 Toluene is nonpolar and is insoluble in water.
Br_2 is nonpolar but because of its size is polarizable and is soluble in water.
KBr is an ionic compound and is very soluble in water.
toluene $< Br_2 <$ KBr   (solubility in H_2O)

11.2 (a) Na^+ has the larger (more negative) hydration energy because the Na^+ ion is smaller than the Cs^+ ion and water molecules can approach more closely and bind more tightly to the Na^+ ion.
(b) Ba^{2+} has the larger (more negative) hydration energy because of its higher charge.

11.3 NaCl, 58.44 amu; 1.00 mol NaCl = 58.44 g
1.00 L H_2O = 1000 mL = 1000 g (assuming a density of 1.00 g/mL)

$$\text{weight \% NaCl} = \frac{58.44 \text{ g}}{1000 \text{ g} + 58.44 \text{ g}} \times 100\% = 5.52 \text{ wt \%}$$

11.4 $$\text{ppm} = \frac{\text{mass of } CO_2}{\text{total mass of solution}} \times 10^6 \text{ ppm}$$

total mass of solution = density x volume = (1.3 g/L)(1.0 L) = 1.3 g

$$35 \text{ ppm} = \frac{\text{mass of } CO_2}{1.3 \text{ g}} \times 10^6 \text{ ppm}$$

$$\text{mass of } CO_2 = \frac{(35 \text{ ppm})(1.3 \text{ g})}{10^6 \text{ ppm}} = 4.6 \times 10^{-5} \text{ g } CO_2$$

11.5 Assume 1.00 L of sea water.
mass of 1.00 L = (1000 mL)(1.025 g/mL) = 1025 g

$$\frac{\text{mass NaCl}}{1025 \text{ g}} \times 100\% = 3.50 \text{ wt \%}; \qquad \text{mass NaCl} = \frac{1025 \text{ g} \times 3.50}{100} = 35.88 \text{ g}$$

There are 35.88 g NaCl per 1.00 L of solution.

$$M = \frac{(35.88 \text{ g NaCl})\left(\dfrac{1 \text{ mol NaCl}}{58.44 \text{ g NaCl}}\right)}{1.00 \text{ L}} = 0.614 \text{ M}$$

11.6 $C_{27}H_{46}O$, 386.7 amu; $CHCl_3$, 119.4 amu; $\quad 40.0 \text{ g} \times \dfrac{1 \text{ kg}}{1000 \text{ g}} = 0.0400 \text{ kg}$

$$\text{molality} = \frac{\text{mol } C_{27}H_{46}O}{\text{kg } CHCl_3} = \frac{\left(0.385 \text{ g} \times \dfrac{1 \text{ mol}}{386.7 \text{ g}}\right)}{0.0400 \text{ kg}} = 0.0249 \text{ } m$$

$$X_{C_{27}H_{46}O} = \frac{\text{mol } C_{27}H_{46}O}{\text{mol } C_{27}H_{46}O + \text{mol } CHCl_3}$$

$$X_{C_{27}H_{46}O} = \frac{\left(0.385 \text{ g} \times \dfrac{1 \text{ mol}}{386.7 \text{ g}}\right)}{\left[\left(0.385 \text{ g} \times \dfrac{1 \text{ mol}}{386.7 \text{ g}}\right) + \left(40.0 \text{ g} \times \dfrac{1 \text{ mol}}{119.4 \text{ g}}\right)\right]} = 2.96 \times 10^{-3}$$

11.7 CH_3CO_2Na, 82.03 amu

$$\text{kg } H_2O = (0.150 \text{ mol } CH_3CO_2Na)\left(\frac{1 \text{ kg } H_2O}{0.500 \text{ mol } CH_3CO_2Na}\right) = 0.300 \text{ kg } H_2O$$

$$(0.150 \text{ mol } CH_3CO_2Na)\left(\frac{82.03 \text{ g } CH_3CO_2Na}{1 \text{ mol } CH_3CO_2Na}\right) = 12.3 \text{ g } CH_3CO_2Na$$

mass of solution needed = 300 g + 12.3 g = 312 g

11.8   Assume you have a solution with 1.000 kg (1000 g) of H_2O. If this solution is 0.258 m, then it must also contain 0.258 mol glucose.

$$\text{mass of glucose} = 0.258 \text{ mol} \times \frac{180.2 \text{ g}}{1 \text{ mol}} = 46.5 \text{ g glucose}$$

mass of solution = 1000 g + 46.5 g = 1046.5 g
density = 1.0173 g/mL

$$\text{volume of solution} = 1046.5 \text{ g} \times \frac{1 \text{ mL}}{1.0173 \text{ g}} = 1028.7 \text{ mL}$$

$$\text{volume} = 1028.7 \text{ mL} \times \frac{1 \text{ L}}{1000 \text{ mL}} = 1.029 \text{ L}; \qquad \text{molarity} = \frac{0.258 \text{ mol}}{1.029 \text{ L}} = 0.251 \text{ M}$$

11.9   Assume 1.00 L of solution.
mass of 1.00 L = (1.0042 g/mL)(1000 mL) = 1004.2 g of solution

$$0.500 \text{ mol } CH_3COOH \times \frac{60.05 \text{ g } CH_3COOH}{1 \text{ mol } CH_3COOH} = 30.02 \text{ g } CH_3COOH$$

$$1004.2 \text{ g} - 30.02 \text{ g} = 974.2 \text{ g} = 0.9742 \text{ kg of } H_2O; \quad \text{molality} = \frac{0.500 \text{ mol}}{0.9742 \text{ kg}} = 0.513 \text{ } m$$

11.10   Assume you have 100.0 g of seawater.
mass NaCl = (0.0350)(100.0 g) = 3.50 g NaCl
mass H_2O = 100.0 g – 3.50 g = 96.5 g H_2O

NaCl, 58.44 amu; $\text{mol NaCl} = 3.50 \text{ g} \times \dfrac{1 \text{ mol}}{58.44 \text{ g}} = 0.0599 \text{ mol NaCl}$

$$\text{mass } H_2O = 96.5 \text{ g} \times \frac{1 \text{ kg}}{1000 \text{ g}} = 0.0965 \text{ kg } H_2O; \quad \text{molality} = \frac{0.0599 \text{ mol}}{0.0965 \text{ kg}} = 0.621 \text{ } m$$

11.11 $M = k \cdot P;\ k = \dfrac{M}{P} = \dfrac{3.2 \times 10^{-2}\ M}{1.0\ atm} = 3.2 \times 10^{-2}\ mol/(L \cdot atm)$

11.12 (a) $M = k \cdot P = [3.2 \times 10^{-2}\ mol/(L \cdot atm)](2.5\ atm) = 0.080\ M$

(b) $M = k \cdot P = [3.2 \times 10^{-2}\ mol/(L \cdot atm)](4.0 \times 10^{-4}\ atm) = 1.3 \times 10^{-5}\ M$

11.13 $C_7H_6O_2$, 122.1 amu; C_2H_6O, 46.07 amu

$$X_{solv} = \frac{mol\ C_2H_6O}{mol\ C_2H_6O + mol\ C_7H_6O_2} = \frac{\left(100\ g \times \dfrac{1\ mol}{46.07\ g}\right)}{\left(100\ g \times \dfrac{1\ mol}{46.07\ g}\right) + \left(5.00\ g \times \dfrac{1\ mol}{122.1\ g}\right)} = 0.981$$

$P_{soln} = P_{solv} \cdot X_{solv} = (100.5\ mm\ Hg)(0.981) = 98.6\ mm\ Hg$

11.14 $P_{soln} = P_{solv} \cdot X_{solv};\qquad X_{solv} = \dfrac{P_{soln}}{P_{solv}} = \dfrac{(55.3 - 1.30)\ mm\ Hg}{55.3\ mm\ Hg} = 0.976$

NaBr dissociates into two ions in aqueous solution.

$$X_{solv} = \frac{mol\ H_2O}{mol\ H_2O + mol\ Na^+ + mol\ Br^-}$$

$$X_{solv} = 0.976 = \frac{\left(250\ g \times \dfrac{1\ mol}{18.02\ g}\right)}{\left(250\ g \times \dfrac{1\ mol}{18.02\ g}\right) + x\ mol\ Na^+ + x\ mol\ Br^-}$$

$0.976 = \dfrac{13.9\ mol}{13.9\ mol + 2x\ mol};\qquad$ solve for x.

$x = 0.171\ mol\ Na^+ = 0.171\ mol\ Br^- = 0.171\ mol\ NaBr$

NaBr, 102.9 amu; mass NaBr $= 0.171\ mol \times \dfrac{102.9\ g}{1\ mol} = 17.6\ g\ NaBr$

11.15 C_2H_6O, 46.07 amu; H_2O, 18.02 amu

(a) $25.0\ g\ C_2H_6O \times \dfrac{1\ mol\ C_2H_6O}{46.07\ g\ C_2H_6O} = 0.5426\ mol\ C_2H_6O$

$100.0\ g\ H_2O \times \dfrac{1\ mol\ H_2O}{18.02\ g\ H_2O} = 5.549\ mol\ H_2O$

$X_{C_2H_6O} = \dfrac{0.5426\ mol}{0.5426\ mol + 5.549\ mol} = 0.08907$

$X_{H_2O} = \dfrac{5.549\ mol}{0.5426\ mol + 5.549\ mol} = 0.9109$

$P_{soln} = X_{C_2H_6O}P^{\circ}_{C_2H_6O} + X_{H_2O}P^{\circ}_{H_2O}$

$P_{soln} = (0.08907)(61.2\ mm\ Hg) + (0.9109)(23.8\ mm\ Hg) = 27.1\ mm\ Hg$

(b) $100 \text{ g } C_2H_6O \times \dfrac{1 \text{ mol } C_2H_6O}{46.07 \text{ g } C_2H_6O} = 2.171 \text{ mol } C_2H_6O$

$25.0 \text{ g } H_2O \times \dfrac{1 \text{ mol } H_2O}{18.02 \text{ g } H_2O} = 1.387 \text{ mol } H_2O$

$X_{C_2H_6O} = \dfrac{2.171 \text{ mol}}{2.171 \text{ mol } + 1.387 \text{ mol}} = 0.6102$

$X_{H_2O} = \dfrac{1.387 \text{ mol}}{2.171 \text{ mol } + 1.387 \text{ mol}} = 0.3898$

$P_{soln} = X_{C_2H_6O} P^{\circ}_{C_2H_6O} + X_{H_2O} P^{\circ}_{H_2O}$

$P_{soln} = (0.6102)(61.2 \text{ mm Hg}) + (0.3898)(23.8 \text{ mm Hg}) = 46.6 \text{ mm Hg}$

11.16 $C_9H_8O_4$, 180.2 amu; $CHCl_3$ is the solvent. For $CHCl_3$, $K_b = 3.63 \dfrac{^{\circ}C \cdot kg}{mol}$

$75.00 \text{ g } \times \dfrac{1 \text{ kg}}{1000 \text{ g}} = 0.075 \ 00 \text{ kg}$

$\Delta T_b = K_b \cdot m = \left(3.63 \dfrac{^{\circ}C \cdot kg}{mol} \right) \dfrac{\left(1.50 \text{ g} \times \dfrac{1 \text{ mol}}{180.2 \text{ g}} \right)}{(0.07500 \text{ kg})} = 0.40^{\circ}C$

solution boiling point $= 61.7^{\circ}C + \Delta T_b = 61.7^{\circ}C + 0.40^{\circ}C = 62.1^{\circ}C$

11.17 K_2SO_4, 174.3 amu; there are 3 ions (solute particles)/K_2SO_4

$\Delta T_f = K_f \cdot (3 \cdot m) = \left(1.86 \dfrac{^{\circ}C \cdot kg}{mol} \right)(3) \dfrac{\left(7.40 \text{ g} \times \dfrac{1 \text{ mol}}{174.3 \text{ g}} \right)}{0.110 \text{ kg}} = 2.15^{\circ}C$

Solution freezing point $= 0.00^{\circ}C - \Delta T_f = 0.00^{\circ}C - 2.15^{\circ}C = -2.15^{\circ}C$

11.18 There are 2 ions/KBr. $\qquad \Delta T_f = K_f \cdot 2 \cdot m$
freezing point $= -2.95^{\circ}C = 0.00^{\circ}C - \Delta T_f; \Delta T_f = 2.95^{\circ}C$

$m = \dfrac{\Delta T_f}{K_f \cdot 2} = \dfrac{2.95^{\circ}C}{\left(1.86 \dfrac{^{\circ}C \cdot kg}{mol} \right)(2)} = 0.793 \ m$

11.19 For $CaCl_2$ there are 3 ions (solute particles)/$CaCl_2$
$\Pi = MRT$; For $CaCl_2$, $\Pi = 3MRT$

$\Pi = (3)(0.125 \text{ mol/L}) \left(0.082 \ 06 \dfrac{L \cdot atm}{mol \cdot K} \right) (310 \text{ K}) = 9.54 \text{ atm}$

11.20 $\Pi = MRT;$ $M = \dfrac{\Pi}{RT} = \dfrac{(3.85\text{ atm})}{\left(0.082\ 06\ \dfrac{L \cdot atm}{mol \cdot K}\right)(300\text{ K})} = 0.156\text{ M}$

11.21 $\Delta T_f = K_f \cdot m;$ $m = \dfrac{\Delta T_f}{K_f} = \dfrac{2.10°C}{37.7\ \dfrac{°C \cdot kg}{mol}} = 0.0557\ m$

$0.0557\ \dfrac{mol}{kg} = \dfrac{\left(0.250\text{ g}\ \times\ \dfrac{1}{molar\ mass}\right)}{0.035\ 00\text{ kg}};$ solve for the molar mass.

molar mass = 128 g/mol

11.22 $\Pi = MRT;$ $M = \dfrac{\Pi}{RT} = \dfrac{\left(149\text{ mm Hg}\ \times\ \dfrac{1\text{ atm}}{760\text{ mm Hg}}\right)}{\left(0.08206\ \dfrac{L \cdot atm}{mol \cdot K}\right)(298\text{ K})} = 8.02 \times 10^{-3}\ \dfrac{mol}{L}$

molarity = $8.02 \times 10^{-3}\ \dfrac{mol}{L} = \dfrac{\left(0.822\text{ g}\ \times\ \dfrac{1}{molar\ mass}\right)}{(0.300\text{ L})}$

Solve for the molar mass.   molar mass = 342 g/mol

11.23   (a) bp ≈ 107°C   (b) $X_{toluene}$ ≈ 0.64 and $X_{benzene}$ ≈ 0.36

Understanding Key Concepts

11.24   (a) < (b) < (c)

11.25   The surface area of a solid plays an important role in determining how rapidly a solid dissolves. The larger the surface area, the more solid–solvent interactions, and the more rapidly the solid will dissolve. Powdered NaCl has a much larger surface area than a large block of NaCl, and it will dissolve more rapidly.

11.26   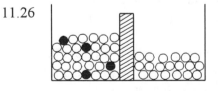   Assume that only the blue (open) spheres (solvent) can pass through the semipermeable membrane. There will be a net transfer of solvent from the right compartment (pure solvent) to the left compartment (solution) to achieve equilibrium.

11.27   NaCl is a nonvolatile solute. Methyl alcohol is a volatile solute. When NaCl is added to water, the vapor pressure of the solution is decreased, which means that the boiling point of the solution will increase. When methyl alcohol is added to water, the vapor pressure of the solution is increased which means that the boiling point of the solution will decrease.

11.28 K_f for snow (H_2O) is 1.86 $\dfrac{°C \cdot kg}{mol}$. Reasonable amounts of salt are capable of lowering the freezing point (ΔT_f) of the snow below an air temperature of $-2°C$. Reasonable amounts of salt, however, are not capable of causing a ΔT_f of more than $30°C$ which would be required if it is to melt snow when the air temperature is $-30°C$.

11.29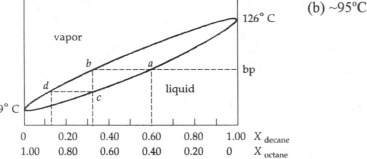

(b) ~95°C

11.30 ΔH_{soln} for HBr is exothermic. The HBr solution will be warm to touch.
ΔH_{soln} for $AgNO_3$ is endothermic. The $AgNO_3$ solution will be cool to touch.

11.31   The vapor pressure of the NaCl solution is lower than that of pure H_2O. More H_2O molecules will go into the vapor from the pure H_2O than from the NaCl solution. More H_2O vapor molecules will go into the NaCl solution than into pure H_2O. The result is represented by (b).

Additional Problems
Solutions and Energy Changes

11.32   Homogeneous mixtures can be classified according to the size of their constituent particles.
Colloids contain particles with diameters in the range 2 - 1000 nm.
Suspensions contain particles that are greater than about 1000 nm in diameter.
Solutions contain particles the size of a typical ion or covalent molecule, 0.2 - 2 nm in diameter.

11.34   Substances tend to dissolve when the solute and solvent have the same type and magnitude of intermolecular forces; thus the rule of thumb "like dissolves like."

11.36   Energy is required to overcome intermolecular forces holding solute particles together in the crystal. For an ionic solid, this is the lattice energy. Substances with higher lattice energies tend to be less soluble than substances with lower lattice energies.

11.38   Ethyl alcohol and water are both polar with small dispersion forces. They both can hydrogen bond, and are miscible.
Pentyl alcohol is slightly polar and can hydrogen bond. It has, however, a relatively large dispersion force because of its size, which limits its water solubility.

11.40 $CaCl_2$, 110.98 amu
For a 1.00 m solution:
heat released = 81,300 J
mass of solution = 1000 g H_2O + 110.98 g $CaCl_2$ = 1110.98 g

$$\Delta T = \frac{q}{(\text{specific heat})(\text{mass of solution})} = \frac{81,300 \text{ J}}{[4.18 \text{ J/(K} \cdot \text{g)}](1110.98 \text{ g})} = 17.5 \text{ K} = 17.5°C$$

Final temperature = 25.0°C + 17.5°C = 42.5°C

Units of Concentration

11.42 $\text{molarity} = \dfrac{\text{moles of solute}}{\text{liters of solution}}$; $\text{molality} = \dfrac{\text{moles of solute}}{\text{kg of solvent}}$

11.44 (a) Dissolve 1.50 mol of glucose in water; dilute to 1.00 L.
(b) Dissolve 1.135 mol of KBr in 1.00 kg of H_2O.
(c) Mix together 0.15 mol of CH_3OH with 0.85 mol of H_2O.

11.46 $C_7H_6O_2$, 122.12 amu, 165 mL = 0.165 L

$$\text{mol } C_7H_6O_2 = 0.165 \text{ L x } \frac{0.0268 \text{ mol}}{1.00 \text{ L}} = 0.004\ 42 \text{ mol}$$

$$\text{mass } C_7H_6O_2 = 0.004\ 42 \text{ mol x } \frac{122.12 \text{ g}}{1 \text{ mol}} = 0.540 \text{ g}$$

Dissolve 4.42 x 10^{-3} mol (0.540 g) of $C_7H_6O_2$ in enough $CHCl_3$ to make 165 mL of solution.

11.48 (a) KCl, 74.6 amu
A 0.500 M KCl solution contains 37.3 g of KCl per 1.00 L of solution.
A 0.500 wt % KCl solution contains 5.00 g of KCl per 995 g of water.
The 0.500 M KCl solution is more concentrated (that is, it contains more solute per amount of solvent).
(b) Both solutions contain the same amount of solute. The 1.75 M solution contains less solvent than the 1.75 m solution. The 1.75 M solution is more concentrated.

11.50 (a) $C_6H_8O_7$, 192.12 amu

$$0.655 \text{ mol } C_6H_8O_7 \text{ x } \frac{192.12 \text{ g } C_6H_8O_7}{1 \text{ mol } C_6H_8O_7} = 126 \text{ g } C_6H_8O_7$$

$$\text{weight \% } C_6H_8O_7 = \frac{126 \text{ g}}{126 \text{ g } + 1000 \text{ g}} \text{ x } 100\% = 11.2 \text{ wt \%}$$

(b) 0.135 mg = 0.135 x 10^{-3} g
(5.00 mL H_2O)(1.00 g/mL) = 5.00 g H_2O

$$\text{weight \% KBr} = \frac{0.135 \text{ x } 10^{-3} \text{ g}}{(0.135 \text{ x } 10^{-3} \text{ g}) + 5.00 \text{ g}} \text{ x } 100\% = 0.002\ 70 \text{ wt \% KBr}$$

(c) weight % aspirin $= \dfrac{5.50\ \text{g}}{5.50\ \text{g} + 145\ \text{g}} \times 100\% = 3.65$ wt % aspirin

11.52　$P_{O_3} = P_{total} \cdot X_{O_3};$ 　　$X_{O_3} = \dfrac{P_{O_3}}{P_{total}} = \dfrac{1.6 \times 10^{-9}\ \text{atm}}{1.3 \times 10^{-2}\ \text{atm}} = 1.2 \times 10^{-7}$

Assume one mole of air (29 g/mol)

mol $O_3 = n_{air} \cdot X_{O_3} = (1\ \text{mol})(1.2 \times 10^{-7}) = 1.2 \times 10^{-7}$ mol O_3

O_3, 48.00 amu;　　　mass $O_3 = 1.2 \times 10^{-7}$ mol $\times \dfrac{48.0\ \text{g}}{1\ \text{mol}} = 5.8 \times 10^{-6}$ g O_3

ppm $O_3 = \dfrac{5.8 \times 10^{-6}\ \text{g}}{29\ \text{g}} \times 10^6 = 0.20$ ppm

11.54　(a) H_2SO_4, 98.08 amu;　　molality $= \dfrac{\left(25.0\ \text{g} \times \dfrac{1\ \text{mol}}{98.08\ \text{g}} \right)}{1.30\ \text{kg}} = 0.196\ m$

(b) $C_{10}H_{14}N_2$, 162.23 amu; CH_2Cl_2, 84.93 amu

2.25 g $C_{10}H_{14}N_2 \times \dfrac{1\ \text{mol}\ C_{10}H_{14}N_2}{162.23\ \text{g}\ C_{10}H_{14}N_2} = 0.0139$ mol $C_{10}H_{14}N_2$

80.0 g $CH_2Cl_2 \times \dfrac{1\ \text{mol}\ CH_2Cl_2}{84.93\ \text{g}\ CH_2Cl_2} = 0.942$ mol CH_2Cl_2

$X_{C_{10}H_{14}N_2} = \dfrac{0.0139\ \text{mol}}{0.942\ \text{mol} + 0.0139\ \text{mol}} = 0.0145$

$X_{CH_2Cl_2} = \dfrac{0.942\ \text{mol}}{0.942\ \text{mol} + 0.0139\ \text{mol}} = 0.985$

11.56　16.0 wt % $= \dfrac{16.0\ \text{g}\ H_2SO_4}{16.0\ \text{g}\ H_2SO_4 + 84.0\ \text{g}\ H_2O}$

H_2SO_4, 98.08 amu;　　density = 1.1094 g/mL

volume of solution = 100.0 g $\times \dfrac{1\ \text{mL}}{1.1094\ \text{g}} = 90.14$ mL $= 0.090\ 14$ L

molarity $= \dfrac{\left(16.0\ \text{g} \times \dfrac{1\ \text{mol}}{98.08\ \text{g}} \right)}{0.090\ 14\ \text{L}} = 1.81$ M

11.58　molality $= \dfrac{\left(40.0\ \text{g} \times \dfrac{1\ \text{mol}}{62.07\ \text{g}} \right)}{0.0600\ \text{kg}} = 10.7\ m$

11.60 $C_{19}H_{21}NO_3$, 311.34 amu; 1.5 mg = 1.5×10^{-3} g

$$1.3 \times 10^{-3} \frac{mol}{kg} = \frac{\left(1.5 \times 10^{-3} g \times \frac{1\ mol}{311.34\ g}\right)}{kg\ of\ solvent};\quad \text{solve for kg of solvent.}$$

kg of solvent = 0.0037 kg

Because the solution is very dilute, kg of solvent ≈ kg of solution.

$$g\ of\ solution = (0.0037\ kg)\left(\frac{1000\ g}{1\ kg}\right) = 3.7\ g$$

11.62 $C_6H_{12}O_6$, 180.16 amu; H_2O, 18.02 amu;   Assume 1.00 L of solution.

mass of solution = (1000 mL)(1.0624 g/mL) = 1062.4 g

$$mass\ of\ solute = 0.944\ mol \times \frac{180.16\ g}{1\ mol} = 170.1\ g\ C_6H_{12}O_6$$

mass of H_2O = 1062.4 g – 170.1 g = 892.3 g H_2O

$$mol\ C_6H_{12}O_6 = 0.944\ mol;\quad mol\ H_2O = 892.3\ g \times \frac{1\ mol}{18.02\ g} = 49.5\ mol$$

(a) $$X_{C_6H_{12}O_6} = \frac{mol\ C_6H_{12}O_6}{mol\ C_6H_{12}O_6 + mol\ H_2O} = \frac{0.944\ mol}{0.944\ mol + 49.5\ mol} = 0.0187$$

(b) $$wt\ \% = \frac{mass\ C_6H_{12}O_6}{total\ mass\ of\ solution} \times 100\% = \frac{170.1\ g}{1062.4\ g} \times 100\% = 16.0\%$$

(c) $$molality = \frac{mol\ C_6H_{12}O_6}{kg\ H_2O} = \frac{0.944\ mol}{0.8923\ kg} = 1.06\ m$$

Solubility and Henry's Law

11.64 $$M = k \cdot P = (0.091\ \frac{mol}{L \cdot atm})(0.75\ atm) = 0.068\ M$$

11.66 $M = k \cdot P$

Calculate k: $$k = \frac{M}{P} = \frac{2.21 \times 10^{-3}\ mol/L}{1.00\ atm} = 2.21 \times 10^{-3}\ \frac{mol}{L \cdot atm}$$

Convert 4 mg/L to mol/L:

4 mg = 4×10^{-3} g

$$O_2\ molarity = \frac{\left(4 \times 10^{-3}\ g \times \frac{1\ mol}{32.00\ g}\right)}{1.00\ L} = 1.25 \times 10^{-4}\ M$$

$$P_{O_2} = \frac{M}{k} = \frac{1.25 \times 10^{-4}\ \frac{mol}{L}}{2.21 \times 10^{-3}\ \frac{mol}{L \cdot atm}} = 0.06\ atm$$

11.68   [Xe] = 10 mmol/L = 0.010 M at STP

$$M = k \cdot P; \quad k = \frac{M}{P} = \frac{0.010\,M}{1.0\,atm} = 0.010\,mol/(L \cdot atm)$$

Colligative Properties

11.70   The difference in entropy between a solution and a pure solvent is responsible for colligative properties.

11.72   Osmotic pressure is the amount of pressure that needs to be applied to cause osmosis to stop.

11.74

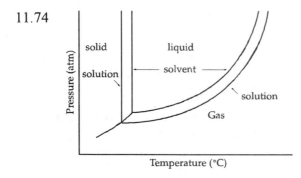

11.76   (a) CH_4N_2O, 60.06 amu; H_2O, 18.02 amu

$$10.0\,g\,CH_4N_2O \times \frac{1\,mol\,CH_4N_2O}{60.06\,g\,CH_4N_2O} = 0.167\,mol\,CH_4N_2O$$

$$150.0\,g\,H_2O \times \frac{1\,mol\,H_2O}{18.02\,g\,H_2O} = 8.32\,mol\,H_2O$$

$$X_{H_2O} = \frac{8.32\,mol}{8.32\,mol + 0.167\,mol} = 0.980$$

$$P_{soln} = P^{o}_{H_2O} \cdot X_{H_2O} = (71.93\,mm\,Hg)(0.980) = 70.5\,mm\,Hg$$

(b) LiCl, 42.39 amu; $\quad 10.0\,g\,LiCl \times \frac{1\,mol\,LiCl}{42.39\,g\,LiCl} = 0.236\,mol\,LiCl$

LiCl dissociates into Li^+(aq) and Cl^-(aq) in H_2O.
mol Li^+ = mol Cl^- = mol LiCl = 0.236 mol

$$150.0\,g\,H_2O \times \frac{1\,mol\,H_2O}{18.02\,g\,H_2O} = 8.32\,mol\,H_2O$$

$$X_{H_2O} = \frac{8.32\,mol}{8.32\,mol + 0.236\,mol + 0.236\,mol} = 0.946$$

$$P_{soln} = P^{o}_{H_2O} \cdot X_{H_2O} = (71.93\,mm\,Hg)(0.946) = 68.0\,mm\,Hg$$

11.78   For H_2O, $K_b = 0.51 \dfrac{°C \cdot kg}{mol}$;   150.0 g = 0.1500 kg

(a) $\Delta T_b = K_b \cdot m = \left(0.51 \dfrac{°C \cdot kg}{mol} \right) \left(\dfrac{0.167\ mol}{0.1500\ kg} \right) = 0.57°C$

solution boiling point = 100.00°C + ΔT_b = 100.00°C + 0.57°C = 100.57°C

(b) $\Delta T_b = K_b \cdot m = \left(0.51 \dfrac{°C \cdot kg}{mol} \right) \left(\dfrac{2(0.236\ mol)}{0.1500\ kg} \right) = 1.6°C$

solution boiling point = 100.00°C + ΔT_b = 100.00°C + 1.6°C = 101.6°C

11.80   Acetone, C_3H_6O, 58.08 amu, $P^o_{C_3H_6O} = 285$ mm Hg

Ethyl acetate, $C_4H_8O_2$, 88.11 amu, $P^o_{C_4H_8O_2} = 118$ mm Hg

$25.0\ g\ C_3H_6O\ \times\ \dfrac{1\ mol\ C_3H_6O}{58.08\ g\ C_3H_6O} = 0.430\ mol\ C_3H_6O$

$25.0\ g\ C_4H_8O_2\ \times\ \dfrac{1\ mol\ C_4H_8O_2}{88.11\ g\ C_4H_8O_2} = 0.284\ mol\ C_4H_8O_2$

$X_{C_3H_6O} = \dfrac{0.430\ mol}{0.430\ mol\ +\ 0.284\ mol} = 0.602$; $X_{C_4H_8O_2} = \dfrac{0.284\ mol}{0.430\ mol\ +\ 0.284\ mol} = 0.398$

$P_{soln} = P^o_{C_3H_6O} \cdot X_{C_3H_6O} + P^o_{C_4H_8O_2} \cdot X_{C_4H_8O_2}$

$P_{soln} = (285\ mm\ Hg)(0.602) + (118\ mm\ Hg)(0.398) = 219\ mm\ Hg$

11.82   In the liquid, $X_{acetone} = 0.602$ and $X_{ethyl\ acetate} = 0.398$
In the vapor, $P_{Total} = 219$ mm Hg

$P_{acetone} = P^o_{acetone} \cdot X_{acetone} = (285\ mm\ Hg)(0.602) = 172\ mm\ Hg$

$P_{ethyl\ acetate} = P^o_{ethyl\ acetate} \cdot X_{ethyl\ acetate} = (118\ mm\ Hg)(0.398) = 47\ mm\ Hg$

$X_{acetone} = \dfrac{P_{acetone}}{P_{total}} = \dfrac{172\ mm\ Hg}{219\ mm\ Hg} = 0.785$

$X_{ethyl\ acetate} = \dfrac{P_{ethyl\ acetate}}{P_{total}} = \dfrac{47\ mm\ Hg}{219\ mm\ Hg} = 0.215$

11.84 $C_9H_8O_4$, 180.16 amu;   215 g = 0.215 kg

$\Delta T_b = K_b \cdot m = 0.47\ °C$; $K_b = \dfrac{\Delta T_b}{m} = \dfrac{0.47°C}{\left(\dfrac{5.00\ g\ \times\ \dfrac{1\ mol}{180.16\ g}}{0.215\ kg} \right)} = 3.6\ \dfrac{°C \cdot kg}{mol}$

11.86 $\Delta T_b = K_b \cdot m = 1.76\ °C;$ $\qquad m = \dfrac{\Delta T_b}{K_b} = \dfrac{1.76\ °C}{3.07\ \dfrac{°C \cdot kg}{mol}} = 0.573\ m$

11.88 $\Pi = MRT$

(a) NaCl 58.44 amu; 350.0 mL = 0.3500 L

There are 2 moles of ions/mole of NaCl

$\Pi = (2)\left(\dfrac{5.00\ g \times \dfrac{1mol}{58.44\ g}}{0.3500\ L}\right)\left(0.082\ 06\ \dfrac{L \cdot atm}{mol \cdot K}\right)(323\ K) = 13.0\ atm$

(b) CH_3CO_2Na, 82.03 amu; 55.0 mL = 0.0550 L

There are 2 moles of ions/mole of CH_3CO_2Na

$\Pi = (2)\left(\dfrac{6.33\ g \times \dfrac{1\ mol}{82.03\ g}}{0.0550\ L}\right)\left(0.082\ 06\ \dfrac{L \cdot atm}{mol \cdot K}\right)(283K) = 65.2\ atm$

11.90 $\Pi = MRT;$ $\qquad M = \dfrac{\Pi}{RT} = \dfrac{4.85\ atm}{\left(0.082\ 06\ \dfrac{L \cdot atm}{mol \cdot K}\right)(300\ K)} = 0.197\ M$

Uses of Colligative Properties

11.92 Osmotic pressure is most often used for the determination of molecular weight because, of the four colligative properties, osmotic pressure gives the largest colligative property change per mole of solute.

11.94 $\Pi = 407.2\ mm\ Hg \times \dfrac{1\ atm}{760\ mm\ Hg} = 0.5358\ atm$

$\Pi = MRT;\ M = \dfrac{\Pi}{RT} = \dfrac{0.5358\ atm}{\left(0.082\ 06\ \dfrac{L \cdot atm}{mol \cdot K}\right)(298.15\ K)} = 0.021\ 90\ M = 0.021\ 90\ mol/L$

$M = 0.021\ 90\ mol/L = \dfrac{\left(\dfrac{1.500\ g}{molar\ mass}\right)}{0.2000\ L}$; solve for the molar mass

molar mass of cellobiose = 342.5 g/mol; molecular weight = 342.5 amu

11.96 HCl is a strong electrolyte in H_2O and completely dissociates into two solute particles per each HCl.

HF is a weak electrolyte in H_2O. Only a few percent of the HF molecules dissociates into ions.

11.98 First, determine the empirical formula:
Assume 100.0 g of β-carotene.

10.51% H $10.51 \text{ g H} \times \dfrac{1 \text{ mol H}}{1.008 \text{ g H}} = 10.43 \text{ mol H}$

89.49% C $89.49 \text{ g C} \times \dfrac{1 \text{ mol C}}{12.01 \text{ g C}} = 7.45 \text{ mol C}$

$C_{7.45}H_{10.43}$;   Divide each subscript by the smaller, 7.45.
$C_{7.45/7.45}H_{10.43/7.45}$
$CH_{1.4}$
Multiply each subscript by 5 to obtain integers. Empirical formula is C_5H_7, 67.1 amu.
Second, calculate the molecular weight:

$\Delta T_f = K_f \cdot m$; $m = \dfrac{\Delta T_f}{K_f} = \dfrac{1.17^\circ\text{C}}{37.7 \dfrac{^\circ\text{C} \cdot \text{kg}}{\text{mol}}} = 0.0310 \dfrac{\text{mol}}{\text{kg}}$

$m = 0.0310 \text{ mol/kg} = \dfrac{\left(\dfrac{0.0250 \text{ g}}{\text{molar mass}}\right)}{0.001\ 50 \text{ kg}}$; solve for the molar mass.

molar mass of β–carotene = 538 g/mol; molecular weight = 538 amu

Finally, determine the molecular formula:
Divide the molecular weight by the empirical formula weight.
$\dfrac{538 \text{ amu}}{67.1 \text{ amu}} = 8$;   molecular formula is $C_{(8 \times 5)}H_{(8 \times 7)}$, or $C_{40}H_{56}$

11.100 The mixture will begin to boil at ~ 55°C.

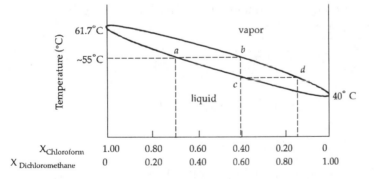

General Problems

11.102 $C_2H_6O_2$, 62.07 amu; $\Delta T_f = 22.0°C$

$\Delta T_f = K_f \cdot m$; $m = \dfrac{\Delta T_f}{K_f} = \dfrac{22.0^\circ\text{C}}{1.86 \dfrac{^\circ\text{C} \cdot \text{kg}}{\text{mol}}} = 11.8\ m$

$$m = 11.8 \ \frac{\text{mol}}{\text{kg}} = \frac{\left((\text{mass of ethylene glycol}) \ \times \ \dfrac{1 \ \text{mol}}{62.07 \ \text{g}} \right)}{3.55 \ \text{kg}}$$

Solve for the mass of ethylene glycol.   mass of ethylene glycol $= 2600 \ \text{g} = 2.60 \ \times \ 10^3 \ \text{g}$

11.104   For KBr, the solute particle molality is $2 \times 0.665 \ m = 1.33 \ m$
For $CaCl_2$, the solute particle molality is $3 \times 0.400 \ m = 1.20 \ m$
The $CaCl_2$ solution has the smaller solute particle molality, and its freezing point depression (ΔT_f) will not be as large. The $CaCl_2$ solution will have the higher freezing point.

11.106   When solid $CaCl_2$ is added to liquid water, the temperature rises because ΔH_{soln} for $CaCl_2$ is exothermic.
When solid $CaCl_2$ is added to ice at $0°C$, some of the ice will melt and the temperature will fall because the $CaCl_2$ lowers the freezing point of an ice/water mixture.

11.108 $C_{10}H_8$, 128.17 amu; 150.0 g = 0.1500 kg; $\Delta T_f = 0.35°C$

$$\Delta T_f = K_f \cdot m; \ m = \frac{\Delta T_f}{K_f} = \frac{0.35° \ C}{5.12 \ \dfrac{°C \cdot kg}{\text{mol}}} = 0.0684 \ \frac{\text{mol}}{\text{kg}}$$

$$m = 0.0684 \ \frac{\text{mol}}{\text{kg}} = \frac{\left((\text{mass of } C_{10}H_8) \ \times \ \dfrac{1 \ \text{mol}}{128.17 \ \text{g}} \right)}{0.1500 \ \text{kg}}$$

Solve for the mass of $C_{10}H_8$. mass of $C_{10}H_8 = 1.3 \ \text{g}$

11.110   NaCl, 58.44 amu; there are 2 ions/NaCl
A 3.5 wt % aqueous solution of NaCl contains 3.5 g NaCl and 96.5 g H_2O.

$$\text{molality} = \frac{\left(3.5 \ \text{g} \ \times \ \dfrac{1 \ \text{mol}}{58.44 \ \text{g}} \right)}{0.0965 \ \text{kg}} = 0.62 \ m$$

$$\Delta T_f = K_f \cdot 2 \cdot m = \left(1.86 \ \frac{°C \cdot kg}{\text{mol}} \right)(2)(0.62 \ \text{mol/kg}) = 2.3°C$$

freezing point $= 0.0°C - \Delta T_f = 0.0°C - 2.3°C = -2.3°C$

$$\Delta T_b = K_b \cdot 2 \cdot m = \left(0.51 \ \frac{°C \cdot kg}{\text{mol}} \right)(2)(0.62 \ \text{mol/kg}) = 0.63°C$$

boiling point $= 100.00°C + \Delta T_b = 100.00°C + 0.63°C = 100.63°C$

11.112   (a) 90 wt % isopropyl alcohol $= \dfrac{10.5 \ \text{g}}{10.5 \ \text{g} \ + \ \text{mass of} \ H_2O} \ \times \ 100\%$

Solve for the mass of H_2O.

$$\text{mass of } H_2O = \left(10.5 \text{ g } \times \frac{100}{90}\right) - 10.5 \text{ g} = 1.2 \text{ g}$$

mass of solution = 10.5 g + 1.2 g = 11.7 g
11.7 g of rubbing alcohol contains 10.5 g of isopropyl alcohol.

(b) C_3H_8O, 60.10 amu
mass C_3H_8O = (0.90)(50.0 g) = 45 g

$$45 \text{ g } C_3H_8O \times \frac{1 \text{ mol } C_3H_8O}{60.10 \text{ g } C_3H_8O} = 0.75 \text{ mol } C_3H_8O$$

11.114 First, determine the empirical formula.
3.47 mg = 3.47 x 10^{-3} g sample
10.10 mg = 10.10 x 10^{-3} g CO_2
2.76 mg = 2.76 x 10^{-3} g H_2O

$$\text{mass C} = 10.10 \text{ x } 10^{-3} \text{ g } CO_2 \times \frac{12.01 \text{ g C}}{44.01 \text{ g } CO_2} = 2.76 \text{ x } 10^{-3} \text{ g C}$$

$$\text{mass H} = 2.76 \text{ x } 10^{-3} \text{ g } H_2O \times \frac{2 \text{ x } 1.008 \text{ g H}}{18.02 \text{ g } H_2O} = 3.09 \text{ x } 10^{-4} \text{ g H}$$

mass O = 3.47 x 10^{-3} g – 2.76 x 10^{-3} g C – 3.09 x 10^{-4} g H = 4.01 x 10^{-4} g O

$$2.76 \text{ x } 10^{-3} \text{ g C} \times \frac{1 \text{ mol C}}{12.01 \text{ g C}} = 2.30 \text{ x } 10^{-4} \text{ mol C}$$

$$3.09 \text{ x } 10^{-4} \text{ g H} \times \frac{1 \text{ mol H}}{1.008 \text{ g H}} = 3.07 \text{ x } 10^{-4} \text{ mol H}$$

$$4.01 \text{ x } 10^{-4} \text{ g O} \times \frac{1 \text{ mol O}}{16.00 \text{ g O}} = 2.51 \text{ x } 10^{-5} \text{ mol O} = 0.251 \text{ x } 10^{-4} \text{ mol O}$$

To simplify the empirical formula, divide each mol quantity by 10^{-4}.
$C_{2.30}H_{3.07}O_{0.251}$; Divide all subscripts by the smallest, 0.251.
$C_{2.30 / 0.251}H_{3.07 / 0.251}O_{0.251 / 0.251}$
$C_{9.16}H_{12.23}O$, empirical formula is $C_9H_{12}O$ (136 amu)

Second, determine the molecular weight.

$$7.55 \text{ mg} = 7.55 \text{ x } 10^{-3} \text{ g estradiol}; \quad 0.500 \text{ g} \times \frac{1 \text{ kg}}{1000 \text{ g}} = 5.00 \text{ x } 10^{-4} \text{ kg camphor}$$

$$\Delta T = K_f \cdot m; \quad m = \frac{\Delta T}{K_f} = \frac{2.10 °C}{37.7 \text{ } (°C \cdot kg)/mol} = 0.0557 \text{ } m$$

$$m = \frac{\text{mol estradiol}}{\text{kg solvent}}$$

mol estradiol = m x (kg solvent) = (0.0557 mol/kg)(5.00 x 10^{-4} kg) = 2.79 x 10^{-5} mol

$$\text{molar mass} = \frac{7.55 \text{ x } 10^{-3} \text{ g estradiol}}{2.79 \text{ x } 10^{-5} \text{ mol estradiol}} = 271 \text{ g/mol}; \text{ molecular weight} = 271 \text{ amu}$$

Finally, determine the molecular formula:

Divide the molecular weight by the empirical formula weight.

$\dfrac{271 \text{ amu}}{136 \text{ amu}} = 2;$ molecular formula is $C_{(2 \times 9)}H_{(2 \times 12)}O_{(2 \times 1)}$, or $C_{18}H_{24}O_2$

11.116 (a) H_2SO_4, 98.08 amu; $2.238 \text{ mol } H_2SO_4 \times \dfrac{98.08 \text{ g } H_2SO_4}{1 \text{ mol } H_2SO_4} = 219.50 \text{ g } H_2SO_4$

mass of 2.238 m solution = 219.50 g H_2SO_4 + 1000 g H_2O = 1219.50 g

volume of 2.238 m solution = $1219.50 \text{ g} \times \dfrac{1.0000 \text{ mL}}{1.1243 \text{ g}} = 1084.68 \text{ mL} = 1.0847 \text{ L}$

molarity of 2.238 m solution = $\dfrac{2.238 \text{ mol}}{1.0847 \text{ L}} = 2.063 \text{ M}$

The molarity of the H_2SO_4 solution is less than the molarity of the $BaCl_2$ solution. Because equal volumes of the two solutions are mixed, H_2SO_4 is the limiting reactant and the number of moles of H_2SO_4 determines the number of moles of $BaSO_4$ produced as the white precipitate.

$(0.05000 \text{ L}) \times (2.063 \text{ mol } H_2SO_4/\text{L}) \times \dfrac{1 \text{ mol } BaSO_4}{1 \text{ mol } H_2SO_4} \times \dfrac{233.39 \text{ g } BaSO_4}{1 \text{ mol } BaSO_4} = 24.07 \text{ g } BaSO_4$

(b) More precipitate will form because of the excess $BaCl_2$ in the solution.

11.118 Let $x = X_{H_2O}$ and $y = X_{CH_3OH}$ and assume $n_{total} = 1.00 \text{ mol}$

(14.5 mm Hg)x + (82.5 mm Hg)y = 39.4 mm Hg

(26.8 mm Hg)x + (140.3 mm Hg)y = 68.2 mm Hg

$x = \dfrac{68.2 - 140.3y}{26.8}$

$\dfrac{14.5(68.2 - 140.3y)}{26.8} + 82.5y = 39.4$

$\dfrac{(988.9 - 2034.35y)}{26.8} + 82.5y = 39.4$

$36.90 - 75.91y + 82.5y = 39.4;\ 6.59 = 2.5;\ y = \dfrac{2.5}{6.59} = 0.3794$

$x = \dfrac{[68.2 - 140.3(0.3794)]}{26.8} = 0.5586$

$X_{LiCl} = 1 - X_{H_2O} - X_{CH_3OH} = 1 - .5586 - 0.3794 = 0.0620$

The mole fraction equals the number of moles of each component because $n_{total} = 1.00 \text{ mol}$.

mass LiCl = $0.0620 \text{ mol LiCl} \times \dfrac{42.39 \text{ g LiCl}}{1 \text{ mol LiCl}} = 2.6 \text{ g LiCl}$

mass H_2O = $0.5588 \text{ mol } H_2O \times \dfrac{18.02 \text{ g } H_2O}{1 \text{ mol } H_2O} = 10.1 \text{ g } H_2O$

$$\text{mass } CH_3OH = 0.3794 \text{ mol } CH_3OH \times \frac{32.04 \text{ g } CH_3OH}{1 \text{ mol } CH_3OH} = 12.2 \text{ g } CH_3OH$$

total mass = 2.6 g + 10.1 g + 12.2 g = 24.9 g

$$\text{wt \% LiCl} = \frac{2.6 \text{ g}}{24.9 \text{ g}} \times 100\% = 10\%; \quad \text{wt \% } H_2O = \frac{10.1 \text{ g}}{24.9 \text{ g}} \times 100\% = 41\%$$

$$\text{wt \% } CH_3OH = \frac{12.2 \text{ g}}{24.9 \text{ g}} \times 100\% = 49\%$$

12 Chemical Kinetics

12.1 $2 N_2O_5(g) \rightarrow 4 NO_2(g) + O_2(g)$

time	$[N_2O_5]$	$[O_2]$
200 s	0.0142 M	0.0029 M
300 s	0.0120 M	0.0040 M

Rate of decomposition of $N_2O_5 = -\dfrac{\Delta[N_2O_5]}{\Delta t} = -\dfrac{0.0120\,M - 0.0142\,M}{300\,s - 200\,s} = 2.2 \times 10^{-5}\,M/s$

Rate of formation of $O_2 = \dfrac{\Delta[O_2]}{\Delta t} = \dfrac{0.0040\,M - 0.0029\,M}{300\,s - 200\,s} = 1.1 \times 10^{-5}\,M/s$

12.2 $3 I^-(aq) + H_3AsO_4(aq) + 2 H^+(aq) \rightarrow I_3^-(aq) + H_3AsO_3(aq) + H_2O(l)$

(a) $-\dfrac{\Delta[I^-]}{\Delta t} = 4.8 \times 10^{-4}\,M/s$

$\dfrac{\Delta[I_3^-]}{\Delta t} = \dfrac{1}{3}\left(-\dfrac{\Delta[I^-]}{\Delta t}\right) = \left(\dfrac{1}{3}\right)(4.8 \times 10^{-4}\,M/s) = 1.6 \times 10^{-4}\,M/s$

(b) $-\dfrac{\Delta[H^+]}{\Delta t} = 2\left(\dfrac{\Delta[I_3^-]}{\Delta t}\right) = (2)(1.6 \times 10^{-4}\,M/s) = 3.2 \times 10^{-4}\,M/s$

12.3 Rate $= k[BrO_3^-][Br^-][H^+]^2$
1st order in BrO_3^-, 1st order in Br^-, 2nd order in H^+, 4th order overall
Rate $= k[H_2][I_2]$, 1st order in H_2, 1st order in I_2, 2nd order overall
Rate $= k[CH_3CHO]^{3/2}$, 3/2 order in CH_3CHO, 3/2 order overall

12.4 $H_2O_2(aq) + 3 I^-(aq) + 2 H^+(aq) \rightarrow I_3^-(aq) + 2 H_2O(l)$

Rate $= \dfrac{\Delta[I_3^-]}{\Delta t} = k[H_2O_2]^m[I^-]^n$

(a) $\dfrac{Rate_3}{Rate_1} = \dfrac{2.30 \times 10^{-4}\,M/s}{1.15 \times 10^{-4}\,M/s} = 2 \qquad \dfrac{[H_2O_2]_3}{[H_2O_2]_1} = \dfrac{0.200\,M}{0.100\,M} = 2$

Since both ratios are the same, m = 1.

$\dfrac{Rate_2}{Rate_1} = \dfrac{2.30 \times 10^{-4}\,M/s}{1.15 \times 10^{-4}\,M/s} = 2 \qquad \dfrac{[I^-]_2}{[I^-]_1} = \dfrac{0.200\,M}{0.100\,M} = 2$

Since both ratios are the same, n = 1.
The rate law is: Rate $= k[H_2O_2][I^-]$

(b) $k = \dfrac{\text{Rate}}{[H_2O_2][I^-]}$

Using data from Experiment 1: $k = \dfrac{1.15 \times 10^{-4}\,M/s}{(0.100\,M)(0.100\,M)} = 1.15 \times 10^{-2}\,/(M \cdot s)$

(c) $\text{Rate} = k[H_2O_2][I^-] = [1.15 \times 10^{-2}/(M \cdot s)](0.300\,M)(0.400\,M) = 1.38 \times 10^{-3}\,M/s$

12.5   <u>Rate Law</u>   <u>Units of k</u>

Rate Law	Units of k
$\text{Rate} = k[(CH_3)_3CBr]$	$1/s$
$\text{Rate} = k[Br_2]$	$1/s$
$\text{Rate} = k[BrO_3^-][Br^-][H^+]^2$	$1/(M^3 \cdot s)$
$\text{Rate} = k[H_2][I_2]$	$1/(M \cdot s)$
$\text{Rate} = [CH_3CHO]^{3/2}$	$1/(M^{1/2} \cdot s)$

12.6   (a) $\log \dfrac{[Co(NH_3)_5Br^{2+}]_t}{[Co(NH_3)_5Br^{2+}]_0} = \dfrac{-kt}{2.303}$

$k = 6.3 \times 10^{-6}/s; \qquad t = 10\,h \times \dfrac{3600\,s}{1\,h} = 36{,}000\,s$

$\log[Co(NH_3)_5Br^{2+}]_t = \dfrac{-kt}{2.303} + \log[Co(NH_3)_5Br^{2+}]_0$

$\log[Co(NH_3)_5Br^{2+}]_t = \dfrac{-(6.3 \times 10^{-6}/s)(36{,}000\,s)}{2.303} + \log(0.100)$

$\log[Co(NH_3)_5Br^{2+}]_t = -1.0985; \qquad$ After 10 h, $[Co(NH_3)_5Br^{2+}] = 10^{-1.0985} = 0.080\,M$

(b) $[Co(NH_3)_5Br^{2+}]_0 = 0.100\,M$
If 75% of the $Co(NH_3)_5Br^{2+}$ reacts then 25% remains.
$[Co(NH_3)_5Br^{2+}]_t = (0.25)(0.100\,M) = 0.025\,M$

$\log \dfrac{[Co(NH_3)_5Br^{2+}]_t}{[Co(NH_3)_5Br^{2+}]_0} = \dfrac{-kt}{2.303}; \qquad t = \dfrac{(2.303)\log \dfrac{[Co(NH_3)_5Br^{2+}]_t}{[Co(NH_3)_5Br^{2+}]_0}}{-k}$

$t = \dfrac{(2.303)\log\left(\dfrac{0.025}{0.100}\right)}{-(6.3 \times 10^{-6}/s)} = 2.2 \times 10^5\,s; \quad t = 2.2 \times 10^5\,s \times \dfrac{1\,h}{3600\,s} = 61\,h$

12.7

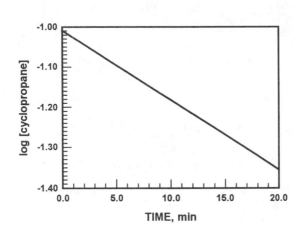

Slope $= -0.0173/\text{min} = -2.9 \times 10^{-4}/\text{s}$ and $k = -2.303(\text{slope})$

A plot of log[cyclopropane] versus time is linear, indicating that the data fit the equation for a first-order reaction. $k = 6.7 \times 10^{-4}/\text{s}$ (0.040/min)

12.8 (a) $k = 1.8 \times 10^{-5}/\text{s}$

$$t_{1/2} = \frac{0.693}{k} = \frac{0.693}{1.8 \times 10^{-5}/\text{s}} = 38,500 \text{ s}; \qquad t_{1/2} = 38,500 \text{ s} \times \frac{1 \text{ h}}{3600 \text{ s}} = 11 \text{ h}$$

(b) $0.30 \text{ M} \xrightarrow{t_{1/2}} 0.15 \text{ M} \xrightarrow{t_{1/2}} 0.075 \text{ M} \xrightarrow{t_{1/2}} 0.0375 \text{ M} \xrightarrow{t_{1/2}} 0.019 \text{ M}$

(c) Since 25% of the initial concentration corresponds to 1/4 or $(1/2)^2$ of the initial concentration, the time required is two half-lives: $t = 2t_{1/2} = 2(11 \text{ h}) = 22 \text{ h}$

12.9

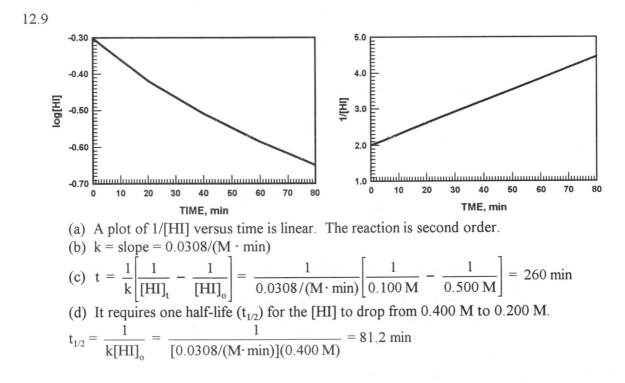

(a) A plot of 1/[HI] versus time is linear. The reaction is second order.

(b) $k = \text{slope} = 0.0308/(\text{M} \cdot \text{min})$

(c) $t = \frac{1}{k}\left[\frac{1}{[\text{HI}]_t} - \frac{1}{[\text{HI}]_o}\right] = \frac{1}{0.0308/(\text{M} \cdot \text{min})}\left[\frac{1}{0.100 \text{ M}} - \frac{1}{0.500 \text{ M}}\right] = 260 \text{ min}$

(d) It requires one half-life ($t_{1/2}$) for the [HI] to drop from 0.400 M to 0.200 M.

$$t_{1/2} = \frac{1}{k[\text{HI}]_o} = \frac{1}{[0.0308/(\text{M} \cdot \text{min})](0.400 \text{ M})} = 81.2 \text{ min}$$

12.10 (a)

$$NO_2(g) + F_2(g) \rightarrow NO_2F(g) + F(g)$$
$$\underline{F(g) + NO_2(g) \rightarrow NO_2F(g)}$$

Overall reaction $\quad 2\,NO_2(g) + F_2(g) \rightarrow 2\,NO_2F(g)$

Because F(g) is produced in the first reaction and consumed in the second, it is a reaction intermediate.

(b) In each reaction there are two reactants, so each elementary reaction is bimolecular.

12.11 (a) Rate = $k[O_3][O]$ (b) Rate = $k[Br]^2[Ar]$ (c) Rate = $k[Co(CN)_5(H_2O)^{2-}]$

12.12

$$Co(CN)_5(H_2O)^{2-}(aq) \rightarrow Co(CN)_5^{2-}(aq) + H_2O(l) \qquad \text{(slow)}$$
$$\underline{Co(CN)_5^{2-}(aq) + I^-(aq) \rightarrow Co(CN)_5I^{3-}(aq)} \qquad \text{(fast)}$$

Overall reaction $\quad Co(CN)_5(H_2O)^{2-}(aq) + I^-(aq) \rightarrow Co(CN)_5I^{3-}(aq) + H_2O(l)$

The overall reaction is the stoichiometric reaction.

The rate law for the first (slow) elementary reaction is: $\quad$ Rate = $k[Co(CN)_5(H_2O)^{2-}]$

12.13 (a) $\log\left(\dfrac{k_2}{k_1}\right) = \left(\dfrac{-E_a}{2.303\,R}\right)\left(\dfrac{1}{T_2} - \dfrac{1}{T_1}\right)$

$k_1 = 3.7 \times 10^{-5}/s, \ T_1 = 25°C = 298 \ K$

$k_2 = 1.7 \times 10^{-3}/s, \ T_2 = 55°C = 328 \ K$

$E_a = -\dfrac{[\log k_2 - \log k_1](2.303)(R)}{\left(\dfrac{1}{T_2} - \dfrac{1}{T_1}\right)}$

$E_a = -\dfrac{[\log(1.7 \times 10^{-3}) - \log(3.7 \times 10^{-5})](2.303)[8.314 \times 10^{-3} \ kJ/(K \cdot mol)]}{\left(\dfrac{1}{328 \ K} - \dfrac{1}{298 \ K}\right)} = 104 \ kJ/mol$

(b) $k_1 = 3.7 \times 10^{-5}/s, \ T_1 = 25°C = 298 \ K$

solve for $k_2, \ T_2 = 35°C = 308 \ K$

$\log k_2 = \left(\dfrac{-E_a}{2.303\,R}\right)\left(\dfrac{1}{T_2} - \dfrac{1}{T_1}\right) + \log k_1$

$\log k_2 = \left(\dfrac{-104 \ kJ/mol}{(2.303)[8.314 \times 10^{-3} \ kJ/(K \cdot mol)]}\right)\left(\dfrac{1}{308 \ K} - \dfrac{1}{298 \ K}\right) + \log(3.7 \times 10^{-5})$

$\log k_2 = -3.84; \qquad k_2 = 10^{-3.84} = 1.4 \times 10^{-4}/s$

Understanding Key Concepts

12.14 Because Rate = $k[A][B]$, the rate is proportional to the product of the number of A molecules and the number of B molecules. The relative rates of the reaction in vessels (a) – (d) are 2 : 1 : 4 : 2.

12.15 Because the same reaction takes place in each vessel, the k's are all the same.

12.16 (a) Because Rate = k[A], the rate is proportional to the number of A molecules in each reaction vessel. The relative rates of the reaction are 2 : 4 : 3.
(b) For a first-order reaction, half-lives are independent of concentration. The half-lives are the same.
(c) Concentrations will double, rates will double, and half-lives will be unaffected.

12.17 (a) For the first-order reaction, half of the A molecules are converted to B molecules each minute.

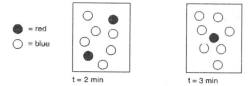

(b) Because half of the A molecules are converted to B molecules in 1 min, the half-life is 1 min.

12.18 (a) Because the half-life is inversely proportional to the concentration of A molecules, the reaction is second order in A.
(b) Rate = k[A]^2
(c) The second box represents the passing of one half-life, and the third box represents the passing of a second half-life for a second-order reaction. A relative value of k can be calculated.

$$k = \frac{1}{t_{1/2}[A]} = \frac{1}{(1)(16)} = 0.0625$$

$t_{1/2}$ in going from box 3 to box 4 is: $t_{1/2} = \dfrac{1}{k[A]} = \dfrac{1}{(0.0625)(4)} = 4 \text{ min}$

(For fourth box, t = 7 min)

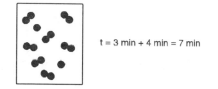

t = 3 min + 4 min = 7 min

12.19 (a) bimolecular   (b) unimolecular   (c) termolecular

12.20 (a) Rate = k[B$_2$][C]
(b) B$_2$ + C → CB + B   (slow)
      CB + A → AB + C   (fast)
(c) C is a catalyst. C does not appear in the chemical equation because it is consumed in the first step and regenerated in the second step.

12.21 (a) BC + D → B + CD
(b) 1. B–C + D (reactants), A (catalyst); 2. B---C---A (transition state), D (reactant); 3. A–C (intermediate), B (product), D (reactant); 4. A---C---D (transition state),

B (product); 5. A (catalyst), C–D + B (products)
(c) The first step is rate determining because the first maximum in the potential energy curve is greater than the second (relative) maximum; Rate = k[A][BC]
(d) Endothermic

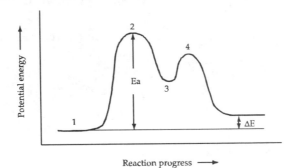

Additional Problems
Reaction Rates

12.22 M/s or $\dfrac{mol}{L \cdot s}$

12.24 (a) Rate $= \dfrac{-\Delta[\text{cyclopropane}]}{\Delta t} = -\dfrac{0.080\ M - 0.098\ M}{5.0\ min - 0.0\ min} = 3.6 \times 10^{-3}$ M/min

Rate $= 3.6 \times 10^{-3}\ \dfrac{M}{min} \times \dfrac{1\ min}{60\ s} = 6.0 \times 10^{-5}$ M/s

(b) Rate $= \dfrac{-\Delta[\text{cyclopropane}]}{\Delta t} = -\dfrac{0.044\ M - 0.054\ M}{20.0\ min - 15.0\ min} = 2.0 \times 10^{-3}$ M/min

Rate $= 2.0 \times 10^{-3}\ \dfrac{M}{min} \times \dfrac{1\ min}{60\ s} = 3.3 \times 10^{-5}$ M/s

12.26

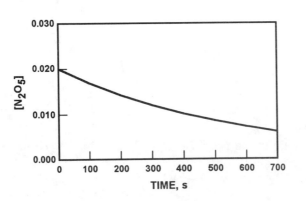

(a) The instantaneous rate of decomposition of N_2O_5 at t = 200 s is determined from the slope of the curve at t = 200 s.

$$\text{Rate} = -\frac{\Delta[N_2O_5]}{\Delta t} = -\text{ slope} = -\frac{(1.20 \times 10^{-2}\ M) - (1.69 \times 10^{-2}\ M)}{300\ s - 100\ s} = 2.4 \times 10^{-5}\ M/s$$

(b) The initial rate of decomposition of N_2O_5 is determined from the slope of the curve at $t = 0$ s. This is equivalent to the slope of the curve from 0 s to 100 s because in this time interval the curve is almost linear.

$$\text{Initial rate} = -\frac{\Delta[N_2O_5]}{\Delta t} = -\text{ slope} = -\frac{(1.69 \times 10^{-2}\ M) - (2.00 \times 10^{-2}\ M)}{100\ s - 0\ s} = 3.1 \times 10^{-5}\ M/s$$

12.28 (a) $-\dfrac{\Delta[H_2]}{\Delta t} = -3\,\dfrac{\Delta[N_2]}{\Delta t}$; The rate of consumption of H_2 is 3 times faster.

(b) $\dfrac{\Delta[NH_3]}{\Delta t} = -2\,\dfrac{\Delta[N_2]}{\Delta t}$; The rate of formation of NH_3 is 2 times faster.

12.30 $N_2(g) + 3\,H_2(g) \rightarrow 2\,NH_3(g)$; $\quad -\dfrac{\Delta[N_2]}{\Delta t} = -\dfrac{1}{3}\,\dfrac{\Delta[H_2]}{\Delta t} = \dfrac{1}{2}\,\dfrac{\Delta[NH_3]}{\Delta t}$

Rate Laws

12.32 Rate $= k[NO]^2[Br_2]$; 2nd order in NO; 1st order in Br_2; 3rd order overall

12.34 Rate $= k[H_2][ICl]$; units for k are $\dfrac{L}{mol \cdot s}$ or $1/(M \cdot s)$

12.36 (a) Rate $= k[CH_3COCH_3]^m$

$$m = \frac{\log\left(\dfrac{\text{Rate}_2}{\text{Rate}_1}\right)}{\log\left(\dfrac{[CH_3COCH_3]_2}{[CH_3COCH_3]_1}\right)} = \frac{\log\left(\dfrac{7.8 \times 10^{-5}}{5.2 \times 10^{-5}}\right)}{\log\left(\dfrac{9.0 \times 10^{-3}}{6.0 \times 10^{-3}}\right)} = 1; \quad \text{Rate} = k[CH_3COCH_3]$$

(b) From Experiment 1: $k = \dfrac{\text{Rate}}{[CH_3COCH_3]} = \dfrac{5.2 \times 10^{-5}\ M/s}{6.0 \times 10^{-3}\ M} = 8.7 \times 10^{-3}/s$

(c) Rate $= k[CH_3COCH_3] = (8.7 \times 10^{-3}/s)(1.8 \times 10^{-3}M) = 1.6 \times 10^{-5}\ M/s$

12.38 (a) Rate $= k[NH_4^+]^m[NO_2^-]^n$

$$m = \frac{\log\left(\dfrac{\text{Rate}_2}{\text{Rate}_1}\right)}{\log\left(\dfrac{[NH_4^+]_2}{[NH_4^+]_1}\right)} = \frac{\log\left(\dfrac{3.6 \times 10^{-6}}{7.2 \times 10^{-6}}\right)}{\log\left(\dfrac{0.12}{0.24}\right)} = 1; \quad n = \frac{\log\left(\dfrac{\text{Rate}_3}{\text{Rate}_2}\right)}{\log\left(\dfrac{[NO_2^-]_3}{[NO_2^-]_2}\right)} = \frac{\log\left(\dfrac{5.4 \times 10^{-6}}{3.6 \times 10^{-6}}\right)}{\log\left(\dfrac{0.15}{0.10}\right)} = 1$$

Rate $= k[NH_4^+][NO_2^-]$

(b) From Experiment 1: $k = \dfrac{\text{Rate}}{[NH_4^+][NO_2^-]} = \dfrac{7.2 \times 10^{-6}\ M/s}{(0.24\ M)(0.10\ M)} = 3.0 \times 10^{-4}/(M \cdot s)$

(c) Rate $= k[NH_4^+][NO_2^-] = [3.0 \times 10^{-4}/(M \cdot s)](0.39\ M)(0.052\ M) = 6.1 \times 10^{-6}\ M/s$

Integrated Rate Law; Half-Life

12.40 $\log \dfrac{[C_3H_6]_t}{[C_3H_6]_0} = \dfrac{-kt}{2.303}$, $k = 6.7 \times 10^{-4}/s$

(a) $t = 30\ \text{min} \times \dfrac{60\ s}{1\ \text{min}} = 1800\ s$

$\log[C_3H_6]_t = \dfrac{-kt}{2.303} + \log[C_3H_6]_0 = \dfrac{-(6.7 \times 10^{-4}/s)(1800\ s)}{2.303} + \log(0.0500) = -1.825$

$[C_3H_6]_t = 10^{-1.825} = 0.015\ M$

(b) $t = \dfrac{(2.303)\log \dfrac{[C_3H_6]_t}{[C_3H_6]_0}}{-k} = \dfrac{(2.303)\log\left(\dfrac{0.0100}{0.0500}\right)}{-(6.7 \times 10^{-4}/s)} = 2403\ s$

$t = 2403\ s \times \dfrac{1\ \text{min}}{60\ s} = 40\ \text{min}$

(c) $[C_3H_6]_0 = 0.0500\ M$;   If 25% of the C_3H_6 reacts then 75% remains.
$[C_3H_6]_t = (0.75)(0.0500\ M) = 0.0375\ M$.

$t = \dfrac{(2.303)\log \dfrac{[C_3H_6]_t}{[C_3H_6]_0}}{-k} = \dfrac{(2.303)\log\left(\dfrac{0.0375}{0.0500}\right)}{-(6.7 \times 10^{-4}/s)} = 429\ s$

$t = 429\ s \times \dfrac{1\ \text{min}}{60\ s} = 7.2\ \text{min}$

12.42 $t_{1/2} = \dfrac{0.693}{k} = \dfrac{0.693}{6.7 \times 10^{-4}/s} = 1034\ s = 17\ \text{min}$

$t = \left(\dfrac{-2.303}{k}\right)\log\dfrac{[C_3H_6]_t}{[C_3H_6]_0} = \left(\dfrac{-2.303}{6.7 \times 10^{-4}/s}\right)\log\dfrac{(0.0625)(0.0500)}{(0.0500)} = 4140\ s$

$t = 4140\ s \times \dfrac{1\ \text{min}}{60\ s} = 69\ \text{min}$

This is also 4 half-lives. $100 \to 50 \to 25 \to 12.5 \to 6.25$

12.44 $kt = \dfrac{1}{[C_4H_6]_t} - \dfrac{1}{[C_4H_6]_0}$, $k = 4.0 \times 10^{-2}/(M \cdot s)$

(a) $t = 1.00\ h \times \dfrac{60\ \text{min}}{1\ hr} \times \dfrac{60\ s}{1\ \text{min}} = 3600\ s$

$$\frac{1}{[C_4H_6]_t} = kt + \frac{1}{[C_4H_6]_0} = (4.0 \times 10^{-2}/(M \cdot s))(3600\ s) + \frac{1}{0.0200\ M}$$

$$\frac{1}{[C_4H_6]_t} = 194/M \quad \text{and} \quad [C_4H_6] = 5.2 \times 10^{-3}\ M$$

(b) $\quad t = \frac{1}{k}\left[\frac{1}{[C_4H_6]_t} - \frac{1}{[C_4H_6]_0}\right]$

$$t = \frac{1}{4.0 \times 10^{-2}/(M \cdot s)}\left[\frac{1}{(0.0020\ M)} - \frac{1}{(0.0200\ M)}\right] = 11{,}250\ s$$

$$t = 11{,}250\ s \times \frac{1\ min}{60\ s} \times \frac{1\ hr}{60\ min} = 3.1\ h$$

12.46 $\quad t_{1/2} = \dfrac{1}{k[C_4H_6]_0} = \dfrac{1}{[4.0 \times 10^{-2}/(M \cdot s)](0.0200\ M)} = 1250\ s = 21\ min$

$t = t_{1/2} = \dfrac{1}{k[C_4H_6]_0} = \dfrac{1}{[4.0 \times 10^{-2}/(M \cdot s)](0.0100\ M)} = 2500\ s = 42\ min$

12.48

time (min)	[N₂O]	log[N₂O]	1/[N₂O]
0	0.250	−0.602	4.00
60	0.218	−0.662	4.59
90	0.204	−0.690	4.90
120	0.190	−0.721	5.26
180	0.166	−0.780	6.02

A plot of log[N₂O] versus time is linear. The reaction is first order in N₂O.

$k = -2.303(\text{slope}) = -2.303(-9.88 \times 10^{-4}/min) = 2.28 \times 10^{-3}/min$

$k = 2.28 \times 10^{-3}/min \times \dfrac{1\ min}{60\ s} = 3.79 \times 10^{-5}/s$

12.50 $\quad k = \dfrac{0.693}{t_{1/2}} = \dfrac{0.693}{248\ s} = 2.79 \times 10^{-3}/s$

Reaction Mechanisms

12.52 There is no relationship between the coefficients in a balanced chemical equation for an overall reaction and the exponents in the rate law unless the overall reaction occurs in a single elementary step, in which case the coefficients in the balanced equation are the exponents in the rate law.

12.54 (a)

$$H_2(g) + ICl(g) \rightarrow HI(g) + HCl(g)$$
$$\underline{HI(g) + ICl(g) \rightarrow I_2(g) + HCl(g)}$$

Overall reaction $\quad H_2(g) + 2 ICl(g) \rightarrow I_2(g) + 2 HCl(g)$

(b) Because HI(g) is produced in the first step and consumed in the second step, it is a reaction intermediate.

(c) In each reaction there are two reactant molecules, so each elementary reaction is bimolecular.

12.56 (a) bimolecular, Rate = k[O₃][Cl]   (b) unimolecular, Rate = k[NO₂]
(c) bimolecular, Rate = k[ClO][O]   (d) termolecular, Rate = k[Cl]²[N₂]

12.58 (a)

$$NO_2Cl(g) \rightarrow NO_2(g) + Cl(g)$$
$$\underline{Cl(g) + NO_2Cl(g) \rightarrow NO_2(g) + Cl_2(g)}$$

Overall reaction $\quad 2 NO_2Cl(g) \rightarrow 2 NO_2(g) + Cl_2(g)$

(b) 1. unimolecular; 2. bimolecular
(c) Rate = k[NO₂Cl]

12.60 $H_2(g) + ICl(g) \rightarrow HI(g) + HCl(g)$ $\qquad$ (slow)
$\qquad HI(g) + ICl(g) \rightarrow I_2(g) + HCl(g)$ $\qquad$ (fast)

The Arrhenius Equation

12.62 Very few collisions involve a collision energy greater than or equal to the activation energy, and only a fraction of those have the proper orientation for reaction.

12.64 Plot log k versus 1/T to determine the activation energy, E_a

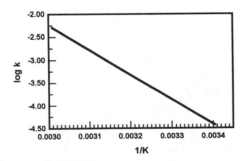

Slope = −5411 K
$E_a = -2.303R(\text{slope}) = -(2.303)(8.314 \times 10^{-3} \text{ kJ/(K} \cdot \text{mol}))(-5411 \text{ K}) = 104 \text{ kJ/mol}$

12.66 (a) $\log\left(\dfrac{k_2}{k_1}\right) = \left(\dfrac{-E_a}{2.303\,R}\right)\left(\dfrac{1}{T_2} - \dfrac{1}{T_1}\right)$

$k_1 = 1.3/(M \cdot s),\ T_1 = 700\ K$

$k_2 = 23.0/(M \cdot s),\ T_2 = 800\ K$

$E_a = -\dfrac{[\log k_2 - \log k_1](2.303)(R)}{\left(\dfrac{1}{T_2} - \dfrac{1}{T_1}\right)}$

$E_a = -\dfrac{[\log(23.0) - \log(1.3)](2.303)[8.314 \times 10^{-3}\ kJ/(K \cdot mol)]}{\left(\dfrac{1}{800\ K} - \dfrac{1}{700\ K}\right)} = 134\ kJ/mol$

(b) $k_1 = 1.3/(M \cdot s),\ T_1 = 700\ K$

solve for $k_2,\ T_2 = 750\ K$

$\log k_2 = \left(\dfrac{-E_a}{2.303\,R}\right)\left(\dfrac{1}{T_2} - \dfrac{1}{T_1}\right) + \log k_1$

$\log k_2 = \left(\dfrac{-133.8\ kJ/mol}{(2.303)[8.314 \times 10^{-3}\ kJ/(K \cdot mol)]}\right)\left(\dfrac{1}{750\ K} - \dfrac{1}{700\ K}\right) + \log(1.3) = 0.779$

$k_2 = 10^{0.779} = 6.0/(M \cdot s)$

12.68 (a) 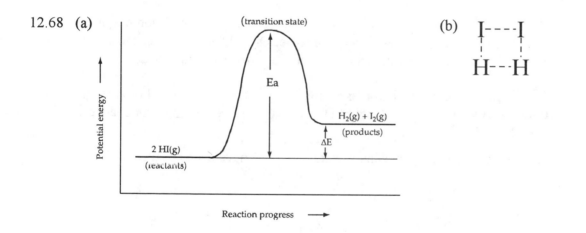 (b)

$$\underset{\text{H--H}}{\text{I----I}}$$

Catalysis

12.70 A catalyst does participate in the reaction, but it is not consumed because it reacts in one step of the reaction and is regenerated in a subsequent step.

12.72 A catalyst increases the rate of a reaction by changing the reaction mechanism and lowering the activation energy.

12.74 (a) $O_3(g) + O(g) \rightarrow 2\,O_2(g)$ (b) Cl acts as a catalyst.
(c) ClO is a reaction intermediate.
(d) A catalyst reacts in one step and is regenerated in a subsequent step. A reaction intermediate is produced in one step and consumed in another.

12.76 (a) $NH_2NO_2(aq) + OH^-(aq) \rightarrow NHNO_2^-(aq) + H_2O(l)$
$\underline{NHNO_2^-(aq) \rightarrow N_2O(g) + OH^-(aq)}$
Overall reaction $NH_2NO_2(aq) \rightarrow N_2O(g) + H_2O(l)$
(b) OH^- acts as a catalyst because it is used in the first step and regenerated in the second. $NHNO_2^-$ is a reaction intermediate because it is produced in the first step and consumed in the second.
(c) The rate will decrease because added acid decreases the concentration of OH^-, which appears in the rate law since it is a catalyst.

General Problems

12.78 The first maximum represents the potential energy of the transition state for the first step. The second maximum represents the potential energy of the transition state for the second step. The saddle point between the two maxima represents the potential energy of the intermediate products.

12.80 $A + B \rightarrow AB$ (slow)
$AB + B \rightarrow AB_2$ (fast)

12.82 As the temperature of a gas is raised by 10°C, even though the collision frequency increases by only ~2%, the reaction rate increases by 100% or more because there is an exponential increase in the fraction of the collisions that leads to products.

12.84 (a) From the data in the table for Experiment 1, we see that 0.20 mol of A reacts with 0.10 mol of B to produce 0.10 mol of D. The balanced equation for the reaction is:
$2\,A + B \rightarrow D$

(b) From the data in the table, initial Rates $= -\dfrac{\Delta A}{\Delta t}$ have been calculated.

For example, from Experiment 1:
$$\text{Initial rate} = -\frac{\Delta A}{\Delta t} = -\frac{(4.80\,M - 5.00\,M)}{60\,s} = 3.33 \times 10^{-3}\,M/s$$
Initial concentrations and initial rate data have been collected in the table below.

EXPT	$[A]_o$ (M)	$[B]_o$ (M)	$[C]_o$ (M)	Initial Rate (M/s)
1	5.00	2.00	1.00	3.33×10^{-3}
2	10.00	2.00	1.00	6.66×10^{-3}
3	5.00	4.00	1.00	3.33×10^{-3}
4	5.00	2.00	2.00	6.66×10^{-3}

Rate $= k[A]^m[B]^n[C]^p$

From Expts 1 and 2, [A] doubles and the initial rate doubles; therefore m = 1.
From Expts 1 and 3, [B] doubles but the initial rate does not change; therefore n = 0.
From Expts 1 and 4, [C] doubles and the initial rate doubles; therefore p = 1.
The reaction is: first order in A; zero order in B; first order in C; second order overall.

(c) Rate = k[A][C]

(d) C is a catalyst. C appears in the rate law, but it is not consumed in the reaction.

(e) $A + C \rightarrow AC$ (slow)
    $AC + B \rightarrow AB + C$ (fast)
    $A + AB \rightarrow D$ (fast)

(f) From data in Expt 1:

$$k = \frac{Rate}{[A][C]} = \frac{\Delta D/\Delta t}{[A][C]} = \frac{0.10\ M/60\ s}{(5.00\ M)(1.00\ M)} = 3.4 \times 10^{-4}/(M \cdot s)$$

12.86 $\log\left(\dfrac{k_2}{k_1}\right) = \left(\dfrac{-E_a}{2.303\ R}\right)\left(\dfrac{1}{T_2} - \dfrac{1}{T_1}\right)$

$k_2 = 2.5 k_1$
$k_1 = 1.0, \quad T_1 = 20°C = 293\ K$
$k_2 = 2.5, \quad T_2 = 30°C = 303\ K$

$E_a = -\dfrac{[\log k_2 - \log k_1](2.303)(R)}{\left(\dfrac{1}{T_2} - \dfrac{1}{T_1}\right)}$

$E_a = -\dfrac{[\log(2.5) - \log(1.0)](2.303)[8.314 \times 10^{-3}\ kJ/(K\cdot mol)]}{\left(\dfrac{1}{303\ K} - \dfrac{1}{293\ K}\right)} = 68\ kJ/mol$

$k_1 = 1.0, \quad T_1 = 120°C = 393\ K$
$k_2 = ?, \quad\quad T_2 = 130°C = 403\ K$
Solve for k_2.

$\log k_2 = \dfrac{-E_a}{2.303\ R}\left(\dfrac{1}{T_2} - \dfrac{1}{T_1}\right) + \log k_1$

$\log k_2 = \dfrac{-68\ kJ/mol}{(2.303)[8.314 \times 10^{-3}\ kJ/(K\cdot mol)]}\left(\dfrac{1}{403\ K} - \dfrac{1}{393\ K}\right) + \log(1.0) = 0.224$

$k_2 = 10^{0.224} = 1.7;$ The rate increases by a factor of 1.7.

12.88 (a) $2\ NO(g) \rightleftarrows N_2O_2(g)$ (fast)
                          $N_2O_2(g) + H_2(g) \rightarrow N_2O(g) + H_2O(g)$ (slow)
                          $\underline{N_2O(g) + H_2(g) \rightarrow N_2(g) + H_2O(g)}$ (fast)
Overall reaction $2\ NO(g) + 2\ H_2(g) \rightarrow N_2(g) + 2\ H_2O(g)$

(b) N_2O_2 and N_2O are reaction intermediates because they are produced in one step of the reaction and used up in a subsequent step.

(c) Rate = $k_2[N_2O_2][H_2]$

(d) Because the forward and reverse rates in step 1 are equal, $k_1[NO]^2 = k_{-1}[N_2O_2]$.

Solving for $[N_2O_2]$ and substituting into the rate law for the second step gives

$$\text{Rate} = k_2[N_2O_2][H_2] = \frac{k_1 k_2}{k_{-1}}[NO]^2[H_2]$$

Because the rate law for the overall reaction is equal to the rate law for the rate-determining step, the rate law for the overall reaction is

$$\text{Rate} = k[NO]^2[H_2] \quad \text{where } k = \frac{k_1 k_2}{k_{-1}}$$

12.90 $2\,N_2O(g) \rightarrow 2\,N_2(g) + O_2(g)$

P_{O_2} (in exit gas) = 1.0 mm Hg; P_{total} = 1.50 atm = 1140 mm Hg

From the reaction stoichiometry:

P_{N_2} (in exit gas) = 2 P_{O_2} = 2.0 mm Hg

P_{N_2O} (in exit gas) = $P_{total} - P_{N_2} - P_{O_2}$ = 1140 − 2.0 − 1.0 = 1137 mm Hg

Assume P_{N_2O} (initial) = P_{total} = 1140 mm Hg (In assuming a constant total pressure in the tube, we are neglecting the slight change in pressure due to the reaction.)

Volume of tube = $\pi r^2 l = \pi(1.25 \text{ cm})^2(20 \text{ cm})$ = 98.2 cm³ = 0.0982 L

Time, t, gases are in the tube = $\dfrac{\text{volume of tube}}{\text{flow rate}}$ x $\dfrac{0.0982 \text{ L}}{0.75 \text{ L/min}}$ x $\dfrac{60 \text{ s}}{1 \text{ min}}$ = 7.86 s

At time t, $\dfrac{[N_2O]_t}{[N_2O]_0} = \dfrac{P_{N_2O} \text{ (in exit gas)}}{P_{N_2O} \text{(initial)}} = \dfrac{1137 \text{ mm Hg}}{1140 \text{ mm Hg}}$ = 0.997 37

Because $k = A\,e^{-\frac{E_a}{RT}}$ and A = 4.2 x 10⁹ s⁻¹, k has units of s⁻¹. Therefore, this is a first-order reaction and the appropriate integrated rate law is $\log\dfrac{[N_2O]_t}{[N_2O]_0} = \dfrac{-kt}{2.303}$.

$$k = \frac{-2.303}{t}\log\frac{[N_2O]_t}{[N_2O]_0} = \frac{-2.303}{7.86 \text{ s}}\log(0.997\,37) = 3.35 \times 10^{-4} \text{ s}^{-1}$$

From the Arrhenius equation, $\log k = \log A - \dfrac{E_a}{2.303\,RT}$

$$T = \frac{E_a}{2.303\,R[\log A - \log k]} = \frac{222 \text{ kJ/mol}}{(2.303)(8.314 \times 10^{-3} \text{ kJ/(K}\cdot\text{mol}))[9.623 - (3.475)]} = 885 \text{ K}$$

12.92 (a) 1 → 1/2 → 1/4 → 1/8

After three half-lives, 1/8 of the strontium-90 will remain.

(b) $k = \dfrac{0.693}{t_{1/2}} = \dfrac{0.693}{29 \text{ y}}$ = 0.0239/y = 0.024/y

(c) $t = \left(\dfrac{-2.303}{k}\right)\log\dfrac{(\text{Sr-90})_t}{(\text{Sr-90})_o} = \left(\dfrac{-2.303}{0.0239/\text{y}}\right)\log\dfrac{(0.01)}{(1)}$ = 193 y

Chemical Equilibrium

13.1 (a) $K_c = \dfrac{[SO_3]^2}{[SO_2]^2[O_2]}$ (b) $K_c = \dfrac{[SO_2]^2[O_2]}{[SO_3]^2}$

13.2 (a) $K_c = \dfrac{[SO_3]^2}{[SO_2]^2[O_2]} = \dfrac{(5.0 \times 10^{-2})^2}{(3.0 \times 10^{-3})^2(3.5 \times 10^{-3})} = 7.9 \times 10^4$

(b) $K_c = \dfrac{[SO_2]^2[O_2]}{[SO_3]^2} = \dfrac{(3.0 \times 10^{-3})^2(3.5 \times 10^{-3})}{(5.0 \times 10^{-2})^2} = 1.3 \times 10^{-5}$

13.3 $K_p = \dfrac{(P_{CO_2})(P_{H_2})}{(P_{CO})(P_{H_2O})} = \dfrac{(6.12)(20.3)}{(1.31)(10.0)} = 9.48$

13.4 $2\,NO(g) + O_2 \rightleftharpoons 2\,NO_2(g);$ $\Delta n = 2 - 3 = -1$
$K_p = K_c(RT)^{\Delta n},$ $K_c = K_p(1/RT)^{\Delta n}$
at 500 K: $K_p = (6.9 \times 10^5)[(0.0821)(500)]^{-1} = 1.7 \times 10^4$

at 1000 K: $K_c = (1.3 \times 10^{-2})\left(\dfrac{1}{(0.0821)(1000)}\right)^{-1} = 1.1$

13.5 (a) $K_c = \dfrac{[H_2]^3}{[H_2O]^3},$ $K_p = \dfrac{(P_{H_2})^3}{(P_{H_2O})^3}$ (b) $K_c = [H_2]^2[O_2],$ $K_p = (P_{H_2})^2(P_{O_2})$

13.6 $K_c = 1.2 \times 10^{-42}$. Since K_c is very small, the equilibrium mixture contains mostly H_2 molecules. H is in periodic group 1A. A very small value of K_c is consistent with strong bonding between 2 H atoms, each with one valence electron.

13.7 The container volume of 5.0 L must be included to calculate molar concentrations.

(a) $Q_c = \dfrac{[NO_2]_t^2}{[NO]_t^2[O_2]_t} = \dfrac{(0.80\ \text{mol}/5.0\ \text{L})^2}{(0.060\ \text{mol}/5.0\ \text{L})^2(1.0\ \text{mol}/5.0\ \text{L})} = 890$

Because $Q_c < K_c$, the reaction is not at equilibrium. The reaction will proceed to the right to reach equilibrium.

(b) $Q_c = \dfrac{[NO_2]_t^2}{[NO]_t^2[O_2]_t} = \dfrac{(4.0\ \text{mol}/5.0\ \text{L})^2}{(5.0 \times 10^{-3}\ \text{mol}/5.0\ \text{L})^2(0.20\ \text{mol}/5.0\ \text{L})} = 1.6 \times 10^7$

Because $Q_c > K_c$, the reaction is not at equilibrium. The reaction will proceed to the left to reach equilibrium.

13.8 $K_c = \dfrac{[H]^2}{[H_2]} = 1.2 \times 10^{-42}$

(a) $[H] = \sqrt{K_c[H_2]} = \sqrt{(1.2 \times 10^{-42})(0.10)} = 3.5 \times 10^{-22}$ M

(b) H atoms = $(3.5 \times 10^{-22}$ mol/L$)(1.0$ L$)(6.022 \times 10^{23}$ atoms/mol$) = 210$ H atoms

H_2 molecules = $(0.10$ mol/L$)(1.0$ L$)(6.022 \times 10^{23}$ molecules/mol$) = 6.0 \times 10^{22}$ H_2 molecules

13.9

	CO(g)	+	$H_2O(g)$	$\rightleftarrows$	$CO_2(g)$	+	$H_2(g)$
initial (M)	0.150		0.150		0		0
change (M)	–x		–x		+x		+x
equil (M)	0.150 – x		0.150 – x		x		x

$K_c = 4.24 = \dfrac{[CO_2][H_2]}{[CO][H_2O]} = \dfrac{x^2}{(0.150-x)^2}$

Take the square root of both sides and solve for x.

$\sqrt{4.24} = \sqrt{\dfrac{x^2}{(0.150-x)^2}}$; $2.06 = \dfrac{x}{0.150-x}$; $x = 0.101$

At equilibrium, $[CO_2] = [H_2] = x = 0.101$ M

$[CO] = [H_2O] = 0.150 - x = 0.150 - 0.101 = 0.049$ M

13.10

	$N_2O_4(g)$	$\rightleftarrows$	$2\ NO_2(g)$
initial (M)	0.0500		0
change (M)	–x		+2x
equil (M)	0.0500 – x		2x

$K_c = 4.64 \times 10^{-3} = \dfrac{[NO_2]^2}{[N_2O_4]} = \dfrac{(2x)^2}{(0.0500-x)}$

$4x^2 + (4.64 \times 10^{-3})x - (2.32 \times 10^{-4}) = 0$

Use the quadratic formula to solve for x.

$x = \dfrac{-(4.64 \times 10^{-3}) \pm \sqrt{(4.64 \times 10^{-3})^2 - 4(4)(-2.32 \times 10^{-4})}}{2(4)} = \dfrac{-0.00464 \pm 0.06110}{8}$

$x = -0.008\ 22$ and $0.007\ 06$

Discard the solution that uses the negative square root ($-0.008\ 22$) because it will lead to negative concentrations and that is impossible.

$[N_2O_4] = 0.0500 - x = 0.0500 - 0.007\ 06 = 0.0429$ M

$[NO_2] = 2x = 2(0.007\ 06) = 0.0141$ M

13.11 $N_2O_4(g) \rightleftarrows 2\ NO_2(g)$

$Q_c = \dfrac{[NO_2]_t^2}{[N_2O_4]_t} = \dfrac{(0.0300\ \text{mol/L})^2}{(0.0200\ \text{mol/L})} = 0.0450$; $Q_c > K_c$

The reaction will approach equilibrium by going from right to left.

$$N_2O_4(g) \rightleftharpoons 2\ NO_2(g)$$

initial (M)	0.0200	0.0300
change (M)	+x	−2x
equil (M)	0.0200 + x	0.0300 − 2x

$$K_c = 4.64 \times 10^{-3} = \frac{[NO_2]^2}{[N_2O_4]} = \frac{(0.0300 - 2x)^2}{(0.0200 + x)}$$

$$4x^2 - 0.1246x - (8.072 \times 10^{-4}) = 0$$

Use the quadratic formula to solve for x.

$$x = \frac{-(-0.1246) \pm \sqrt{(-0.1246)^2 - 4(4)(-8.072 \times 10^{-4})}}{2(4)} = \frac{0.1246 \pm 0.05119}{8}$$

x = 0.0220 and 0.009 18

Discard the solution that uses the positive square root (0.0220) because it will lead to negative concentration of NO_2, and that is impossible.

$[N_2O_4] = 0.0200 + x = 0.0200 + 0.009\ 18 = 0.0292$ M

$[NO_2] = 0.0300 - 2x = 0.0300 - 2(0.009\ 18) = 0.0116$ M

13.12   (a) CO(reactant) added, H_2 concentration increases.
(b) CO_2 (product) added, H_2 concentration decreases.
(c) H_2O (reactant) removed, H_2 concentration decreases.
(d) CO_2 (product) removed, H_2 concentration increases.

At equilibrium, $Q_c = K_c = \dfrac{[CO_2][H_2]}{[CO][H_2O]}$. If some CO_2 is removed from the

equilibrium mixture, the numerator in Q_c is decreased, which means that $Q_c < K_c$ and the reaction will shift to the right, increasing the H_2 concentration.

13.13   (a) Because there are 2 mol of gas on both sides of the balanced equation, the composition of the equilibrium mixture is unaffected by a change in pressure. The number of moles of reaction products remains the same.
(b) Because there are 2 mol of gas on the left side and 1 mol of gas on the right side of the balanced equation, the stress of an increase in pressure is relieved by a shift in the reaction to the side with fewer moles of gas (in this case, to products). The number of moles of reaction products increases.
(c) Because there is 1 mol of gas on the left side and 2 mol of gas on the right side of the balanced equation, the stress of an increase in pressure is relieved by a shift in the reaction to the side with fewer moles of gas (in this case, to reactants). The number of moles of reaction product decreases.

13.14   Le Châtelier's principle predicts that a stress of added heat will be relieved by net reaction in the direction that absorbs the heat. Since the reaction is endothermic, the equilibrium will shift from left to right (K_c will increase) with an increase in temperature. Therefore, the equilibrium mixture will contain more of the offending NO, the higher the temperature.

13.15 The reaction is exothermic. As the temperature is increased the reaction shifts from right to left. The amount of ethyl acetate decreases.

$$K_c = \frac{[CH_3COOC_2H_5][H_2O]}{[CH_3COOH][C_2H_5OH]}$$

As the temperature is decreased, the reaction shifts from left to right. The product concentrations increase, and the reactant concentrations decrease. This corresponds to an increase in K_c.

13.16 (a) A catalyst does not affect the equilibrium composition. The amount of CO remains the same.

(b) The reaction is exothermic. An increase in temperature shifts the reaction toward reactants. The amount of CO increases.

(c) Because there are 3 mol of gas on the left side and 2 mol of gas on the right side of the balanced equation, the stress of an increase in pressure is relieved by a shift in the reaction to the side with fewer moles of gas (in this case, to products). The amount of CO decreases.

(d) An increase in pressure as a result of the addition of an inert gas (with no volume change) does not affect the equilibrium composition. The amount of CO remains the same.

(e) Adding O_2 increases the O_2 concentration and shifts the reaction toward products. The amount of CO decreases.

13.17 (a) Since $k_f \gg k_r$, K_c will be large and the equilibrium mixture will have a larger concentration of products than reactants.

(b) $K_c = \dfrac{k_f}{k_r} = \dfrac{1.32 \times 10^{-4} M^{-1}s^{-1}}{1.22 \times 10^{-6} M^{-1}s^{-1}} = 108$

Understanding Key Concepts

13.18 (a) (1) and (3) since the number of A and B's are the same in the third and fourth box.

(b) $K_c = \dfrac{[B]}{[A]} = \dfrac{6}{4} = 1.5$

(c) Because the same number of molecules appear on both sides of the equation, the volume terms in K_c all cancel. Therefore, we can calculate K_c without including the volume.

13.19 (a) $A_2 + C_2 \rightleftharpoons 2\,AC$ (most product molecules)

(b) $A_2 + B_2 \rightleftharpoons 2\,AB$ (fewest product molecules)

13.20 (a) (2) (b) (1), reverse; (3), forward

13.21 (a) $A_2 + 2B \rightleftharpoons 2\,AB$

(b) The number of AB molecules will increase, because as the volume is decreased at constant temperature, the pressure will increase and the reaction will shift to the side of fewer molecules to reduce the pressure.

13.22　When the stopcock is opened, the reaction will go in the reverse direction because there will be initially an excess of AB molecules.

13.23　As the temperature is raised, the reaction proceeds in the reverse direction. This is consistent with an exothermic reaction where "heat" can be considered as a product.

13.24　(a) AB $\rightarrow$ A + B
(b) The reaction is endothermic because a stress of added heat (higher temperature) shifts the AB $\rightleftharpoons$ A + B equilibrium to the right.
(c) If the volume is increased, the pressure is decreased. The stress of decreased pressure will be relieved by a shift in the equilibrium from left to right, thus increasing the number of A atoms.

13.25　Heat + $BaCO_3(s)$ $\rightleftharpoons$ $BaO(s)$ + $CO_2(g)$

(a)　　　　　　　　　(b)

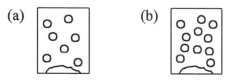

Additional Problems
Equilibrium Expressions and Equilibrium Constants

13.26　(a) $K_c = \dfrac{[PCl_3][Cl_2]}{[PCl_5]}$　　(b) $K_c = \dfrac{[ClNO]^2}{[NO]^2[Cl_2]}$　　(c) $K_c = \dfrac{[NO]^2}{[N_2][O_2]}$

13.28　(a) $K_p = \dfrac{(P_{PCl_3})(P_{Cl_2})}{(P_{PCl_5})}$, $\Delta n = 1$ and $K_c = K_p\left(\dfrac{1}{RT}\right)$

(b) $K_p = \dfrac{(P_{ClNO})^2}{(P_{NO})^2(P_{Cl_2})}$, $\Delta n = -1$ and $K_c = K_p\left(\dfrac{1}{RT}\right)^{-1}$

(c) $K_p = \dfrac{(P_{NO})^2}{(P_{N_2})(P_{O_2})}$, $\Delta n = 0$ and $K_c = K_p$

13.30　$K_c = \dfrac{[C_2H_5OC_2H_5][H_2O]}{[C_2H_5OH]^2}$

13.32　$K_c = \dfrac{[\text{malic acid}]}{[\text{fumaric acid}]}$

13.34 The two reactions are the reverse of each other.

$$K_c(\text{reverse}) = \frac{1}{K_c(\text{forward})} = \frac{1}{7.5 \times 10^{-9}} = 1.3 \times 10^8$$

13.36 $\quad K_c = \dfrac{[PCl_3][Cl_2]}{[PCl_5]} = \dfrac{(1.5 \times 10^{-2})(3.2 \times 10^{-2})}{(8.3 \times 10^{-3})} = 0.058$

13.38 The container volume of 2.00 L must be included to calculate molar concentrations.
Initial $[HI] = 9.30 \times 10^{-3}$ mol/2.00 L $= 4.65 \times 10^{-3}$ M $= 0.004\ 65$ M

	$H_2(g)$	+	$I_2(g)$	$\rightleftarrows$	$2\ HI(g)$
initial (M)	0		0		0.004 65
change (M)	+x		+x		−2x
equil (M)	x		x		0.004 65 − 2x

$x = [H_2] = [I_2] = 6.29 \times 10^{-4}$ M $= 0.000\ 629$ M
$[HI] = 0.004\ 65 - 2x = 0.004\ 65 - 2(0.000\ 629) = 0.003\ 39$ M

$$K_c = \frac{[HI]^2}{[H_2][I_2]} = \frac{(0.003\ 39)^2}{(0.000\ 629)^2} = 29.0$$

13.40 (a) $\quad K_c = \dfrac{[CH_3COOC_2H_5][H_2O]}{[CH_3COOH][C_2H_5OH]}$

(b) $\quad\quad\quad\quad\quad CH_3COOH(\text{soln}) + C_2H_5OH(\text{soln}) \rightleftarrows CH_3COOC_2H_5(\text{soln}) + H_2O(\text{soln})$

initial (mol)	1.00	1.00	0	0
change (mol)	−x	−x	+x	+x
equil (mol)	1.00 − x	1.00 − x	x	x

$x = 0.65$ mol; $1.00 - x = 0.35$ mol; $\quad K_c = \dfrac{(0.65)^2}{(0.35)^2} = 3.4$

Because there are the same number of molecules on both sides of the equation, the volume terms in K_c cancel. Therefore, we can calculate K_c without including the volume.

13.42 Because $\Delta n = 0$, $K_p = K_c = 4.24$

13.44 $\quad K_p = P_{H_2O} = 0.0313$ atm; $\Delta n = 1$

$$K_c = K_p\left(\frac{1}{RT}\right) = (0.0313)\left(\frac{1}{(0.0821)(298)}\right) = 1.28 \times 10^{-3}$$

13.46 (a) $K_c = \dfrac{[CO_2]^3}{[CO]^3}$, $\quad K_p = \dfrac{(P_{CO_2})^3}{(P_{CO})^3}$ $\quad\quad$ (b) $K_c = \dfrac{1}{[O_2]^3}$, $\quad K_p = \dfrac{1}{(P_{O_2})^3}$

$\quad\quad$ (c) $K_c = [SO_3]$, $\quad K_p = P_{SO_3}$ $\quad\quad\quad\quad\quad\quad$ (d) $K_c = [Ba^{2+}][SO_4^{2-}]$

Using the Equilibrium Constant

13.48   (a) Because K_c is very large, the equilibrium mixture contains mostly product.
(b) Because K_c is very small, the equilibrium mixture contains mostly reactant.

13.50   (a) Because K_c is very small, the equilibrium mixture contains mostly reactant.
(b) Because K_c is very large, the equilibrium mixture contains mostly product.
(c) Because $K_c = 1.8$, the equilibrium mixture contains an appreciable concentration of both reactants and products.

13.52 $K_c = 1.2 \times 10^{82}$ is very large. When equilibrium is reached, very little if any ethanol will remain because the reaction goes to completion.

13.54   The container volume of 10.0 L must be included to calculate molar concentrations.

$$Q_c = \frac{[CS_2]_t[H_2]_t^4}{[CH_4]_t[H_2S]_t^2} = \frac{(3.0 \text{ mol/}10.0 \text{ L})(3.0 \text{ mol/}10.0 \text{ L})^4}{(2.0 \text{ mol/}10.0 \text{ L})(4.0 \text{ mol/}10.0 \text{ L})^2} = 7.6 \times 10^{-2}; \quad K_c = 2.5 \times 10^{-3}$$

The reaction is not at equilibrium because $Q_c > K_c$. The reaction will proceed from right to left to reach equilibrium.

13.56 $K_c = \dfrac{[NH_3]^2}{[N_2][H_2]^3} = 1.7 \times 10^2;$ At equilibrium, $[N_2] = 0.020$ M and $[H_2] = 0.18$ M

$[NH_3] = \sqrt{[N_2] \times [H_2]^3 \times K_c} = \sqrt{(0.020)(0.18)^3(1.7 \times 10^2)} = 0.14$ M

13.58

	$N_2(g)$	$+$	$O_2(g)$	$\rightleftharpoons$	$2\,NO(g)$
initial (M)	1.40		1.40		0
change (M)	$-x$		$-x$		$+2x$
equil (M)	$1.40 - x$		$1.40 - x$		$2x$

$K_c = 1.7 \times 10^{-3} = \dfrac{[NO]^2}{[N_2][O_2]} = \dfrac{(2x)^2}{(1.40 - x)^2}$

Take the square root of both sides and solve for x.

$\sqrt{1.7 \times 10^{-3}} = \sqrt{\dfrac{(2x)^2}{(1.40 - x)^2}}; \quad 4.1 \times 10^{-2} = \dfrac{2x}{1.40 - x}; \quad x = 2.8 \times 10^{-2}$

At equilibrium, $[NO] = 2x = 2(2.8 \times 10^{-2}) = 0.056$ M
$[N_2] = [O_2] = 1.40 - x = 1.40 - (2.8 \times 10^{-2}) = 1.37$ M

13.60

	$PCl_5(g)$	$\rightleftharpoons$	$PCl_3(g)$	$+$	$Cl_2(g)$
initial (M)	0.160		0		0
change (M)	$-x$		$+x$		$+x$
equil (M)	$0.160 - x$		x		x

$K_c = \dfrac{[PCl_3][Cl_2]}{[PCl_5]} = 5.8 \times 10^{-2} = \dfrac{x^2}{0.160 - x}$

$x^2 + (5.8 \times 10^{-2})x - 0.00928 = 0$

Use the quadratic formula to solve for x.

$$x = \frac{(-5.8 \times 10^{-2}) \pm \sqrt{(5.8 \times 10^{-2})^2 - 4(1)(-0.00928)}}{2(1)} = \frac{(-5.8 \times 10^{-2}) \pm 0.20}{2}$$

$x = 0.071$ and -0.129

Discard the solution that uses the negative square root ($x = -0.129$) because it gives negative concentrations of PCl_3 and Cl_2 and that is impossible.

$[PCl_3] = [Cl_2] = x = 0.071$ M; $\qquad [PCl_5] = 0.160 - x = 0.160 - 0.071 = 0.089$ M

13.62 (a) $K_c = \dfrac{[CH_3COOC_2H_5][H_2O]}{[CH_3COOH][C_2H_5OH]} = 3.4 = \dfrac{(x)(12.0)}{(4.0)(6.0)}$; $\quad x = 6.8$ moles $CH_3COOC_2H_5$

Note that the volume cancels because the same number of molecules appear on both sides of the chemical equation.

(b) $\qquad CH_3COOH(soln) + C_2H_5OH(soln) \rightleftarrows CH_3COOC_2H_5(soln) + H_2O(soln)$

initial (mol)	1.00	10.00	0	0
change (mol)	$-x$	$-x$	$+x$	$+x$
equil (mol)	$1.00 - x$	$10.00 - x$	x	x

$K_c = 3.4 = \dfrac{x^2}{(1.00 - x)(10.00 - x)}$

$2.4x^2 - 37.4x + 34 = 0$

Use the quadratic formula to solve for x.

$$x = \frac{-(-37.4) \pm \sqrt{(-37.4)^2 - 4(2.4)(34)}}{2(2.4)} = \frac{37.4 \pm 32.75}{4.8}$$

$x = 0.969$ and 14.6

Discard the solution that uses the positive square root ($x = 14.6$) because it leads to negative concentrations and that is impossible.

mol $CH_3COOH = 1.00 - x = 1.00 - 0.969 = 0.03$ mol

mol $C_2H_5OH = 10.00 - x = 10.00 - 0.969 = 9.03$ mol

mol $CH_3COOC_2H_5 =$ mol $H_2O = x = 0.97$ mol

Le Châtelier's Principle

13.64 (a) Cl^- (reactant) added, $AgCl(s)$ increases

(b) Ag^+ (reactant) added, $AgCl(s)$ increases

(c) Ag^+ (reactant) removed, $AgCl(s)$ decreases

(d) Cl^- (reactant) removed, $AgCl(s)$ decreases

Disturbing the equilibrium by decreasing $[Cl^-]$ increases Q_c ($Q_c = \dfrac{1}{[Ag^+]_t[Cl^-]_t}$) to a value greater than K_c. To reach a new state of equilibrium, Q_c must decrease, which means that the denominator must increase; that is, the reaction must go from right to left, thus decreasing the amount of solid AgCl.

13.66   (a) Because there are 2 mol of gas on the left side and 3 mol of gas on the right side of the balanced equation, the stress of an increase in pressure is relieved by a shift in the reaction to the side with fewer moles of gas (in this case, to reactants). The number of moles of reaction products decreases.
(b) Because there are 2 mol of gas on both sides of the balanced equation, the composition of the equilibrium mixture is unaffected by a change in pressure. The number of moles of reaction product remains the same.
(c) Because there are 2 mol of gas on the left side and 1 mol of gas on the right side of the balanced equation, the stress of an increase in pressure is relieved by a shift in the reaction to the side with fewer moles of gas (in this case, to products). The number of moles of reaction products increases.

13.68 $CO(g) + H_2O(g) \rightleftharpoons CO_2(g) + H_2(g)$ $\Delta H° = -41.2$ kJ
The reaction is exothermic. $[H_2]$ decreases when the temperature is increased.
As the temperature is decreased, the reaction shifts to the right. $[CO_2]$ and $[H_2]$ increase, $[CO]$ and $[H_2O]$ decrease, and K_c increases.

13.70   (a) I_2 (reactant) added; the HI concentration increases.
(b) H_2 (reactant) removed; the HI concentration decreases.
(c) A catalyst does not affect the composition of the equilibrium mixture; no change in the HI concentration.
(d) An exothermic reaction shifts to the left as the temperature increases; the HI concentration decreases.

13.72   (a) The reaction is exothermic. The amount of CH_3OH (product) decreases as the temperature increases.
(b) When the volume decreases, the reaction shifts to the side with fewer gas molecules. The amount of CH_3OH increases.
(c) Addition of an inert gas (He) does not affect the equilibrium composition. There is no change.
(d) Addition of CO (reactant) shifts the reaction toward product. The amount of CH_3OH increases.
(e) Addition or removal of a catalyst does not affect the equilibrium composition. There is no change.

Chemical Equilibrium and Chemical Kinetics

13.74 $A + B \rightleftharpoons C$
rate$_f$ = $k_f[A][B]$ and rate$_r$ = $k_r[C]$;   at equilibrium, rate$_f$ = rate$_r$
$$k_f[A][B] = k_r[C]; \qquad \frac{k_f}{k_r} = \frac{[C]}{[A][B]} = K_c$$

13.76 $K_c = \dfrac{k_f}{k_r} = \dfrac{0.13}{6.2 \times 10^{-4}} = 210$

Chapter 13 – Chemical Equilibrium

General Problems

13.78 $Hb + O_2 \rightleftharpoons Hb(O_2)$

If CO binds to Hb, Hb is removed from the reaction and the reaction will shift to the left resulting in O_2 being released from $Hb(O_2)$. This will decrease the effectiveness of Hb for carrying O_2.

13.80 (a) $[N_2O_4] = \dfrac{0.500 \text{ mol}}{4.00 \text{ L}} = 0.125 \text{ M}$

	$N_2O_4(g)$	$\rightleftharpoons$	$2 NO_2(g)$
initial (M)	0.125		0
change (M)	$-(0.793)(0.125)$		$+(2)(0.793)(0.125)$
equil (M)	$0.125 - (0.793)(0.125)$		$(2)(0.793)(0.125)$

At equilibrium, $[N_2O_4] = 0.125 - (0.793)(0.125) = 0.0259 \text{ M}$
$[NO_2] = (2)(0.793)(0.125) = 0.198 \text{ M}$

$K_c = \dfrac{[NO_2]^2}{[N_2O_4]} = \dfrac{(0.198)^2}{(0.0259)} = 1.51$

$\Delta n = 2 - 1 = 1$ and $K_p = K_c(RT)^{\Delta n};$ $K_p = K_c(RT) = (1.51)(0.0821)(400) = 49.6$

(b)

13.82 $K_c = \dfrac{[NH_3]^2}{[N_2][H_2]^3} = 0.291$

At equilibrium, $[N_2] = 1.0 \times 10^{-3} \text{ M}$ and $[H_2] = 2.0 \times 10^{-3} \text{ M}$

$[NH_3] = \sqrt{[N_2] \times [H_2]^3 \times K_c} = \sqrt{(1.0 \times 10^{-3})(2.0 \times 10^{-3})^3(0.291)} = 1.5 \times 10^{-6} \text{ M}$

13.84 $2 HI(g) \rightleftharpoons H_2(g) + I_2(g)$

Calculate K_c. $K_c = \dfrac{[H_2][I_2]}{[HI]^2} = \dfrac{(0.13)(0.70)}{(2.1)^2} = 0.0206$

$[HI] = \dfrac{0.20 \text{ mol}}{0.5000 \text{ L}} = 0.40 \text{ M}$

	$2 HI(g)$	$\rightleftharpoons$	$H_2(g)$	+	$I_2(g)$
initial (M)	0.40		0		0
change (M)	$-2x$		$+x$		$+x$
equil (M)	$0.40 - 2x$		x		x

$K_c = 0.0206 = \dfrac{[H_2][I_2]}{[HI]^2} = \dfrac{x^2}{(0.40 - 2x)^2}$

Take the square root of both sides, and solve for x.

$$\sqrt{0.0206} = \sqrt{\frac{x^2}{(0.40 - 2x)^2}}; \quad 0.144 = \frac{x}{0.40 - 2x}; \quad x = 0.045$$

At equilibrium, $[H_2] = [I_2] = x = 0.045$ M

$[HI] = 0.40 - 2x = 0.40 - 2(0.045) = 0.31$ M

13.86 $[H_2O] = \dfrac{6.00 \text{ mol}}{5.00 \text{ L}} = 1.20$ M

	C(s)	+	H_2O(g)	⇌	CO(g)	+	H_2(g)
initial (M)			1.20		0		0
change (M)			−x		+x		+x
equil (M)			1.20 − x		x		x

$$K_c = \frac{[CO][H_2]}{[H_2O]} = 3.0 \times 10^{-2} = \frac{x^2}{1.20 - x}$$

$x^2 + (3.0 \times 10^{-2})x - 0.036 = 0$

Use the quadratic formula to solve for x.

$$x = \frac{-(0.030) \pm \sqrt{(0.030)^2 - 4(-0.036)}}{2(1)} = \frac{-0.030 \pm 0.381}{2}$$

x = 0.176 and −0.206

Discard the solution that uses the negative square root (x = −0.206) because it leads to negative concentrations and that is impossible.

$[CO] = [H_2] = x = 0.18$ M; $[H_2O] = 1.20 - x = 1.20 - 0.18 = 1.02$ M

13.88 A decrease in volume (a) and the addition of reactants (c) will affect the composition of the equilibrium mixture, but leave the value of K_c unchanged.

A change in temperature (b) affects the value of K_c.

Addition of a catalyst (d) or an inert gas (e) affects neither the composition of the equilibrium mixture nor the value of K_c.

13.90 2 monomer ⇌ dimer

(a) In benzene, $K_c = 1.51 \times 10^2$

	2 monomer	⇌	dimer
initial (M)	0.100		0
change (M)	−2x		+x
equil (M)	0.100 − 2x		x

$$K_c = \frac{[dimer]}{[monomer]^2} = 1.51 \times 10^2 = \frac{x}{(0.100 - 2x)^2}$$

$604x^2 - 61.4x + 1.51 = 0$

Use the quadratic formula to solve for x.

$$x = \frac{-(-61.4) \pm \sqrt{(-61.4)^2 - (4)(604)(1.51)}}{2(604)} = \frac{61.4 \pm 11.04}{1208}$$

x = 0.0600 and 0.0417

Discard the solution that uses the positive square root (x = 0.0600) because it gives a negative concentration of the monomer and that is impossible.

[monomer] = 0.100 − 2x = 0.100 − 2(0.0417) = 0.017 M;   [dimer] = x = 0.0417 M

$$\frac{[dimer]}{[monomer]} = \frac{0.0417\,M}{0.017\,M} = 2.5$$

(b) In H_2O, $K_c = 3.7 \times 10^{-2}$

	2 monomer	⇌	dimer
initial (M)	0.100		0
change (M)	−2x		+x
equil (M)	0.100 − 2x		x

$$K_c = \frac{[dimer]}{[monomer]^2} = 3.7 \times 10^{-2} = \frac{x}{(0.100 - 2x)^2}$$

$0.148x^2 - 1.0148x + 0.000\,37 = 0$

Use the quadratic formula to solve for x.

$$x = \frac{-(-1.0148) \pm \sqrt{(-1.0148)^2 - (4)(0.148)(0.00037)}}{2(0.148)} = \frac{1.0148 \pm 1.0147}{0.296}$$

$x = 6.86$ and 3.7×10^{-4}

Discard the solution that uses the positive square root (x = 6.86) because it gives a negative concentration of the monomer and that is impossible.

[monomer] = 0.100 − 2x = 0.100 − 2(3.7 x 10⁻⁴) = 0.099 M;   [dimer] = x = 3.7 × 10⁻⁴ M

$$\frac{[dimer]}{[monomer]} = \frac{3.7 \times 10^{-4}\,M}{0.099\,M} = 0.0038$$

(c) K_c for the water solution is so much smaller than K_c for the benzene solution because H_2O can hydrogen bond with acetic acid, thus preventing acetic acid dimer formation. Benzene cannot hydrogen bond with acetic acid.

13.92 (a) $K_c = \dfrac{[C_2H_6][C_2H_4]}{[C_4H_{10}]}$ $K_p = \dfrac{(P_{C_2H_6})(P_{C_2H_4})}{P_{C_4H_{10}}}$

(b) $K_p = 12$; $\Delta n = 1$; $K_c = K_p\left(\dfrac{1}{RT}\right) = (12)\left(\dfrac{1}{(0.0821)(773)}\right) = 0.19$

(c)

	$C_4H_{10}(g)$	⇌	$C_2H_6(g)$	+	$C_2H_4(g)$
initial (atm)	50		0		0
change (atm)	−x		+x		+x
equil (atm)	50 − x		x		x

$$K_p = 12 = \frac{x^2}{50 - x}; \qquad x^2 + 12x - 600 = 0$$

Use the quadratic formula to solve for x.

$$x = \frac{(-12) \pm \sqrt{(12)^2 - 4(1)(-600)}}{2(1)} = \frac{-12 \pm 50.44}{2}$$

$x = -31.22$ and 19.22

Discard the solution that uses the negative square root (x = −31.22) because it leads to

170

negative concentrations and that is impossible.

% C_4H_{10} converted $= \dfrac{19.22}{50}$ x 100% = 38%

$P_{total} = P_{C_4H_{10}} + P_{C_2H_6} + P_{C_2H_4} = (50 - x) + x + x = (50 - 19) + 19 + 19 = 69$ atm

(d) A decrease in volume would decrease the % conversion of C_4H_{10}.

13.94 (a) $K_p = 3.45$; $\Delta n = 1$; $K_c = K_p\left(\dfrac{1}{RT}\right) = (3.45)\left(\dfrac{1}{(0.0821)(500)}\right) = 0.0840$

(b) $[(CH_3)_3CCl] = 1.00$ mol/5.00 L = 0.200 M

	$(CH_3)_3CCl(g)$	$\rightleftarrows$	$(CH_3)_2C{=}CH_2(g)$	+	$HCl(g)$
initial (M)	0.200		0		0
change (M)	−x		+x		+x
equil (M)	0.200 − x		x		x

$K_c = 0.0840 = \dfrac{x^2}{0.200 - x}$; $x^2 + 0.0840x - 0.0168 = 0$

Use the quadratic formula to solve for x.

$x = \dfrac{(-0.0840) \pm \sqrt{(0.0840)^2 - 4(1)(-0.0168)}}{2(1)} = \dfrac{-0.0840 \pm 0.272}{2}$

x = −0.178 and 0.094

Discard the solution that uses the negative square root (x = −0.178) because it leads to negative concentrations and that is impossible.

$[(CH_3)_2C{=}CCH_2] = [HCl] = x = 0.094$ M

$[(CH_3)_3CCl] = 0.200 - x = 0.200 - 0.094 = 0.106$ M

(c) $K_p = 3.45$

	$(CH_3)_3CCl(g)$	$\rightleftarrows$	$(CH_3)_2C{=}CH_2(g)$	+	$HCl(g)$
initial (atm)	0		0.400		0.600
change (atm)	+x		−x		−x
equil (atm)	x		0.400 − x		0.600 − x

$K_p = 3.45 - \dfrac{(0.400 - x)(0.600 - x)}{x}$

$x^2 - 4.45x + 0.240 = 0$

Use the quadratic formula to solve for x.

$x = \dfrac{-(-4.45) \pm \sqrt{(-4.45)^2 - 4(1)(0.240)}}{2(1)} = \dfrac{4.45 \pm 4.34}{2}$

x = 0.055 and 4.40

Discard the solution that uses the positive square root (x = 4.40) because it leads to a negative partial pressures and that is impossible.

$P_{t-butyl\ chloride} = x = 0.055$ atm; $P_{isobutylene} = 0.400 - x = 0.400 - 0.055 = 0.345$ atm

$P_{HCl} = 0.600 - x = 0.600 - 0.055 = 0.545$ atm

13.96 The activation energy (E_a) is positive, and for an exothermic reaction, $E_{a,r} > E_{a,f}$.

$$k_f = A_f\, e^{-E_{a,f}/RT}, \quad k_r = A_r\, e^{-E_{a,r}/RT}$$

$$K_c = \frac{k_f}{k_r} = \frac{A_f\, e^{-E_{a,f}/RT}}{A_r\, e^{-E_{a,r}/RT}} = \frac{A_f}{A_r}\, e^{(E_{a,r}-E_{a,f})/RT}$$

($E_{a,r} - E_{a,f}$) is positive, so the exponent is always positive. As the temperature increases, the exponent, $(E_{a,r} - E_{a,f})/RT$, decreases and the value for K_c decreases as well.

13.98 (a) CO_2, 44.01 amu; CO, 28.01 amu

$$79.2 \text{ g } CO_2 \times \frac{1 \text{ mol } CO_2}{44.01 \text{ g } CO_2} = 1.80 \text{ mol } CO_2$$

	$CO_2(g)$	+	C(s)	⇌	2 CO(g)
initial (mol)	1.80				0
change (mol)	–x				+2x
equil (mol)	1.80 – x				2x

total mass of gas in flask = (16.3 g/L)(5.00 L) = 81.5 g
81.5 = (1.80 – x)(44.01) + (2x)(28.01)
81.5 = 79.22 – 44.01x + 56.02x; 2.28 = 12.01x; x = 2.28/12.01 = 0.19
n_{CO_2} = 1.80 – x = 1.80 – 0.19 = 1.61 mol CO_2; n_{CO} = 2x = 2(0.19) = 0.38 mol CO

$$P_{CO_2} = \frac{nRT}{V} = \frac{(1.61 \text{ mol})\left(0.082\,06\,\frac{L\cdot atm}{mol\cdot K}\right)(1000 \text{ K})}{5.0 \text{ L}} = 26.4 \text{ atm}$$

$$P_{CO} = \frac{nRT}{V} = \frac{(0.38 \text{ mol})\left(0.082\,06\,\frac{L\cdot atm}{mol\cdot K}\right)(1000 \text{ K})}{5.0 \text{ L}} = 6.24 \text{ atm}$$

$$K_p = \frac{(P_{CO})^2}{(P_{CO_2})} = \frac{(6.24)^2}{(26.4)} = 1.47$$

(b) At 1100K, the total mass of gas in flask = (16.9 g/L)(5.00 L) = 84.5 g
84.5 = (1.80 – x)(44.01) + (2x)(28.01)
84.5 = 79.22 – 44.01x + 56.02x; 5.28 = 12.01x; x = 5.28/12.01 = 0.44
n_{CO_2} = 1.80 – x = 1.80 – 0.44 = 1.36 mol CO_2; n_{CO} = 2x = 2(0.44) = 0.88 mol CO

$$P_{CO_2} = \frac{nRT}{V} = \frac{(1.36 \text{ mol})\left(0.082\,06\,\frac{L\cdot atm}{mol\cdot K}\right)(1100 \text{ K})}{5.0 \text{ L}} = 24.6 \text{ atm}$$

$$P_{CO} = \frac{nRT}{V} = \frac{(0.88 \text{ mol})\left(0.082\,06\,\frac{L\cdot atm}{mol\cdot K}\right)(1100 \text{ K})}{5.0 \text{ L}} = 15.9 \text{ atm}$$

$$K_p = \frac{(P_{CO})^2}{(P_{CO_2})} = \frac{(15.9)^2}{(24.6)} = 10.3$$

(c) In agreement with Le Châtelier's principle, the reaction is endothermic because K_p increases with increasing temperature.

13.100 The atmosphere is 21% (0.21) O_2; $P_{O_2} = (0.21)\left(720 \text{ mm Hg} \times \dfrac{1 \text{ atm}}{760 \text{ mm Hg}} \right) = 0.199$ atm

$$2 \ O_3(g) \ \rightleftharpoons \ 3 \ O_2(g)$$

$$K_p = \frac{(P_{O_2})^3}{(P_{O_3})^2}; \quad P_{O_3} = \sqrt{\frac{(P_{O_2})^3}{K_p}} = \sqrt{\frac{(0.199)^3}{1.3 \times 10^{57}}} = 2.46 \times 10^{-30} \text{ atm}$$

$$\text{vol} = 10 \times 10^6 \text{ m}^3 \times \left(\frac{100 \text{ cm}}{1 \text{ m}} \right)^3 \times \frac{1 \text{ L}}{1000 \text{ cm}^3} = 1.0 \times 10^{10} \text{ L}$$

$$n_{O_3} = \frac{PV}{RT} = \frac{(2.46 \times 10^{-30} \text{ atm})(1.0 \times 10^{10} \text{ L})}{\left(0.082\,06 \ \dfrac{\text{L} \cdot \text{atm}}{\text{mol} \cdot \text{K}} \right)(298 \text{ K})} = 1.0 \times 10^{-21} \text{ mol } O_3$$

$$O_3 \text{ molecules} = 1.0 \times 10^{-21} \text{ mol } O_3 \times \frac{6.022 \times 10^{23} \ O_3 \text{ molecules}}{1 \text{ mol } O_3} = 6.0 \times 10^2 \ O_3 \text{ molecules}$$

Hydrogen, Oxygen, and Water

14.1 $PV = nRT; \quad PV = \dfrac{g}{molar\ mass} RT$

$d_{H_2} = \dfrac{g}{V} = \dfrac{P(molar\ mass)}{RT} = \dfrac{(1.00\ atm)(2.016\ g/mol)}{\left(0.08206\ \dfrac{L \cdot atm}{mol \cdot K}\right)(298\ K)} = 0.0824\ g/L$

$1\ L = 1000\ mL = 1000\ cm^3$

$d_{H_2} = 0.0824\ g/1000\ cm^3 = 8.24 \times 10^{-5}\ g/cm^3$

$\dfrac{d_{air}}{d_{H_2}} = \dfrac{1.185 \times 10^{-3}\ g/cm^3}{8.24 \times 10^{-5}\ g/cm^3} = 14.4;$ Air is 14 times more dense than H_2.

14.2   For every 100.0 g, there are:
61.4 g O, 22.9 g C, 10.0 g H, 2.6 g N, and 3.1 g other

$22.9\ g\ C \times \dfrac{1\ mol\ C}{12.011\ g\ C} = 1.907\ mol\ C$

$10.0\ g\ H \times \dfrac{1\ mol\ H}{1.008\ g\ H} = 9.921\ mol\ H$

Assume the sample contains 1.907 mol ^{13}C and 9.921 mol D.

$mass\ ^{13}C = 1.907\ mol\ ^{13}C \times \dfrac{13.0034\ g\ ^{13}C}{1\ mol\ ^{13}C} = 24.8\ g\ ^{13}C$

$mass\ D = 9.921\ mol\ D \times \dfrac{2.0141\ g\ D}{1\ mol\ D} = 20.0\ g\ D$

(a) Total mass if all H is D is:
61.4 g O + 22.9 C + 20.0 g D + 2.6 g N + 3.1 g other = 110.0 g

$mass\ \%\ D = \dfrac{20.0\ g\ D}{110.0\ g} \times 100\% = 18.2\%\ D$

(b) Total mass if all C is ^{13}C is:
61.4 g O + 24.8 g ^{13}C + 10.0 g H + 2.6 g N + 3.1 g other = 101.9 g

$mass\ \%\ ^{13}C = \dfrac{24.8\ g\ ^{13}C}{101.9\ g} \times 100\% = 24.3\%\ ^{13}C$

(c) The isotope effect for H is larger than that for C because D is two times the mass of ^1H while ^{13}C is only about 8% heavier than ^{12}C.

14.3 $2\ Ga(s) + 6\ H^+(aq) \rightarrow 3\ H_2(g) + 2\ Ga^{3+}(aq)$

14.4 $CaH_2(s) + 2\ H_2O(l) \rightarrow 2\ H_2(g) + Ca^{2+}(aq) + 2\ OH^-(aq)$
CaH_2, 42.09 amu; 25°C = 298 K

$$PV = nRT; \qquad n_{H_2} = \frac{PV}{RT} = \frac{(1.00 \text{ atm})(2.0 \times 10^5 \text{ L})}{\left(0.082\ 06 \dfrac{\text{L} \cdot \text{atm}}{\text{mol} \cdot \text{K}}\right)(298 \text{ K})} = 8.18 \times 10^3 \text{ mol H}_2$$

$$8.18 \times 10^3 \text{ mol H}_2 \times \frac{1 \text{ mol CaH}_2}{2 \text{ mol H}_2} \times \frac{42.09 \text{ g CaH}_2}{1 \text{ mol CaH}_2} \times \frac{1 \text{ kg}}{1000 \text{ g}} = 1.7 \times 10^2 \text{ kg CaH}_2$$

14.5   Assume 12.0 g of Pd with a volume of 1.0 cm³.

$$V_{H_2} = 935 \text{ cm}^3 = 935 \text{ mL} = 0.935 \text{ L}$$

$$PV = nRT; \qquad n_{H_2} = \frac{PV}{RT} = \frac{(1.00 \text{ atm})(0.935 \text{ L})}{\left(0.082\ 06 \dfrac{\text{L} \cdot \text{atm}}{\text{mol} \cdot \text{K}}\right)(273 \text{ K})} = 0.0417 \text{ mol H}_2$$

$$n_H = 2 n_{H_2} = 0.0834 \text{ mol H}$$

$$12.0 \text{ g Pd} \times \frac{1 \text{ mol Pd}}{106.42 \text{ g Pd}} = 0.113 \text{ mol Pd}$$

$Pd_{0.113}H_{0.0834}$

$Pd_{0.113 / 0.113}H_{0.0834 / 0.113}$

$PdH_{0.74}$

$$g\ H = (0.0834 \text{ mol H})(1.008 \text{ g/mol}) = 0.0841 \text{ g H}$$

$$d_H = 0.0841 \text{ g/cm}^3; \qquad M_H = \frac{0.0834 \text{ mol}}{0.001 \text{ L}} = 83.4 \text{ M}$$

14.6 $2 \text{ KMnO}_4(s) \rightarrow \text{K}_2\text{MnO}_4(s) + \text{MnO}_2(s) + \text{O}_2(g)$

       $KMnO_4$, 158.03 amu; 25°C = 298 K

$$\text{mol O}_2 = 0.200 \text{ g KMnO}_4 \times \frac{1 \text{ mol KMnO}_4}{158.03 \text{ g KMnO}_4} \times \frac{1 \text{ mol O}_2}{2 \text{ mol KMnO}_4} = 6.33 \times 10^{-4} \text{ mol O}_2$$

$$PV = nRT; \qquad V = \frac{nRT}{P} = \frac{(6.33 \times 10^{-4} \text{ mol})\left(0.082\ 06 \dfrac{\text{L} \cdot \text{atm}}{\text{mol} \cdot \text{K}}\right)(298 \text{ K})}{1.00 \text{ atm}} = 0.0155 \text{ L}$$

$$V = 0.0155 \text{ L} \times \frac{1000 \text{ mL}}{1 \text{ L}} = 15.5 \text{ mL O}_2$$

14.7   (a) $Li_2O(s) + H_2O(l) \rightarrow 2 \text{ Li}^+(aq) + 2 \text{ OH}^-(aq)$
         (b) $SO_3(l) + H_2O(l) \rightarrow \text{H}^+(aq) + \text{HSO}_4^-(aq)$
         (c) $Cr_2O_3(s) + 6 \text{ H}^+(aq) \rightarrow 2 \text{ Cr}^{3+}(aq) + 3 \text{ H}_2O(l)$
         (d) $Cr_2O_3(s) + 2 \text{ OH}^-(aq) + 3 \text{ H}_2O(l) \rightarrow 2 \text{ Cr(OH)}_4^-(aq)$

14.8   (a) Rb_2O_2 Rb +1, O −1, peroxide   (b) CaO   Ca +2, O −2, oxide
         (c) CsO_2 Cs +1, O −1/2, superoxide   (d) SrO_2 Sr +2, O −1, peroxide
         (e) CO_2 C +4, O −2, oxide

14.9 (a) $Rb_2O_2(s) + H_2O(l) \rightarrow 2\ Rb^+(aq) + HO_2^-(aq) + OH^-(aq)$
(b) $CaO(s) + H_2O(l) \rightarrow Ca^{2+}(aq) + 2\ OH^-(aq)$
(c) $2\ CsO_2(s) + H_2O(l) \rightarrow O_2(g) + 2\ Cs^+(aq) + HO_2^-(aq) + OH^-(aq)$
(d) $SrO_2(s) + H_2O(l) \rightarrow Sr^{2+}(aq) + HO_2^-(aq) + OH^-(aq)$
(e) $CO_2(g) + H_2O(l) \rightarrow H^+(aq) + HCO_3^-(aq)$

14.10 $H-\overset{..}{\underset{..}{O}}-\overset{..}{\underset{..}{O}}-H$ The electron dot structure indicates a single bond (see text Table 14.2) which is consistent with an O–O bond length of 148 pm.

14.11 $PbS(s) + 4\ H_2O_2(aq) \rightarrow PbSO_4(s) + 4\ H_2O(l)$

14.12 (a) $2\ Li(s) + 2\ H_2O(l) \rightarrow H_2(g) + 2\ Li^+(aq) + 2\ OH^-(aq)$
(b) $Sr(s) + 2\ H_2O(l) \rightarrow H_2(g) + Sr^{2+}(aq) + 2\ OH^-(aq)$
(c) $Br_2(l) + H_2O(l) \rightleftharpoons HOBr(aq) + H^+(aq) + Br^-(aq)$

14.13 mass of H_2O = 5.62 g – 3.10 g = 2.52 g H_2O

$$2.52\ g\ H_2O \times \frac{1\ mol\ H_2O}{18.02\ g\ H_2O} = 0.140\ mol\ H_2O$$

$$3.10\ g\ NiSO_4 \times \frac{1\ mol\ NiSO_4}{154.8\ g\ NiSO_4} = 0.0200\ mol\ NiSO_4$$

number of H_2O's in hydrate $= \dfrac{n_{H_2O}}{n_{NiSO_4}} = \dfrac{0.140\ mol}{0.0200\ mol} = 7$

Hydrate formula is $NiSO_4 \cdot 7\ H_2O$

Understanding Key Concepts

14.14 (a) (1) covalent   (2) ionic   (3) covalent   (4) interstitial
(b) (1) H, +1; other element, –3
(2) H, –1; other element, +1
(3) H, +1; other element, –2

14.15 (a) A, NaH; B, PdH_x; C, H_2S; D, HI
(b) NaH (ionic); PdH_x (interstitial); H_2S and HI (covalent)
(c) H_2S and HI (molecular); NaH and PdH_x (3–dimensional crystal)
(d) NaH: Na +1, H –1
H_2S: S –2, H +1
HI: I –1, H +1

14.16 React H_2O with a reducing agent to produce H_2. Ca or Al could be used.

14.17 (a) 6; $^{16}O_2$, $^{17}O_2$, $^{18}O_2$, $^{16}O^{17}O$, $^{16}O^{18}O$, $^{17}O^{18}O$

(b) 18

$^{16}O_3$	$^{16}O_2{}^{17}O$	$^{18}O_3$
$^{17}O_2{}^{16}O$	$^{17}O_3$	$^{16}O_2{}^{18}O$
$^{18}O_2{}^{16}O$	$^{18}O_2{}^{17}O$	$^{17}O_2{}^{18}O$
$^{16}O^{17}O^{16}O$	$^{17}O^{18}O^{17}O$	$^{16}O^{17}O^{18}O$
$^{16}O^{18}O^{16}O$	$^{17}O^{16}O^{17}O$	$^{18}O^{16}O^{18}O$
$^{17}O^{18}O^{16}O$	$^{18}O^{16}O^{17}O$	$^{18}O^{17}O^{18}O$

14.18 (a), (b), and (d) are different kinds of water and have similar properties. (c) and (e) are different kinds of hydrogen peroxide and have similar properties. The properties of (a), (b), and (d) are quite different from those of (c) and (e).

14.19 (a) (1) –2, +4; (2) –2, +6; (3) –2, +2
      (b) (1) covalent; (2) covalent; (3) ionic
      (c) (1) acidic; (2) acidic; (3) basic
      (d) (1) carbon; (2) sulfur

14.20 (a) (1) –2, +2; (2) –2, +1; (3) –2, +5
      (b) (1) three-dimensional; (2) molecular; (3) molecular
      (c) (1) solid; (2) gas or liquid; (3) gas or liquid
      (d) (2) hydrogen; (3) nitrogen

14.21 (a) A, CaO; B, Al_2O_3; C, SO_3; D, SeO_3
      (b) CaO (basic); Al_2O_3 (amphoteric); SO_3 and SeO_3 (acidic)
      (c) CaO (most ionic); SO_3 (most covalent)
      (d) CaO and Al_2O_3 (3–dimensional crystal); SO_3 and SeO_3 (molecular)
      (e) CaO (highest melting point); SO_3 (lowest melting point)

Additional Problems
Chemistry of Hydrogen

14.22

Isotope	Nucleus Composition	Atom %
protium	1 proton	99.9844%
deuterium	1 proton, 1 neutron	0.0156%
tritium	1 proton, 2 neutrons	$\sim 10^{-16}$%

14.24 Quantitative differences in properties that arise from the differences in the masses of the isotopes are known as isotope effects.
      Examples: H_2 and D_2 have different melting and boiling points.
                    H_2O and D_2O have different dissociation constants.

14.26
$$\frac{\text{mass } {}^2D - \text{mass } {}^1H}{\text{mass } {}^1H} \times 100\% = \frac{2.0141 \text{ amu} - 1.0078 \text{ amu}}{1.0078 \text{ amu}} \times 100\% = 98.85\%$$

$$\frac{\text{mass } {}^3H - \text{mass } {}^2H}{\text{mass } {}^2H} \times 100\% = \frac{3.0160 \text{ amu} - 2.0141 \text{ amu}}{2.0141 \text{ amu}} \times 100\% = 49.74\%$$

The differences in properties will be larger for H_2O and D_2O rather than for D_2O and T_2O because of the larger relative difference in mass for H and D versus D and T. This is supported by the data in Table 14.1.

14.28 There are 18 kinds of H_2O

$H_2{}^{16}O$	$H_2{}^{17}O$	$H_2{}^{18}O$
$D_2{}^{16}O$	$D_2{}^{17}O$	$D_2{}^{18}O$
$T_2{}^{16}O$	$T_2{}^{17}O$	$T_2{}^{18}O$
$HD^{16}O$	$HD^{17}O$	$HD^{18}O$
$HT^{16}O$	$HT^{17}O$	$HT^{18}O$
$DT^{16}O$	$DT^{17}O$	$DT^{18}O$

14.30 The steam–hydrocarbon reforming process is the most important industirial preparation of hydrogen.

$$CH_4(g) + H_2O(g) \xrightarrow[\text{Ni catalyst}]{1100°C} CO(g) + 3\,H_2(g)$$

$$CO(g) + H_2O(g) \xrightarrow{400°C} CO_2(g) + H_2(g)$$

$$CO_2(g) + 2\,OH^-(aq) \rightarrow CO_3{}^{2-}(aq) + H_2O(l)$$

14.32 (a) LiH, 7.95 amu; CaH_2, 42.09 amu

$$LiH(s) + H_2O(l) \rightarrow H_2(g) + Li^+(aq) + OH^-(aq)$$
$$CaH_2(s) + 2\,H_2O(l) \rightarrow 2\,H_2(g) + Ca^{2+}(aq) + 2\,OH^-(aq)$$

You obtain $\dfrac{1 \text{ mol } H_2}{7.95 \text{ g LiH}} = 0.126$ mol H_2/g LiH

and $\dfrac{2 \text{ mol } H_2}{42.09 \text{ g CaH}_2} = 0.0475$ mol H_2/g CaH_2;   Therefore, LiH gives more H_2.

(b) 25°C = 298 K

$$PV = nRT; \quad n_{H_2} = \frac{PV}{RT} = \frac{(150 \text{ atm})(100 \text{ L})}{\left(0.082\ 06\ \dfrac{L \cdot atm}{mol \cdot K}\right)(298K)} = 613.4 \text{ mol } H_2$$

mass CaH_2 = 613.4 mol H_2 x $\dfrac{1 \text{ mol } CaH_2}{2 \text{ mol } H_2}$ x $\dfrac{42.09 \text{ g } CaH_2}{1 \text{ mol } CaH_2}$ x $\dfrac{1 \text{ kg}}{1000 \text{ g}}$ = 12.9 kg CaH_2

14.34 Hydrogen is located in Group 1A of the periodic table because it has one valence electron. Hydrogen's chemical behavior is so different from that of the alkali metals because its ionization potential is so high that it prefers to form covalent compounds.

14.36 (a) HCl(g), covalent; HCl(aq), H^+
(b) CaH_2, H^- (c) SiH_4, covalent   (d) RbH, H^-

14.38   Ionic Hydrides   Covalent Hydrides   Metallic Hydrides

              NaH NH_3 $TiH_{1.7}$

              CaH_2 H_2O $ZrH_{1.9}$

              SrH_2 CH_4 UH_3

14.40 H_2S covalent hydride, gas, weak acid in H_2O

        NaH   ionic hydride, solid (salt like), reacts with H_2O to produce H_2

        PdH_x metallic (interstitial) hydride, solid, stores hydrogen

14.42   (a) KH, ionic bonding   (b) PH_3, covalent bonding

14.44   (a)   H—S̈e—H , bent   (b)   H—Äs—H, trigonal pyramidal

                                                    |

                                                    H

        (c)   H   , tetrahedral

                    |

            H—Si—H

                    |

                    H

14.46   A nonstoichiometric compound is a compound whose atomic composition cannot be expressed as a ratio of small whole numbers. An example is PdH_x. The lack of stoichiometry results from the hydrogen occupying holes in the solid state structure.

14.48 TiH_2, 49.90 amu;   Assume 1.0 cm^3 of TiH_2 which has a mass of 3.9 g.

$$3.9 \text{ g TiH}_2 \times \frac{1 \text{ mol TiH}_2}{49.90 \text{ g TiH}_2} = 0.078 \text{ mol TiH}_2$$

$$0.078 \text{ mol TiH}_2 \times \frac{2 \text{ mol H}}{1 \text{ mol TiH}_2} = 0.156 \text{ mol H}$$

$$0.156 \text{ mol H} \times \frac{1.008 \text{ g H}}{1 \text{ mol H}} = 0.157 \text{ g H}$$

$d_H = 0.16$ g/cm^3; the density of H in TiH_2 is about 2.25 times the density of liquid H_2.

$$PV = nRT; \quad V = \frac{nRT}{P} = \frac{\left(0.16 \text{ g} \times \frac{1 \text{ mol}}{2.016 \text{ g}}\right)\left(0.082\,06 \frac{\text{L} \cdot \text{atm}}{\text{mol} \cdot \text{K}}\right)(273 \text{ K})}{1.00 \text{ atm}} = 1.8 \text{ L H}_2$$

$1.8 \text{ L} = 1.8 \times 10^3 \text{ mL} = 1.8 \times 10^3 \text{ cm}^3$

Chemistry of Oxygen

14.50   (a) O_2 is obtained in industry by the fractional distillation of liquid air.

        (b) In the laboratory, O_2 is prepared by the thermal decomposition of $KClO_3(s)$.

$$2 \text{ KClO}_3(s) \xrightarrow[\text{MnO}_2]{\text{heat}} 2 \text{ KCl}(s) + 3 \text{ O}_2(g)$$

14.52 $\overset{\text{catalyst}}{2\,H_2O_2(aq) \;\rightarrow\; 2\,H_2O(l) \;+\; O_2(g)}$

H_2O_2, 34.01 amu; 25°C = 298 K

$$\text{mol } O_2 = 20.4 \text{ g } H_2O_2 \times \frac{1 \text{ mol } H_2O_2}{34.01 \text{ g } H_2O_2} \times \frac{1 \text{ mol } O_2}{2 \text{ mol } H_2O_2} = 0.300 \text{ mol } O_2$$

$$PV = nRT; \quad V = \frac{nRT}{P} = \frac{(0.300 \text{ mol})\left(0.082\ 06 \dfrac{L \cdot atm}{mol \cdot K}\right)(298 \text{ K})}{1.00 \text{ atm}} = 7.34 \text{ L } O_2$$

14.54   :Ö::Ö: The Lewis electron dot structure shows an O=O double bond. It also shows all electrons paired. This is not consistent with the fact that O_2 is paramagnetic.

14.56   Acidic oxides are covalent and are formed by nonmetals. Examples are CO_2 and Cl_2O_7. Basic oxides are ionic and are formed by the active metals on the left side of the periodic table. Examples are CaO and K_2O.
Amphoteric oxides exhibit both acidic and basic properties. Examples are Al_2O_3 and Ga_2O_3.

14.58 $Li_2O \;<\; BeO \;<\; B_2O_3 \;<\; CO_2 \;<\; N_2O_5$ (see Figure 14.6)

14.60 $Cl_2O_7 \;<\; Al_2O_3 \;<\; Na_2O \;<\; Cs_2O$ (see Figure 14.6)

14.62   (a) CrO_3 (higher Cr oxidation state)   (b) N_2O_5 (higher N oxidation state)
(c) SO_3 (higher S oxidation)

14.64   (a) $Cl_2O_7(l) \;+\; H_2O(l) \;\rightarrow\; 2\,H^+(aq) \;+\; 2\,ClO_4^-(aq)$
(b) $K_2O(s) \;+\; H_2O(l) \;\rightarrow\; 2\,K^+(aq) \;+\; 2\,OH^-(aq)$
(c) $SO_3(l) \;+\; H_2O(l) \;\rightarrow\; H^+(aq) \;+\; HSO_4^-(aq)$

14.66   (a) $ZnO(s) \;+\; 2\,H^+(aq) \;\rightarrow\; Zn^{2+}(aq) \;+\; H_2O(l)$
(b) $ZnO(s) \;+\; 2\,OH^-(aq) \;+\; H_2O(l) \;\rightarrow\; Zn(OH)_4^{2-}(aq)$

14.68   A peroxide has oxygen in the −1 oxidation state, for example, H_2O_2. A superoxide has oxygen in the −1/2 oxidation state, for example, KO_2.

14.70   (a) BaO_2 (b) CaO   (c) CsO_2 (d) Li_2O (e) Na_2O_2

14.72

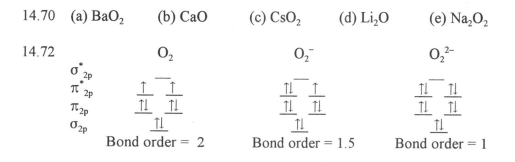

(a) The O–O bond length increases because the bond order decreases. The bond order decreases because of the increased occupancy of antibonding orbitals.

(b) O_2^- has 1 unpaired electron and is paramagnetic. O_2^{2-} has no unpaired electrons and is diamagnetic.

14.74   A reaction in which one substance is both oxidized and reduced is called a disproportionation reaction. Example:

$2 KO_2(s) + H_2O(l) \rightarrow O_2(g) + 2 K^+(aq) + HO_2^-(aq) + OH^-(aq)$

The oxygen in KO_2 undergoes disproportionation.

14.76   (a) $H_2O_2(aq) + 2 H^+(aq) + 2 I^-(aq) \rightarrow I_2(aq) + 2 H_2O(l)$

(b) $3 H_2O_2(aq) + 8 H^+(aq) + Cr_2O_7^{2-}(aq) \rightarrow 2 Cr^{3+}(aq) + 3 O_2(g) + 7 H_2O(l)$

14.78

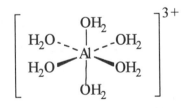

Ozone has two resonance structures which leads to two equivalent O–O bond lengths.

14.80 $3 O_2(g) \xrightarrow{\text{electric discharge}} 2 O_3(g)$

Chemistry of Water

14.82   (a) $2 F_2(g) + 2 H_2O(l) \rightarrow O_2(g) + 4 HF(aq)$

(b) $Cl_2(g) + H_2O(l) \rightleftharpoons HOCl(aq) + H^+(aq) + Cl^-(aq)$

(c) $I_2(s) + H_2O(l) \rightarrow HOI(aq) + H^+(aq) + I^-(aq)$

(d) $Ba(s) + 2 H_2O(l) \rightarrow H_2(g) + Ba^{2+}(aq) + 2 OH^-(aq)$

14.84 $AlCl_3 \cdot 6 H_2O$

$$\left[\begin{array}{c} OH_2 \\ H_2O \cdots \underset{\displaystyle OH_2}{\overset{\displaystyle |}{Al}} \cdots OH_2 \\ H_2O \quad\quad OH_2 \end{array} \right]^{3+}$$

14.86 $CaSO_4 \cdot \frac{1}{2} H_2O$, 145.15 amu; H_2O, 18.02 amu

Assume one mole of $CaSO_4 \cdot \frac{1}{2} H_2O$

$$\text{mass \% } H_2O = \frac{\text{mass } H_2O}{\text{mass hydrate}} \times 100\% = \frac{18.02 \text{ g}}{145.15 \text{ g}} \times 100\% = 12.41\%$$

14.88 $CaSO_4 \cdot \frac{1}{2} H_2O$, 145.15 amu; H_2O, 18.02 amu

mass of H_2O lost = 3.44 g – 2.90 g = 0.54 g H_2O

$$2.90 \text{ g } CaSO_4 \cdot \tfrac{1}{2} H_2O \times \frac{1 \text{ mol}}{145.15 \text{ g}} = 0.020 \text{ mol } CaSO_4 \cdot \tfrac{1}{2} H_2O$$

$$0.54 \text{ g } H_2O \times \frac{1 \text{ mol}}{18.02 \text{ g}} = 0.030 \text{ mol}$$

$$\text{number of } H_2O\text{'s lost} = \frac{0.030 \text{ mol}}{0.020 \text{ mol}} = 1.5 \text{ } H_2O \text{ per } CaSO_4 \cdot \text{½ } H_2O \text{ formed}$$

The mineral gypsum is $CaSO_4 \cdot 2 \text{ } H_2O$; $x = 2$

14.90   Convert mi^3 to cm^3; $(1.0 \text{ mi}^3)\left(\dfrac{1609 \text{ m}}{1 \text{ mi}}\right)^3\left(\dfrac{100 \text{ cm}}{1 \text{ m}}\right)^3 = 4.2 \times 10^{15} \text{ cm}^3$

mass of sea water = volume x density = $(4.2 \times 10^{15} \text{ cm}^3)(1.025 \text{ g/cm}^3) = 4.3 \times 10^{15} \text{ g}$

mass of salts = $(0.035)(4.3 \times 10^{15} \text{ g})\left(\dfrac{1 \text{ kg}}{1000 \text{ g}}\right) = 1.5 \times 10^{11} \text{ kg}$

General Problems

14.92   In drinking water, Cl_2 is used to kill bacteria.
The sedimentation of unsuspended matter in drinking water is accelerated by the addition of CaO and $Al_2(SO_4)_3$. The CaO makes the water slightly basic, which precipitates Al^{3+} as $Al(OH)_3$. The gelatinous $Al(OH)_3$ precipitate slowly settles, carrying with it suspended solid and colloidal material and most of the bacteria.

14.94   Butadiene, C_4H_6, 54.09 amu; 2.7 kg = 2700 g

$$\text{moles } H_2 = 2700 \text{ g } C_4H_6 \times \frac{1 \text{ mol } C_4H_6}{54.09 \text{ g } C_4H_6} \times \frac{2 \text{ mol } H_2}{1 \text{ mol } C_4H_6} = 99.8 \text{ mol } H_2$$

At STP, P = 1.00 atm and T = 273 K

$$PV = nRT; \quad V = \frac{nRT}{P} = \frac{(99.8 \text{ mol})\left(0.082 \text{ 06 } \dfrac{L \cdot atm}{mol \cdot K}\right)(273 \text{ K})}{1.00 \text{ atm}} = 2.2 \times 10^3 \text{ L of } H_2$$

14.96   (a) B_2O_3, diboron trioxide   (b) H_2O_2, hydrogen peroxide
(c) SrH_2, strontium hydride (d) CsO_2, cesium superoxide
(e) $HClO_4$, perchloric acid   (f) BaO_2, barium peroxide

14.98   (a) $2 \text{ } H_2(g) + O_2(g) \rightarrow 2 \text{ } H_2O(l)$
(b) $O_3(g) + 2 \text{ } I^-(aq) + H_2O(l) \rightarrow O_2(g) + I_2(aq) + 2 \text{ } OH^-(aq)$
(c) $H_2O_2(aq) + 2 \text{ } H^+(aq) + 2 \text{ } Br^-(aq) \rightarrow 2 \text{ } H_2O(l) + Br_2(aq)$
(d) $2 \text{ } Na(l) + H_2(g) \rightarrow 2 \text{ } NaH(s)$
(e) $2 \text{ } Na(s) + 2 \text{ } H_2O(l) \rightarrow H_2(g) + 2 \text{ } Na^+(aq) + 2 \text{ } OH^-(aq)$

14.100 K is oxidized by water. F_2 is reduced by water.
Cl_2 and Br_2 disproportionate when treated with water.

14.102 (a) $CO(g) + 2 H_2(g) \rightarrow CH_3OH(l)$
$\Delta H° = \Delta H°_f(CH_3OH) - \Delta H°_f(CO)$
$\Delta H° = (1 \text{ mol})(-238.7 \text{ kJ/mol}) - (1 \text{ mol})(-110.5 \text{ kJ/mol}) = -128.2 \text{ kJ}$
(b) $CO(g) + H_2O(g) \rightarrow CO_2(g) + H_2(g)$
$\Delta H° = \Delta H°_f(CO_2) - [\Delta H°_f(CO) + \Delta H°_f(H_2O)]$
$\Delta H° = (1 \text{ mol})(-393.5 \text{ kJ/mol}) - [(1 \text{ mol})(-110.5 \text{ kJ/mol})$
$\qquad\qquad\qquad + (1 \text{ mol})(-241.8 \text{ kJ/mol})] = -41.2 \text{ kJ}$
(c) $2 KClO_3(s) \rightarrow 2 KCl(s) + 3 O_2(g)$
$\Delta H° = 2 \Delta H°_f(KCl) - 2 \Delta H°_f(KClO_3)$
$\Delta H° = (2 \text{ mol})(-436.7 \text{ kJ/mol}) - (2 \text{ mol})(-397.7 \text{ kJ/mol}) = -78.0 \text{ kJ}$
(d) $6 CO_2(g) + 6 H_2O(l) \rightarrow 6 O_2(g) + C_6H_{12}O_6(s)$
$\Delta H° = \Delta H°_f(C_6H_{12}O_6) - [6 \Delta H°_f(CO_2) + 6 \Delta H°_f(H_2O)]$
$\Delta H° = (1 \text{ mol})(-1260 \text{ kJ/mol}) - [(6 \text{ mol})(-393.5 \text{ kJ/mol})$
$\qquad\qquad\qquad + (6 \text{ mol})(-285.8 \text{ kJ/mol})] = 2816 \text{ kJ}$

14.104 $MH_2(s) + 2 HCl(aq) \rightarrow 2 H_2(g) + M^{2+}(aq) + 2 Cl^-(aq)$
$PV = nRT$

$$n_{H_2} = \frac{PV}{RT} = \frac{\left(750 \text{ mm Hg} \times \dfrac{1.00 \text{ atm}}{760 \text{ mm Hg}}\right)(1.000 \text{ L})}{\left(0.082\ 06 \dfrac{\text{L} \cdot \text{atm}}{\text{mol} \cdot \text{K}}\right)(293 \text{ K})} = 0.0410 \text{ mol } H_2$$

$$0.0410 \text{ mol } H_2 \times \frac{1 \text{ mol } MH_2}{2 \text{ mol } H_2} = 0.0205 \text{ mol } MH_2$$

$$\text{molar mass} = \frac{1.84 \text{ g}}{0.0205 \text{ mol}} = 89.8 \text{ g/mol}$$

mass M = 89.8 – mass of 2 H = 89.8 – 2(1.008) = 87.7 g/mol;   M = Sr; SrH_2

14.106 N_2O_5, 108.01 amu
$N_2O_5(g) + H_2O(l) \rightarrow 2 HNO_3(aq)$
$2 HNO_3(aq) + Zn(s) \rightarrow H_2(g) + Zn(NO_3)_2(aq)$

$$5.4 \text{ g } N_2O_5 \times \frac{1 \text{ mol } N_2O_5}{108.01 \text{ g } N_2O_5} \times \frac{2 \text{ mol } HNO_3}{1 \text{ mol } N_2O_5} \times \frac{1 \text{ mol } H_2}{2 \text{ mol } HNO_3} = 0.050 \text{ mol } H_2$$

$$PV = nRT; \quad P_{H_2} = \frac{nRT}{V} = \frac{(0.050 \text{ mol})\left(0.082\ 06 \dfrac{\text{L} \cdot \text{atm}}{\text{mol} \cdot \text{K}}\right)(298 \text{ K})}{0.500 \text{ L}} = 2.45 \text{ atm}$$

$$P_{H_2} = 2.45 \text{ atm} \times \frac{760 \text{ mm Hg}}{1.00 \text{ atm}} = 1862 \text{ mm Hg}$$

(a) In H there is 0.0156 atom % D.
To get P_{HD} multiply P_{H_2} by the atom % D and then by 2 because H_2 is diatomic.

$$P_{H_2} = (1862 \text{ mm Hg})(0.000156)(2) = 0.58 \text{ mm Hg}$$

(b) $PV = nRT$

$$n_{HD} = \frac{PV}{RT} = \frac{\left(0.58 \text{ mm Hg} \times \frac{1.00 \text{ atm}}{760 \text{ mm Hg}}\right)(0.5000 \text{ L})}{\left(0.082\ 06\ \frac{\text{L}\cdot\text{atm}}{\text{mol}\cdot\text{K}}\right)(298 \text{ K})} = 1.56 \times 10^{-5} \text{ mol HD}$$

$$1.56 \times 10^{-5} \text{ mol HD} \times \frac{6.022 \times 10^{23} \text{ HD molecules}}{1 \text{ mol HD}} = 9.4 \times 10^{18} \text{ HD molecules}$$

(c) $P_{D_2} = (1862 \text{ mm Hg})(0.000156)^2 = 4.53 \times 10^{-5} \text{ mm Hg}$

$PV = nRT$

$$n_{D_2} = \frac{PV}{RT} = \frac{\left(4.53 \times 10^{-5} \text{ mm Hg} \times \frac{1.00 \text{ atm}}{760 \text{ mm Hg}}\right)(0.5000 \text{ L})}{\left(0.082\ 06\ \frac{\text{L}\cdot\text{atm}}{\text{mol}\cdot\text{K}}\right)(298 \text{ K})} = 1.22 \times 10^{-9} \text{ mol D}_2$$

$$1.22 \times 10^{-9} \text{ mol D}_2 \times \frac{6.022 \times 10^{23} \text{ D}_2 \text{ molecules}}{1 \text{ mol D}_2} = 7.3 \times 10^{14} \text{ D}_2 \text{ molecules}$$

15.1 (a) $H_2SO_4(aq) + H_2O(l) \rightleftharpoons H_3O^+(aq) + HSO_4^-(aq)$
              conjugate base

  (b) $HSO_4^-(aq) + H_2O(l) \rightleftharpoons H_3O^+(aq) + SO_4^{2-}(aq)$
              conjugate base

  (c) $H_3O^+(aq) + H_2O(l) \rightleftharpoons H_3O^+(aq) + H_2O(l)$
              conjugate base

15.2 (a) $H_2CO_3(aq) + H_2O(l) \rightleftharpoons H_3O^+(aq) + HCO_3^-(aq)$
    conjugate acid

  (b) $HCO_3^-(aq) + H_2O(l) \rightleftharpoons H_3O^+(aq) + CO_3^{2-}(aq)$
    conjugate acid

  (c) $Base(aq) + H_2O(l) \rightleftharpoons BaseH^+(aq) + OH^-(aq)$
     conjugate acid

15.3 (a) $HF(aq) + NO_3^-(aq) \rightleftharpoons HNO_3(aq) + F^-(aq)$
HNO_3 is a stronger acid than HF, and F^- is a stronger base than NO_3^- (see Table 15.1). Because proton transfer occurs from the stronger acid to the stronger base, the reaction proceeds from right to left.

  (b) $NH_4^+(aq) + CO_3^{2-}(aq) \rightleftharpoons HCO_3^-(aq) + NH_3(aq)$
NH_4^+ is a stronger acid than HCO_3^-, and CO_3^{2-} is a stronger base than NH_3 (see Table 15.1). Because proton transfer occurs from the stronger acid to the stronger base, the reaction proceeds from left to right.

15.4 $[H_3O^+] = \dfrac{K_w}{[OH^-]} = \dfrac{1.0 \times 10^{-14}}{5.0 \times 10^{-6}} = 2.0 \times 10^{-9}$ M
Because $[OH^-] > [H_3O^+]$, the solution is basic.

15.5 $K_w = [H_3O^+][OH^-]$;   In a neutral solution, $[H_3O^+] = [OH^-]$
At 50°C, $[H_3O^+] = [OH^-] = \sqrt{K_w} = \sqrt{5.5 \times 10^{-14}} = 2.3 \times 10^{-7}$ M

15.6 (a) $[H_3O^+] = \dfrac{K_w}{[OH^-]} = \dfrac{1.0 \times 10^{-14}}{1.58 \times 10^{-6}} = 6.3 \times 10^{-9}$ M
$pH = -\log[H_3O^+] = -\log(6.3 \times 10^{-9}) = 8.20$
  (b) $pH = -\log[H_3O^+] = -\log(6.0 \times 10^{-5}) = 4.22$

15.7 $[H_3O^+] = 10^{-pH} = 10^{-7.40} = 4.0 \times 10^{-8}$ M

$$[OH^-] = \frac{K_w}{[H_3O^+]} = \frac{1.0 \times 10^{-14}}{4.0 \times 10^{-8}} = 2.5 \times 10^{-7} \text{ M}$$

15.8 (a) Because $HClO_4$ is a strong acid, $[H_3O^+] = 0.050$ M.
pH = $-\log[H_3O^+] = -\log(0.050) = 1.30$
(b) Because HCl is a strong acid, $[H_3O^+] = 6.0$ M.
pH = $-\log[H_3O^+] = -\log(6.0) = -0.78$
(c) Because KOH is a strong base, $[OH^-] = 0.020$ M.

$$[H_3O^+] = \frac{K_w}{[OH^-]} = \frac{1.0 \times 10^{-14}}{0.020} = 5.0 \times 10^{-13} \text{ M}$$

pH = $-\log[H_3O^+] = -\log(5.0 \times 10^{-13}) = 12.30$
(d) Because $Ba(OH)_2$ is a strong base, $[OH^-] = 2(0.010 \text{ M}) = 0.020$ M.

$$[H_3O^+] = \frac{K_w}{[OH^-]} = \frac{1.0 \times 10^{-14}}{0.020} = 5.0 \times 10^{-13} \text{ M}$$

pH = $-\log[H_3O^+] = -\log(5.0 \times 10^{-13}) = 12.30$

15.9 $BaO(s) + H_2O(l) \rightarrow Ba(OH)_2(aq)$
BaO, 153.33 amu

$$0.25 \text{ g BaO} \times \frac{1 \text{ mol BaO}}{153.33 \text{ g BaO}} \times \frac{1 \text{ mol Ba(OH)}_2}{1 \text{ mol BaO}} \times \frac{2 \text{ mol OH}^-}{1 \text{ mol Ba(OH)}_2} = 3.26 \times 10^{-3} \text{ mol OH}^-$$

$$[OH^-] = \frac{3.26 \times 10^{-3} \text{ mol OH}^-}{0.500 \text{ L}} = 6.52 \times 10^{-3} \text{ M}$$

$$[H_3O^+] = \frac{K_w}{[OH^-]} = \frac{1.0 \times 10^{-14}}{6.52 \times 10^{-3}} = 1.53 \times 10^{-12} \text{ M}$$

pH = $-\log[H_3O^+] = -\log(1.53 \times 10^{-12}) = 11.81$

15.10

	$HOCl(aq) + H_2O(l)$	$\rightleftharpoons$ $H_3O^+(aq)$	$+ OCl^-(aq)$
initial (M)	0.10	~0	0
change (M)	–x	+x	+x
equil (M)	0.10 – x	x	x

$x = [H_3O^+] = 10^{-pH} = 10^{-4.23} = 5.9 \times 10^{-5}$ M
$[OCl^-] = x = 5.9 \times 10^{-5}$ M; $[HOCl] = 0.10 - x = (0.10 - 5.9 \times 10^{-5})$M

$$K_a = \frac{[H_3O^+][OCl^-]}{[HOCl]} = \frac{(5.9 \times 10^{-5})(5.9 \times 10^{-5})}{(0.10 - 5.9 \times 10^{-5})} = 3.5 \times 10^{-8}$$

This value of K_a agrees with the value in Table 15.2.

15.11 (a)

	$CH_3COOH(aq) + H_2O(l)$	$\rightleftharpoons$ $H_3O^+(aq)$	$+ CH_3CO_2^-(aq)$
initial (M)	1.00	~0	0
change (M)	–x	+x	+x
equil (M)	1.00 – x	x	x

$$K_a = \frac{[H_3O^+][CH_3CO_2^-]}{[CH_3COOH]} = 1.8 \times 10^{-5} = \frac{x^2}{1.00-x} \approx \frac{x^2}{1.00}$$

Solve for x. $x = [H_3O^+] = 4.2 \times 10^{-3}$ M

$pH = -\log[H_3O^+] = -\log(4.2 \times 10^{-3}) = 2.38$

$[CH_3CO_2^-] = x = 4.2 \times 10^{-3}$ M; $\qquad [CH_3COOH] = 1.00 - x = 1.00$ M

$$[OH^-] = \frac{K_w}{[H_3O^+]} = \frac{1.0 \times 10^{-14}}{4.2 \times 10^{-3}} = 2.4 \times 10^{-12} \text{ M}$$

(b) $\qquad\qquad$ CH$_3$COOH(aq) + H$_2$O(l) $\rightleftharpoons$ H$_3$O$^+$(aq) + CH$_3$CO$_2^-$(aq)

initial (M)	0.0100	~0	0
change (M)	−x	+x	+x
equil (M)	0.0100 − x	x	x

$$K_a = \frac{[H_3O^+][CH_3CO_2^-]}{[CH_3COOH]} = 1.8 \times 10^{-5} = \frac{x^2}{0.0100-x}$$

$x^2 + (1.8 \times 10^{-5})x - (1.8 \times 10^{-7}) = 0$

Use the quadratic formula to solve for x.

$$x = \frac{-(1.8 \times 10^{-5}) \pm \sqrt{(1.8 \times 10^{-5})^2 - 4(-1.8 \times 10^{-7})}}{2(1)} = \frac{(-1.8 \times 10^{-5}) \pm (8.5 \times 10^{-4})}{2}$$

$x = 4.2 \times 10^{-4}$ and -4.3×10^{-4}

Of the two solutions for x, only the positive value of x has physical meaning because x is the [H$_3$O$^+$].

$x = [H_3O^+] = 4.2 \times 10^{-4}$ M

$pH = -\log[H_3O^+] = -\log(4.2 \times 10^{-4}) = 3.38$

$[CH_3CO_2^-] = x = 4.2 \times 10^{-4}$ M

$[CH_3COOH] = 0.0100 - x = 0.0100 - (4.2 \times 10^{-4}) = 0.0096$ M

$$[OH^-] = \frac{K_w}{[H_3O^+]} = \frac{1.0 \times 10^{-14}}{4.2 \times 10^{-4}} = 2.4 \times 10^{-11} \text{ M}$$

15.12 $\quad$ C$_6$H$_8$O$_6$, 176.13 amu; 250 mg = 0.250 g; 250 mL = 0.250 L

$$[C_6H_8O_6] = \frac{\left(0.250 \text{ g} \times \dfrac{1 \text{ mol}}{176.13 \text{ g}}\right)}{0.250 \text{ L}} = 5.68 \times 10^{-3} \text{ M}$$

$\qquad\qquad\qquad$ C$_6$H$_8$O$_6$(aq) + H$_2$O(l) $\rightleftharpoons$ H$_3$O$^+$(aq) + C$_6$H$_7$O$_6^-$(aq)

initial (M)	5.68 × 10^{-3}	~0	0
change (M)	−x	+x	+x
equil (M)	(5.68 × 10^{-3}) − x	x	x

$$K_a = \frac{[H_3O^+][C_6H_7O_6^-]}{[C_6H_8O_6]} = 8.0 \times 10^{-5} = \frac{x^2}{(5.68 \times 10^{-3})-x}$$

$x^2 + (8.0 \times 10^{-5})x - (4.54 \times 10^{-7}) = 0$

Use the quadratic formula to solve for x.

$$x = \frac{-(8.0 \times 10^{-5}) \pm \sqrt{(8.0 \times 10^{-5})^2 - (4)(-4.54 \times 10^{-7})}}{2(1)} = \frac{(-8.0 \times 10^{-5}) \pm 0.001\ 35}{2}$$

$x = 6.35 \times 10^{-4}$ and -7.15×10^{-4}

Of the two solutions for x, only the positive value of x has physical meaning because x is the $[H_3O^+]$.

$x = [H_3O^+] = 6.35 \times 10^{-4}$ M

$pH = -\log[H_3O^+] = -\log(6.35 \times 10^{-4}) = 3.20$

15.13   (a) From Example 15.9:

$[H_3O^+] = [HF]_{diss} = 4.0 \times 10^{-3}$ M

% dissociation $= \dfrac{[HF]_{diss}}{[HF]_{initial}} \times 100\% = \dfrac{4.0 \times 10^{-3}\ M}{0.050\ M} \times 100\% = 8.0\%$ dissociation

(b)

	HF(aq)	+	H₂O(l)	⇌	H₃O⁺(aq)	+	F⁻(aq)
initial (M)	0.50				~0		0
change (M)	−x				+x		+x
equil (M)	0.50 − x				x		x

$K_a = \dfrac{[H_3O^+][F^-]}{[HF]} = 3.5 \times 10^{-4} = \dfrac{x^2}{0.50 - x}$

$x^2 + (3.5 \times 10^{-4})x - (1.75 \times 10^{-4}) = 0$

Use the quadratic formula to solve for x.

$$x = \frac{-(3.5 \times 10^{-4}) \pm \sqrt{(3.5 \times 10^{-4})^2 - 4(1)(-1.75 \times 10^{-4})}}{2(1)} = \frac{(-3.5 \times 10^{-4}) \pm 0.0265}{2}$$

$x = 0.0131$ and -0.0134

Of the two solutions for x, only the positive value of x has physical meaning, because x is the $[H_3O^+]$.

$[H_3O^+] = [HF]_{diss} = 0.013$ M

% dissociation $= \dfrac{[HF]_{diss}}{[HF]_{initial}} \times 100\% = \dfrac{0.013\ M}{0.50\ M} \times 100\% = 2.6\%$ dissociation

15.14

	H₂SO₃(aq)	+	H₂O(l)	⇌	H₃O⁺(aq)	+	HSO₃⁻(aq)
initial (M)	0.10				~0		0
change (M)	−x				+x		+x
equil (M)	0.10 − x				x		x

$K_{a1} = \dfrac{[H_3O^+][HSO_3^-]}{[H_2SO_3]} = 1.5 \times 10^{-2} = \dfrac{x^2}{0.10 - x}$

$x^2 + 0.015x - 0.0015 = 0$

Use the quadratic formula to solve for x.

$$x = \frac{-(0.015) \pm \sqrt{(0.015)^2 - (4)(-0.0015)}}{2(1)} = \frac{-0.015 \pm 0.079}{2}$$

$x = 0.032$ and -0.047

Of the two solutions for x, only the positive value of x has physical meaning since x is the $[H_3O^+]$.

$x = [H_3O^+] = [HSO_3^-] = 0.032$ M; $[H_2SO_3] = 0.10 - x = 0.10 - 0.032 = 0.07$ M

The second dissociation of H_2SO_3 produces a negligible amount of H_3O^+ compared with that from the first dissociation.

$$HSO_3^-(aq) + H_2O(l) \rightleftharpoons H_3O^+(aq) + SO_3^{2-}(aq)$$

$$K_{a2} = \frac{[H_3O^+][SO_3^{2-}]}{[HSO_3^-]} = 6.3 \times 10^{-8} = \frac{(0.032)[SO_3^{2-}]}{(0.032)}$$

$[SO_3^{2-}] = K_{a2} = 6.3 \times 10^{-8}$ M

$$[OH^-] = \frac{K_w}{[H_3O^+]} = \frac{1.0 \times 10^{-14}}{0.032} = 3.1 \times 10^{-13} \text{ M}$$

$pH = -\log[H_3O^+] = -\log(0.032) = 1.49$

15.15   From the complete dissociation of the first proton, $[H_3O^+] = [HSO_4^-] = 0.50$ M.
For the dissociation of the second proton, the following equilibrium must be considered:

$$HSO_4^-(aq) + H_2O(l) \rightleftharpoons H_3O^+(aq) + SO_4^{2-}(aq)$$

initial (M)	0.50	0.50	0
change (M)	$-x$	$+x$	$+x$
equil (M)	$0.50 - x$	$0.50 + x$	x

$$K_{a2} = \frac{[H_3O^+][SO_4^{2-}]}{[HSO_4^-]} = 1.2 \times 10^{-2} = \frac{(0.50 + x)(x)}{0.50 - x}$$

$x^2 + 0.512x - 0.0060 = 0$

Use the quadratic formula to solve for x.

$$x = \frac{-(0.512) \pm \sqrt{(0.512)^2 - 4(1)(-0.0060)}}{2(1)} = \frac{-0.512 \pm 0.535}{2}$$

$x = 0.011$ and -0.524

Of the two solutions for x, only the positive value of x has physical meaning, since x is the $[SO_4^{2-}]$.

$[H_2SO_4] = 0$ M; $[HSO_4^-] = 0.50 - x = 0.49$ M; $[SO_4^{2-}] = x = 0.011$ M

$[H_3O^+] = 0.50 + x = 0.51$ M

$pH = -\log[H_3O^+] = -\log(0.51) = 0.29$

$$[OH^-] = \frac{K_w}{[H_3O^+]} = \frac{1.0 \times 10^{-14}}{0.51} = 2.0 \times 10^{-14} \text{ M}$$

15.16

$$NH_3(aq) + H_2O(l) \rightleftharpoons NH_4^+(aq) + OH^-(aq)$$

initial (M)	0.40	0	~0
change (M)	$-x$	$+x$	$+x$
equil (M)	$0.40 - x$	x	x

$$K_b = \frac{[NH_4^+][OH^-]}{[NH_3]} = 1.8 \times 10^{-5} = \frac{x^2}{0.40 - x} \approx \frac{x^2}{0.40}$$

Solve for x. $x = [OH^-] = 2.7 \times 10^{-3}$ M

$[NH_4^+] = x = 2.7 \times 10^{-3}$ M; $[NH_3] = 0.40 - x = 0.40$ M

$[H_3O^+] = \dfrac{K_w}{[OH^-]} = \dfrac{1.0 \times 10^{-14}}{2.7 \times 10^{-3}} = 3.7 \times 10^{-12}$ M

$pH = -\log[H_3O^+] = -\log(3.7 \times 10^{-12}) = 11.43$

15.17 $C_{21}H_{22}N_2O_2$, 334.42 amu; 16 mg = 0.016 g

$\text{molarity} = \dfrac{\left(0.016 \text{ g} \times \dfrac{1 \text{ mol}}{334.42 \text{ g}}\right)}{0.100 \text{ L}} = 4.8 \times 10^{-4}$ M

$$C_{21}H_{22}N_2O_2(aq) + H_2O(l) \rightleftharpoons C_{21}H_{23}N_2O_2^+(aq) + OH^-(aq)$$

initial (M)	4.8×10^{-4}	0 ~0
change (M)	$-x$	$+x$ +x
equil (M)	$(4.8 \times 10^{-4}) - x$	x x

$K_b = \dfrac{[C_{21}H_{23}N_2O_2^+][OH^-]}{[C_{21}H_{22}N_2O_2]} = 1.8 \times 10^{-6} = \dfrac{x^2}{(4.8 \times 10^{-4}) - x}$

$x^2 + (1.8 \times 10^{-6})x - (8.6 \times 10^{-10}) = 0$

Use the quadratic formula to solve for x.

$x = \dfrac{-(1.8 \times 10^{-6}) \pm \sqrt{(1.8 \times 10^{-6})^2 - (4)(-8.6 \times 10^{-10})}}{2(1)} = \dfrac{(-1.8 \times 10^{-6}) \pm (5.87 \times 10^{-5})}{2}$

$x = 2.84 \times 10^{-5}$ and -3.02×10^{-5}

Of the two solutions for x, only the positive value of x has physical meaning, because x is the $[OH^-]$.

$[OH^-] = 2.84 \times 10^{-5}$ M

$[H_3O^+] = \dfrac{K_w}{[OH^-]} = \dfrac{1.0 \times 10^{-14}}{2.84 \times 10^{-5}} = 3.52 \times 10^{-10}$ M

$pH = -\log[H_3O^+] = -\log(3.52 \times 10^{-10}) = 9.45$

15.18 (a) $K_a = \dfrac{K_w}{K_b \text{ for } C_5H_{11}N} = \dfrac{1.0 \times 10^{-14}}{1.3 \times 10^{-3}} = 7.7 \times 10^{-12}$

(b) $K_b = \dfrac{K_w}{K_a \text{ for HOCl}} = \dfrac{1.0 \times 10^{-14}}{3.5 \times 10^{-8}} = 2.9 \times 10^{-7}$

15.19 (a) 0.25 M NH_4Br

NH_4^+ is an acidic cation. Br^- is a neutral anion. The salt solution is acidic.

For NH_4^+, $K_a = \dfrac{K_w}{K_b \text{ for } NH_3} = \dfrac{1.0 \times 10^{-14}}{1.8 \times 10^{-5}} = 5.6 \times 10^{-10}$

$$NH_4^+(aq) \ + \ H_2O(l) \ \rightleftharpoons \ H_3O^+(aq) \ + \ NH_3(aq)$$

initial (M)	0.25	~0	0
change (M)	–x	·+x	+x
equil (M)	0.25 – x	x	x

$$K_a = \frac{[H_3O^+][NH_3]}{[NH_4^+]} = 5.6 \times 10^{-10} = \frac{x^2}{0.25 - x} \approx \frac{x^2}{0.25}$$

Solve for x. $x = [H_3O^+] = 1.2 \times 10^{-5}$ M

pH $= -\log[H_3O^+] = -\log(1.2 \times 10^{-5}) = 4.92$

(b) 0.40 M $ZnCl_2$

Zn^{2+} is an acidic cation. Cl^- is a neutral anion. The salt solution is acidic.

$$Zn(H_2O)_6^{2+}(aq) \ + \ H_2O(l) \ \rightleftharpoons \ H_3O^+(aq) \ + \ Zn(H_2O)_5(OH)^+(aq)$$

initial (M)	0.40	~0	0
change (M)	–x	+x	+x
equil(M)	0.40 – x	x	x

$$K_a = \frac{[H_3O^+][Zn(H_2O)_5(OH)^+]}{[Zn(H_2O)_6^{2+}]} = 2.5 \times 10^{-10} = \frac{x^2}{0.40 - x} \approx \frac{x^2}{0.40}$$

Solve for x. $x = [H_3O^+] = 1.0 \times 10^{-5}$ M

pH $= -\log[H_3O^+] = -\log(1.0 \times 10^{-5}) = 5.00$

15.20   For NO_2^-, $K_b = \dfrac{K_w}{K_a \text{ for } HNO_2} = \dfrac{1.0 \times 10^{-14}}{4.6 \times 10^{-4}} = 2.2 \times 10^{-11}$

$$NO_2^-(aq) \ + \ H_2O(l) \ \rightleftharpoons \ HNO_2(aq) \ + \ OH^-(aq)$$

initial (M)	0.20	0	~0
change (M)	–x	+x	+x
equil (M)	0.20 – x	x	x

$$K_b = \frac{[HNO_2][OH^-]}{[NO_2^-]} = 2.2 \times 10^{-11} = \frac{x^2}{0.20 - x} \approx \frac{x^2}{0.20}$$

Solve for x. $x = [OH^-] = 2.1 \times 10^{-6}$ M

$[H_3O^+] = \dfrac{K_w}{[OH^-]} = \dfrac{1.0 \times 10^{-14}}{2.1 \times 10^{-6}} = 4.8 \times 10^{-9}$ M

pH $= -\log[H_3O^+] = -\log(4.8 \times 10^{-9}) = 8.32$

15.21   For NH_4^+, $K_a = \dfrac{K_w}{K_b \text{ for } NH_3} = \dfrac{1.0 \times 10^{-14}}{1.8 \times 10^{-5}} = 5.6 \times 10^{-10}$

For CN^-, $K_b = \dfrac{K_w}{K_a \text{ for } HCN} = \dfrac{1.0 \times 10^{-14}}{4.9 \times 10^{-10}} = 2.0 \times 10^{-5}$

Because $K_b > K_a$, the solution is basic.

15.22   (a) KBr: K^+, neutral cation; Br^-, neutral anion; solution is neutral

(b) $NaNO_2$: Na^+, neutral cation; NO_2^-, basic anion; solution is basic
(c) NH_4Br: NH_4^+, acidic cation; Br^-, neutral anion; solution is acidic
(d) $ZnCl_2$: Zn^{2+}, acidic cation; Cl^-, neutral anion; solution is acidic
(e) NH_4F

For NH_4^+, $K_a = \dfrac{K_w}{K_b \text{ for } NH_3} = \dfrac{1.0 \times 10^{-14}}{1.8 \times 10^{-5}} = 5.6 \times 10^{-10}$

For F^-, $K_b = \dfrac{K_w}{K_a \text{ for } HF} = \dfrac{1.0 \times 10^{-14}}{3.5 \times 10^{-4}} = 2.9 \times 10^{-11}$

Because $K_a > K_b$, the solution is acidic.

15.23 (a) H_2Se is a stronger acid than H_2S because Se is below S in the 6A group and the H–Se bond is weaker than the H–S bond.
(b) HI is a stronger acid than H_2Te because I is to the right of Te in the same row of the periodic table, I is more electronegative than Te, and the H–I bond is more polar.
(c) HNO_3 is a stronger acid than HNO_2 because acid strength increases with increasing oxidation number of N. The oxidation number for N is +5 in HNO_3 and +3 in HNO_2.
(d) H_2SO_3 is a stronger acid than H_2SeO_3 because acid strength increases with increasing electronegativity of the central atom. S is more electronegative than Se.

15.24 (a) Lewis acid, $AlCl_3$; Lewis base, Cl^- (b) Lewis acid, Ag^+; Lewis base, NH_3
(c) Lewis acid, SO_2; Lewis base, OH^- (d) Lewis acid, Cr^{3+}; Lewis base, H_2O

Understanding Key Concepts

15.25 (a) acids, HCO_3^- and H_3O^+; bases, H_2O and CO_3^{2-}
(b) acids, HF and H_2CO_3; bases HCO_3^- and F^-

15.26 (a) X^-, Y^-, Z^- (b) HX < HZ < HY   (c) HY   (d) HX   (e) $(2/10) \times 100\% = 20\%$

15.27 (c) represents a solution of a weak diprotic acid, H_2A. Because K_{a2} is always less than K_{a1}, (a) and (d) represent impossible situations. (b) contains no H_2A.

15.28 (a) $Y^- < Z^- < X^-$
(b) The weakest base, Y^-, has the strongest conjugate acid.
(c) The numbers of HA molecules and OH^- ions are equal because the reaction of A^- with water has a 1:1 stoichiometry: $A^- + H_2O \rightleftharpoons HA + OH^-$

15.29 (a) $A^-(aq) + H_2O(l) \rightleftharpoons HA(aq) + OH^-(aq)$; basic
(b) $M(H_2O)_6^{3+}(aq) + H_2O(l) \rightleftharpoons H_3O^+(aq) + M(H_2O)_5(OH)^{2+}(aq)$; acidic
(c) $2 H_2O(l) \rightleftharpoons H_3O^+(aq) + OH^-(aq)$; neutral
(d) $M(H_2O)_6^{3+}(aq) + A^-(aq) \rightleftharpoons HA(aq) + M(H_2O)_5(OH)^{2+}(aq)$;
acidic because K_a for $M(H_2O)_6^{3+}$ (10^{-4}) is greater than K_b for A^- (10^{-9}).

15.30   The solution is neutral because $K_a = K_b$.

15.31   (a) H_2S, weakest; HBr, strongest. Acid strength for H_nX increases with increasing polarity of the H–X bond and with increasing size of X.
(b) H_2SeO_3, weakest; $HClO_3$, strongest. Acid strength for H_nYO_3 increases with increasing electronegativity of Y.

15.32   (a) Brønsted–Lowry acids: NH_4^+, $H_2PO_4^-$;
   Brønsted–Lowry bases: SO_3^{2-}, OCl^-, $H_2PO_4^-$
(b) Lewis acids: Fe^{3+}, BCl_3
   Lewis bases: SO_3^{2-}, OCl^-, $H_2PO_4^-$

15.33   (a)

$$\ddot{N}H_3 \quad + \quad B(CH_3)_3 \longrightarrow H_3NB(CH_3)_3$$

(b)

$$:\!\ddot{\underset{\cdot\cdot}{Cl}}\!:^- \quad + \quad AlCl_3 \longrightarrow AlCl_4^-$$

(c)

$$:\!\ddot{\underset{\cdot\cdot}{F}}\!:^- \quad + \quad PF_5 \longrightarrow PF_6^-$$

(d)

$$H_2\ddot{O}: \quad + \quad SO_2 \longrightarrow H_2SO_3$$

Additional Problems
Acid–Base Concepts

15.34 NH_3, CN^-, and NO_2^-

15.36   (a) OCl^- (b) PO_4^{3-} (c) OH^- (d) CH_3NH_2 (e) HCO_3^- (f) H^-

15.38   (a) $CH_3COOH(aq) + NH_3(aq) \rightleftarrows NH_4^+(aq) + CH_3COO^-(aq)$
   acid   base ——— acid   base

(b) $CO_3^{2-}(aq) + H_3O^+(aq) \rightleftarrows H_2O(l) + HCO_3^-(aq)$
   base   acid ——— base   acid

(c) $HSO_3^-(aq) + H_2O(l) \rightleftarrows H_3O^+(aq) + SO_3^{2-}(aq)$
   acid   base ——— acid   base

(d) $HSO_3^-(aq) + H_2O(l) \rightleftarrows H_2SO_3(aq) + OH^-(aq)$
   base   acid   acid   base

15.40   From data in Table 15.1:   Strong acids: HNO_3 and H_2SO_4;   Strong bases: H^- and O^{2-}

15.42 (a) left, HCO_3^- is the stronger base   (b) left, F^- is the stronger base
       (c) right, NH_3 is the stronger base   (d) right, CN^- is the stronger base

Dissociation of Water; pH

15.44 $H-\overset{\cdot\cdot}{\underset{\cdot\cdot}{O}}-H$

H_2O as a proton donor (acid): $B: (aq) + H_2O(l) \rightleftharpoons BH^+(aq) + OH^-(aq)$

H_2O as a proton acceptor (base): $HA(aq) + H_2O(l) \rightleftharpoons H_3O^+(aq) + A^-(aq)$

15.46 If $[H_3O^+] > 1.0 \times 10^{-7}$ M, solution is acidic.
       If $[H_3O^+] < 1.0 \times 10^{-7}$ M, solution is basic.
       If $[H_3O^+] = [OH^-] = 1.0 \times 10^{-7}$ M, solution is neutral.
       If $[OH^-] > 1.0 \times 10^{-7}$ M, solution is basic
       If $[OH^-] < 1.0 \times 10^{-7}$ M, solution is acidic.

(a) $[OH^-] = \dfrac{K_w}{[H_3O^+]} = \dfrac{1.0 \times 10^{-14}}{3.4 \times 10^{-9}} = 2.9 \times 10^{-6}$ M, basic

(b) $[H_3O^+] = \dfrac{K_w}{[OH^-]} = \dfrac{1.0 \times 10^{-14}}{0.010} = 1.0 \times 10^{-12}$ M, basic

(c) $[H_3O^+] = \dfrac{K_w}{[OH^-]} = \dfrac{1.0 \times 10^{-14}}{1.0 \times 10^{-10}} = 1.0 \times 10^{-4}$ M, acidic

(d) $[OH^-] = \dfrac{K_w}{[H_3O^+]} = \dfrac{1.0 \times 10^{-14}}{1.0 \times 10^{-7}} = 1.0 \times 10^{-7}$ M, neutral

(e) $[OH^-] = \dfrac{K_w}{[H_3O^+]} = \dfrac{1.0 \times 10^{-14}}{8.6 \times 10^{-5}} = 1.2 \times 10^{-10}$ M, acidic

15.48 (a) $pH = -\log[H_3O^+] = -\log(2.0 \times 10^{-5}) = 4.70$

(b) $[H_3O^+] = \dfrac{K_w}{[OH^-]} = \dfrac{1.0 \times 10^{-14}}{4 \times 10^{-3}} = 2.5 \times 10^{-12}$ M

$pH = -\log[H_3O^+] = -\log(2.5 \times 10^{-12}) = 11.6$
(c) $pH = -\log[H_3O^+] = -\log(3.56 \times 10^{-9}) = 8.449$
(d) $pH = -\log[H_3O^+] = -\log(10^{-3}) = 3$

(e) $[H_3O^+] = \dfrac{K_w}{[OH^-]} = \dfrac{1.0 \times 10^{-14}}{12} = 8.3 \times 10^{-16}$ M

$pH = -\log[H_3O^+] = -\log(8.3 \times 10^{-16}) = 15.08$

15.50 $[H_3O^+] = 10^{-pH}$;   (a) 8×10^{-5} M   (b) 1.5×10^{-11} M   (c) 1.0 M
                                  (d) 5.6×10^{-15} M   (e) 10 M

15.52 $\Delta pH = \log(\Delta[H_3O^+])$;   (a) $\Delta pH = \log(1000) = 3$
       (b) $\Delta pH = \log(1.0 \times 10^5) = 5.00$ (c) $\Delta pH = \log(2.0) = 0.30$

15.54 (a) $pH = -\log[H_3O^+] = -\log(10^{-2}) = 2$

  (b) $pH = -\log[H_3O^+] = -\log(4 \times 10^{-8}) = 7.4$

  (c) $[H_3O^+] = \dfrac{K_w}{[OH^-]} = \dfrac{1.0 \times 10^{-14}}{8 \times 10^{-8}} = 1.25 \times 10^{-7} \text{ M}$

  $pH = -\log[H_3O^+] = -\log(1.25 \times 10^{-7}) = 6.9$

  (d) $[H_3O^+] = \dfrac{K_w}{[OH^-]} = \dfrac{1.0 \times 10^{-14}}{6 \times 10^{-10}} = 1.7 \times 10^{-5} \text{ M}$

  $pH = -\log[H_3O^+] = -\log(1.7 \times 10^{-5}) = 4.8$

  $[H_3O^+] = \dfrac{K_w}{[OH^-]} = \dfrac{1.0 \times 10^{-14}}{2 \times 10^{-6}} = 5 \times 10^{-9} \text{ M}$

  $pH = -\log[H_3O^+] = -\log(5 \times 10^{-9}) = 8.3$

  pH = 4.8 to 8.3

Strong Acids and Strong Bases

15.56 (a) $[H_3O^+] = 0.20 \text{ M};\ pH = -\log[H_3O^+] = -\log(0.20) = 0.70$

  (b) $[OH^-] = 6.3 \times 10^{-3} \text{ M}$

  $[H_3O^+] = \dfrac{K_w}{[OH^-]} = \dfrac{1.0 \times 10^{-14}}{6.3 \times 10^{-3}} = 1.6 \times 10^{-12} \text{ M}$

  $pH = -\log[H_3O^+] = -\log(1.6 \times 10^{-12}) = 11.80$

  (c) $[OH^-] = 2(4.0 \times 10^{-3} \text{ M}) = 8.0 \times 10^{-3} \text{ M}$

  $[H_3O^+] = \dfrac{K_w}{[OH^-]} = \dfrac{1.0 \times 10^{-14}}{8.0 \times 10^{-3}} = 1.25 \times 10^{-12} \text{ M}$

  $pH = -\log[H_3O^+] = -\log(1.25 \times 10^{-12}) = 11.90$

15.58 (a) LiOH, 23.95 amu; 250 mL = 0.250 L

  $\text{molarity of LiOH(aq)} = \dfrac{\left(4.8 \text{ g} \times \dfrac{1 \text{ mol}}{23.95 \text{ g}}\right)}{0.250 \text{ L}} = 0.80 \text{ M}$

  LiOH is a strong base; therefore $[OH^-] = 0.80$ M.

  $[H_3O^+] = \dfrac{K_w}{[OH^-]} = \dfrac{1.0 \times 10^{-14}}{0.80} = 1.25 \times 10^{-14} \text{ M}$

  $pH = -\log[H_3O^+] = -\log(1.25 \times 10^{-14}) = 13.90$

  (b) HCl, 36.46 amu

  $\text{molarity of HCl(aq)} = \dfrac{\left(0.93 \text{ g} \times \dfrac{1 \text{ mol}}{36.46 \text{ g}}\right)}{0.40 \text{ L}} = 0.064 \text{ M}$

  HCl is a strong acid; therefore $[H_3O^+] = 0.064$ M

  $pH = -\log[H_3O^+] = -\log(0.064) = 1.19$

(c) $M_f \cdot V_f = M_i \cdot V_i$

$M_f = \dfrac{M_i \cdot V_i}{V_f} = \dfrac{(0.10\ M)(50\ mL)}{(1000\ mL)} = 5.0 \times 10^{-3}\ M$

$pH = -\log[H_3O^+] = -\log(5.0 \times 10^{-3}) = 2.30$

(d) For HCl, $M_f = \dfrac{M_i \cdot V_i}{V_f} = \dfrac{(2.0 \times 10^{-3}\ M)(100\ mL)}{(500\ mL)} = 4.0 \times 10^{-4}\ M$

For $HClO_4$, $M_f = \dfrac{M_i \cdot V_i}{V_f} = \dfrac{(1.0 \times 10^{-3}\ M)(400\ mL)}{(500\ mL)} = 8.0 \times 10^{-4}\ M$

$[H_3O^+] = (4.0 \times 10^{-4}\ M) + (8.0 \times 10^{-4}\ M) = 1.2 \times 10^{-3}\ M$

$pH = -\log[H_3O^+] = -\log(1.2 \times 10^{-3}) = 2.92$

Weak Acids

15.60 (a) $HClO_2(aq) + H_2O(l) \rightleftharpoons H_3O^+(aq) + ClO_2^-(aq); \quad K_a = \dfrac{[H_3O^+][ClO_2^-]}{[HClO_2]}$

(b) $HOBr(aq) + H_2O(l) \rightleftharpoons H_3O^+(aq) + OBr^-(aq); \quad K_a = \dfrac{[H_3O^+][OBr^-]}{[HOBr]}$

(c) $HCOOH(aq) + H_2O(l) \rightleftharpoons H_3O^+(aq) + HCO_2^-(aq); \quad K_a = \dfrac{[H_3O^+][HCO_2^-]}{[HCOOH]}$

15.62 (a) The larger the K_a, the stronger the acid.

$C_6H_5OH < HOCl < CH_3COOH < HNO_3$

(b) The larger the K_a, the larger the percent dissociation for the same concentration.

$HNO_3 > CH_3COOH > HOCl > C_6H_5OH$

1 M HNO_3, $[H_3O^+] = 1\ M$

1 M CH_3COOH, $[H_3O^+] = \sqrt{K_a} = \sqrt{1.8 \times 10^{-5}} = 4 \times 10^{-3}\ M$

1 M $HOCl$, $[H_3O^+] = \sqrt{K_a} = \sqrt{3.5 \times 10^{-8}} = 2 \times 10^{-4}\ M$

1 M C_6H_5OH, $[H_3O^+] = \sqrt{K_a} = \sqrt{1.3 \times 10^{-10}} = 1 \times 10^{-5}\ M$

15.64

	$HOBr(aq)$	$+$	$H_2O(l)$	$\rightleftharpoons$	$H_3O^+(aq)$	$+$	$OBr^-(aq)$
initial (M)	0.040				~0		0
change (M)	$-x$				$+x$		$+x$
equil (M)	$0.040 - x$				x		x

$x = [H_3O^+] = 10^{-pH} = 10^{-5.05} = 8.9 \times 10^{-6}\ M$

$K_a = \dfrac{[H_3O^+][OBr^-]}{[HOBr]} = \dfrac{x^2}{0.040 - x} = \dfrac{(8.9 \times 10^{-6})^2}{0.040 - (8.9 \times 10^{-6})} = 2.0 \times 10^{-9}$

15.66 $\quad\quad\quad\quad\quad$ $C_6H_5OH(aq)$ + $H_2O(l)$ ⇌ $H_3O^+(aq)$ + $C_6H_5O^-(aq)$

initial (M) $\quad\quad$ 0.10 $\quad\quad\quad\quad\quad\quad\quad$ ~0 $\quad\quad\quad\quad$ 0

change (M) $\quad$ −x $\quad\quad\quad\quad\quad\quad\quad\quad$ +x $\quad\quad\quad$ +x

equil (M) $\quad\quad$ 0.10 − x $\quad\quad\quad\quad\quad\quad$ x $\quad\quad\quad\quad$ x

$$K_a = \frac{[H_3O^+][C_6H_5O^-]}{[C_6H_5OH]} = 1.3 \times 10^{-10} = \frac{x^2}{0.10 - x} \approx \frac{x^2}{0.10}$$

Solve for x. $\quad$ x = 3.6 × 10⁻⁶ M = $[H_3O^+]$ = $[C_6H_5O^-]$

$[C_6H_5OH]$ = 0.10 − x = 0.10 M

pH = −log$[H_3O^+]$ = −log(3.6 × 10⁻⁶) = 5.44

$$[OH^-] = \frac{K_w}{[H_3O^+]} = \frac{1.0 \times 10^{-14}}{3.6 \times 10^{-6}} = 2.8 \times 10^{-9} \text{ M}$$

$$\% \text{ dissociation} = \frac{[C_6H_5OH]_{diss}}{[C_6H_5OH]_{initial}} \times 100\% = \frac{3.6 \times 10^{-6} \text{ M}}{0.10 \text{ M}} \times 100\% = 0.0036\%$$

15.68 $\quad\quad\quad\quad\quad$ $HNO_2(aq)$ + $H_2O(l)$ ⇌ $H_3O^+(aq)$ + $NO_2^-(aq)$

initial (M) $\quad\quad$ 1.5 $\quad\quad\quad\quad\quad\quad\quad\quad$ ~0 $\quad\quad\quad\quad$ 0

change (M) $\quad$ −x $\quad\quad\quad\quad\quad\quad\quad\quad$ +x $\quad\quad\quad$ +x

equil (M) $\quad\quad$ 1.5 − x $\quad\quad\quad\quad\quad\quad\quad$ x $\quad\quad\quad\quad$ x

$$K_a = \frac{[H_3O^+][NO_2^-]}{[HNO_2]} = 4.5 \times 10^{-4} = \frac{x^2}{1.5 - x} \approx \frac{x^2}{1.5}$$

Solve for x. $\quad$ x = 0.026 M = $[H_3O^+]$

pH = −log$[H_3O^+]$ = −log(0.026) = 1.59

$$\% \text{ dissociation} = \frac{[HNO_2]_{diss}}{[HNO_2]_{initial}} \times 100\% = \frac{0.026 \text{ M}}{1.5 \text{ M}} \times 100\% = 1.7\%$$

Polyprotic Acids

15.70 $\quad$ $H_2SeO_4(aq)$ + $H_2O(l)$ ⇌ $H_3O^+(aq)$ + $HSeO_4^-(aq)$; $\quad K_{a1} = \dfrac{[H_3O^+][HSeO_4^-]}{[H_2SeO_4]}$

$\quad\quad$ $HSeO_4^-(aq)$ + $H_2O(l)$ ⇌ $H_3O^+(aq)$ + $SeO_4^{2-}(aq)$; $\quad K_{a2} = \dfrac{[H_3O^+][SeO_4^{2-}]}{[HSeO_4^-]}$

15.72 $\quad\quad\quad\quad\quad$ $H_2CO_3(aq)$ + $H_2O(l)$ ⇌ $H_3O^+(aq)$ + $HCO_3^-(aq)$

initial (M) $\quad\quad$ 0.010 $\quad\quad\quad\quad\quad\quad\quad$ ~0 $\quad\quad\quad\quad$ 0

change (M) $\quad\quad$ −x $\quad\quad\quad\quad\quad\quad\quad\quad$ +x $\quad\quad\quad$ +x

equil (M) $\quad\quad$ 0.010 − x $\quad\quad\quad\quad\quad\quad$ x $\quad\quad\quad\quad$ x

$$K_{a1} = \frac{[H_3O^+][HCO_3^-]}{[H_2CO_3]} = 4.3 \times 10^{-7} = \frac{x^2}{0.010 - x} \approx \frac{x^2}{0.010}$$

Solve for x. $\quad$ x = 6.6 × 10⁻⁵

$[H_3O^+] = [HCO_3^-] = x = 6.6 \times 10^{-5}$ M; $\qquad$ $[H_2CO_3] = 0.010 - x = 0.010$ M

The second dissociation of H_2CO_3 produces a negligible amount of H_3O^+ compared with that from the first dissociation.

$$HCO_3^-(aq) + H_2O(l) \rightleftarrows H_3O^+(aq) + CO_3^{2-}(aq)$$

$$K_{a2} = \frac{[H_3O^+][CO_3^{2-}]}{[HCO_3^-]} = 5.6 \times 10^{-11} = \frac{(6.6 \times 10^{-5})[CO_3^{2-}]}{(6.6 \times 10^{-5})}$$

$[CO_3^{2-}] = K_{a2} = 5.6 \times 10^{-11}$ M

$$[OH^-] = \frac{K_w}{[H_3O^+]} = \frac{1.0 \times 10^{-14}}{6.6 \times 10^{-5}} = 1.5 \times 10^{-10}$ M$$

$pH = -\log[H_3O^+] = -\log(6.6 \times 10^{-5}) = 4.18$

15.74 For the dissociation of the first proton, the following equilibrium must be considered:

$$H_2C_2O_4(aq) + H_2O(l) \rightleftarrows H_3O^+(aq) + HC_2O_4^-(aq)$$

initial (M)	0.20	~0	0
change (M)	–x	+x	+x
equil (M)	0.20 – x	x	x

$$K_{a1} = \frac{[H_3O^+][HC_2O_4^-]}{[H_2C_2O_4]} = 5.9 \times 10^{-2} = \frac{x^2}{0.20 - x}$$

$x^2 + 0.059x - 0.0118 = 0$

Use the quadratic formula to solve for x.

$$x = \frac{-(0.059) \pm \sqrt{(0.059)^2 - 4(1)(-0.0118)}}{2(1)} = \frac{-0.059 \pm 0.225}{2}$$

$x = 0.083$ and -0.142

Of the two solutions for x, only the positive value of x has physical meaning, because x is the $[H_3O^+]$.

$[H_3O^+] = [HC_2O_4^-] = 0.083$ M

For the dissociation of the second proton, the following equilibrium must be considered:

$$HC_2O_4^-(aq) + H_2O(l) \rightleftarrows H_3O^+(aq) + C_2O_4^{2-}(aq)$$

initial (M)	0.083	0.083	0
change (M)	–x	+x	+x
equil (M)	0.083 – x	0.083 + x	x

$$K_{a2} = \frac{[H_3O^+][C_2O_4^{2-}]}{[HC_2O_4^-]} = 6.4 \times 10^{-5} = \frac{(0.083 + x)(x)}{0.083 - x} \approx \frac{(0.083)(x)}{0.083} = x$$

$[H_3O^+] = 0.083 + x = 0.083$ M

$pH = -\log[H_3O^+] = -\log(0.083) = 1.08$

$[C_2O_4^{2-}] = x = 6.4 \times 10^{-5}$ M

Weak Bases; Relation Between K_a and K_b

15.76 (a) $(CH_3)_2NH(aq) + H_2O(l) \rightleftharpoons (CH_3)_2NH_2^+(aq) + OH^-(aq);$ $K_b = \dfrac{[(CH_3)_2NH_2^+][OH^-]}{[(CH_3)_2NH]}$

(b) $C_6H_5NH_2(aq) + H_2O(l) \rightleftharpoons C_6H_5NH_3^+(aq) + OH^-(aq);$ $K_b = \dfrac{[C_6H_5NH_3^+][OH^-]}{[C_6H_5NH_2]}$

(c) $CN^-(aq) + H_2O(l) \rightleftharpoons HCN(aq) + OH^-(aq);$ $K_b = \dfrac{[HCN][OH^-]}{[CN^-]}$

15.78 $[H_3O^+] = 10^{-pH} = 10^{-9.5} = 3.16 \times 10^{-10}$ M

$[OH^-] = \dfrac{K_w}{[H_3O^+]} = \dfrac{1.0 \times 10^{-14}}{3.16 \times 10^{-10}} = 3.16 \times 10^{-5}$ M

$$C_{17}H_{19}NO_3(aq) + H_2O(l) \rightleftharpoons C_{17}H_{20}NO_3^+(aq) + OH^-(aq)$$

	$C_{17}H_{19}NO_3$		$C_{17}H_{20}NO_3^+$	OH^-
initial (M)	7.0×10^{-4}		0	~0
change (M)	$-x$		$+x$	$+x$
equil (M)	$(7.0 \times 10^{-4}) - x$		x	x

$x = [OH^-] = 3.16 \times 10^{-5}$ M

$K_b = \dfrac{[C_{17}H_{20}NO_3^+][OH^-]}{[C_{17}H_{19}NO_3]} = \dfrac{x^2}{(7.0 \times 10^{-4}) - x} =$

$\dfrac{(3.16 \times 10^{-5})^2}{(7.0 \times 10^{-4}) - (3.16 \times 10^{-5})} = 1.49 \times 10^{-6} = 1 \times 10^{-6}$

15.80 (a)

$$CH_3NH_2(aq) + H_2O(l) \rightleftharpoons CH_3NH_3^+(aq) + OH^-(aq)$$

	CH_3NH_2		$CH_3NH_3^+$	OH^-
initial (M)	0.24		0	~0
change (M)	$-x$		$+x$	$+x$
equil (M)	$0.24 - x$		x	x

$K_b = \dfrac{[CH_3NH_3^+][OH^-]}{[CH_3NH_2]} = 3.7 \times 10^{-4} = \dfrac{x^2}{0.24 - x}$

$x^2 + (3.7 \times 10^{-4})x - (8.9 \times 10^{-5}) = 0$

Use the quadratic formula to solve for x.

$x = \dfrac{-(3.7 \times 10^{-4}) \pm \sqrt{(3.7 \times 10^{-4})^2 - (4)(-8.9 \times 10^{-5})}}{2(1)} = \dfrac{(-3.7 \times 10^{-4}) \pm 0.0189}{2}$

$x = 0.0093$ and -0.0096

Of the two solutions for x, only the positive value of x has physical meaning because x is the $[OH^-]$.

$[OH^-] = x = 0.0093$ M

$[H_3O^+] = \dfrac{K_w}{[OH^-]} = \dfrac{1.0 \times 10^{-14}}{0.0093} = 1.1 \times 10^{-12}$ M

$pH = -\log[H_3O^+] = -\log(1.1 \times 10^{-12}) = 11.96$

(b) $\qquad$ $C_5H_5N(aq) + H_2O(l) \rightleftharpoons C_5H_5NH^+(aq) + OH^-(aq)$

initial (M) $\quad$ 0.040 $\qquad\qquad\qquad$ 0 $\qquad$ ~0

change (M) $\quad$ –x $\qquad\qquad\qquad\qquad$ +x $\qquad$ +x

equil (M) $\quad$ 0.040 – x $\qquad\qquad\qquad$ x $\qquad$ x

$$K_b = \frac{[C_5H_5NH^+][OH^-]}{[C_5H_5N]} = 1.8 \times 10^{-9} = \frac{x^2}{0.040 - x} \approx \frac{x^2}{0.040}$$

Solve for x. $\quad x = [OH^-] = 8.5 \times 10^{-6}$ M

$$[H_3O^+] = \frac{K_w}{[OH^-]} = \frac{1.0 \times 10^{-14}}{8.5 \times 10^{-6}} = 1.2 \times 10^{-9} \text{ M}$$

$$pH = -\log[H_3O^+] = -\log(1.2 \times 10^{-9}) = 8.92$$

(c) $\qquad$ $NH_2OH(aq) + H_2O(l) \rightleftharpoons NH_3OH^+(aq) + OH^-(aq)$

initial (M) $\quad$ 0.075 $\qquad\qquad\qquad$ 0 $\qquad$ ~0

change (M) $\quad$ –x $\qquad\qquad\qquad\qquad$ +x $\qquad$ +x

equil (M) $\quad$ 0.075 – x $\qquad\qquad\qquad$ x $\qquad$ x

$$K_b = \frac{[NH_3OH^+][OH^-]}{[NH_2OH]} = 9.1 \times 10^{-9} = \frac{x^2}{0.075 - x} \approx \frac{x^2}{0.075}$$

Solve for x. $\quad x = [OH^-] = 2.6 \times 10^{-5}$ M

$$[H_3O^+] = \frac{K_w}{[OH^-]} = \frac{1.0 \times 10^{-14}}{2.6 \times 10^{-5}} = 3.8 \times 10^{-10} \text{ M}$$

$$pH = -\log[H_3O^+] = -\log(3.8 \times 10^{-10}) = 9.42$$

15.82 $\quad$ (a) $\quad K_a = \dfrac{K_w}{K_b \text{ for } C_3H_7NH_2} = \dfrac{1.0 \times 10^{-14}}{5.1 \times 10^{-4}} = 2.0 \times 10^{-11}$

$\qquad$ (b) $\quad K_a = \dfrac{K_w}{K_b \text{ for } NH_2OH} = \dfrac{1.0 \times 10^{-14}}{9.1 \times 10^{-9}} = 1.1 \times 10^{-6}$

$\qquad$ (c) $\quad K_a = \dfrac{K_w}{K_b \text{ for } C_6H_5NH_2} = \dfrac{1.0 \times 10^{-14}}{4.3 \times 10^{-10}} = 2.3 \times 10^{-5}$

$\qquad$ (d) $\quad K_a = \dfrac{K_w}{K_b \text{ for } C_5H_5N} = \dfrac{1.0 \times 10^{-14}}{1.8 \times 10^{-9}} = 5.6 \times 10^{-6}$

Acid–Base Properties of Salts

15.84 $\quad$ (a) $\quad CH_3NH_3^+(aq) + H_2O(l) \rightleftharpoons H_3O^+(aq) + CH_3NH_2(aq)$

$\qquad\qquad$ acid $\qquad\qquad\quad$ base $\quad$——$\quad$ acid $\qquad\qquad$ base

$\qquad$ (b) $\quad Cr(H_2O)_6^{3+}(aq) + H_2O(l) \rightleftharpoons H_3O^+(aq) + Cr(H_2O)_5(OH)^{2+}(aq)$

$\qquad\qquad$ acid $\qquad\qquad\qquad$ base $\quad$——$\quad$ acid $\qquad\qquad$ base

(c) $CH_3COO^-(aq) + H_2O(l) \rightleftarrows CH_3COOH(aq) + OH^-(aq)$

      base   acid   acid   base

(d) $PO_4^{3-}(aq) + H_2O(l) \rightleftarrows HPO_4^{2-}(aq) + OH^-(aq)$

      base   acid   acid   base

15.86 (a) F^- (conjugate base of a weak acid), basic solution
       (b) Br^- (anion of a strong acid), neutral solution
       (c) NH_4^+ (conjugate acid of a weak base), acidic solution
       (d) $K(H_2O)_6^+$ (neutral cation), neutral solution
       (e) SO_3^{2-} (conjugate base of a weak acid), basic solution
       (f) $Cr(H_2O)_6^{3+}$ (acidic cation), acidic solution

15.88 (a) $(C_2H_5NH_3)NO_3$: $C_2H_5NH_3^+$, acidic cation; NO_3^-, neutral anion
      $C_2H_5NH_2$, $K_b = 6.4 \times 10^{-4}$

$$C_2H_5NH_3^+, \quad K_a = \frac{K_w}{K_b \text{ for } C_2H_5NH_2} = \frac{1.0 \times 10^{-14}}{6.4 \times 10^{-4}} = 1.56 \times 10^{-11}$$

                $C_2H_5NH_3^+(aq) + H_2O(l) \rightleftarrows H_3O^+(aq) + C_2H_5NH_2(aq)$

initial (M)	0.10	~0	0
change (M)	−x	+x	+x
equil (M)	0.10 − x	x	x

$$K_a = \frac{[H_3O^+][C_2H_5NH_2]}{[C_2H_5NH_3^+]} = 1.56 \times 10^{-11} = \frac{x^2}{0.10 - x} \approx \frac{x^2}{0.10}$$

Solve for x. $x = 1.25 \times 10^{-6}$ M $= 1.2 \times 10^{-6}$ M $= [H_3O^+] = [C_2H_5NH_2]$
$pH = -\log[H_3O^+] = -\log(1.25 \times 10^{-6}) = 5.90$
$[C_2H_5NH_3^+] = 0.10 - x = 0.10$ M; $[NO_3^-] = 0.10$ M

$$[OH^-] = \frac{K_w}{[H_3O^+]} = \frac{1.0 \times 10^{-14}}{1.25 \times 10^{-6} \text{ M}} = 8.0 \times 10^{-9}$$

(b) $Na(CH_3CO_2)$: Na^+, neutral cation; $CH_3CO_2^-$, basic anion
CH_3COOH, $K_a = 1.8 \times 10^{-5}$

$$CH_3CO_2^-, \quad K_b = \frac{K_w}{K_a \text{ for } CH_3CO_2H} = \frac{1.0 \times 10^{-14}}{1.8 \times 10^{-5}} = 5.6 \times 10^{-10}$$

                $CH_3CO_2^-(aq) + H_2O(aq) \rightleftarrows CH_3COOH(aq) + OH^-(aq)$

initial (M)	0.10	0	~0
change (M)	−x	+x	+x
equil (M)	0.10 − x	x	x

$$K_b = \frac{[CH_3COOH][OH^-]}{[CH_3CO_2^-]} = 5.6 \times 10^{-10} = \frac{x^2}{0.10 - x} \approx \frac{x^2}{0.10}$$

Solve for x. $x = 7.5 \times 10^{-6}$ M $= [CH_3COOH] = [OH^-]$

$[CH_3CO_2^-] = 0.10 - x = 0.10$ M; $[Na^+] = 0.10$ M

$[H_3O^+] = \dfrac{K_w}{[OH^-]} = \dfrac{1.0 \times 10^{-14}}{7.5 \times 10^{-6}} = 1.3 \times 10^{-9}$ M

$pH = -\log[H_3O^+] = -\log(1.3 \times 10^{-9}) = 8.89$

(c) $NaNO_3$: Na^+, neutral cation; NO_3^-, neutral anion

$[Na^+] = [NO_3^-] = 0.10$ M

$[H_3O^+] = [OH^-] = 1.0 \times 10^{-7}$ M; $pH = 7.00$

Factors That Affect Acid Strength

15.90 (a) $PH_3 < H_2S < HCl$; electronegativity increases from P to Cl
(b) $NH_3 < PH_3 < AsH_3$; X–H bond strength decreases from N to As (down a group)
(c) $HBrO < HBrO_2 < HBrO_3$; acid strength increases with the number of O atoms

15.92 (a) HCl; The strength of a binary acid H_nA increases as A moves from left to right and from top to bottom in the periodic table.
(b) $HClO_3$; The strength of an oxoacid increases with increasing electronegativity and increasing oxidation state of the central atom.
(c) HBr; The strength of a binary acid H_nA increases as A moves from left to right and from top to bottom in the periodic table.

15.94 (a) H_2Te, weaker X–H bond
(b) H_3PO_4, P has higher electronegativity
(c) $H_2PO_4^-$, lower negative charge
(d) NH_4^+, higher positive charge and N is more electronegative than C

Lewis Acids and Bases

15.96 (a) Lewis acid, SiF_4; Lewis base, F^- (b) Lewis acid, Zn^{2+}; Lewis base, NH_3
(c) Lewis acid, $HgCl_2$; Lewis base, Cl^- (d) Lewis acid, CO_2; Lewis base, H_2O

15.98 (a) $2 \; :\!\ddot{F}\!:^- \; + \; SiF_4 \; \longrightarrow \; SiF_6^{2-}$

(b) $4 \; \ddot{N}H_3 \; + \; Zn^{2+} \; \longrightarrow \; Zn(NH_3)_4^{2+}$

(c) $2 \; :\!\ddot{C}l\!:^- \; + \; HgCl_2 \; \longrightarrow \; HgCl_4^{2-}$

(d) $H_2\ddot{O}\!: \; + \; CO_2 \; \longrightarrow \; H_2CO_3$

15.100 (a) CN^-, Lewis base   (b) H^+, Lewis acid   (c) H_2O, Lewis base
(d) Fe^{3+}, Lewis acid   (e) OH^-, Lewis base   (f) CO_2, Lewis acid
(g) $P(CH_3)_3$, Lewis base   (h) $B(CH_3)_3$, Lewis acid

General Problems

15.102 In aqueous solution:
(1) H_2S acts as an acid only.
(2) HS^- can act as both an acid and a base.
(3) S^{2-} can act as a base only.
(4) H_2O can act as both an acid and a base.
(5) H_3O^+ acts as an acid only.
(6) OH^- acts as a base only.

15.104

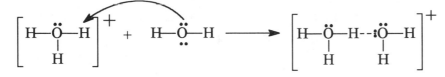

H_3O^+ can hydrogen bond with additional H_2O molecules.

15.106 $HCO_3^-(aq) + Al(H_2O)_6^{3+}(aq) \rightarrow H_2O(l) + CO_2(g) + Al(H_2O)_5(OH)^{2+}(aq)$

15.108 H_2O, 18.02 amu

at 0 °C, $[H_2O] = \dfrac{\left(0.9998 \text{ g} \times \dfrac{1 \text{ mol}}{18.02 \text{ g}}\right)}{0.001 \text{ L}} = 55.48 \text{ M}$

$K_w = [H_3O^+][OH^-]$, for a neutral solution $[H_3O^+] = [OH^-]$

$[H_3O^+] = \sqrt{K_w} = \sqrt{1.14 \times 10^{-15}} = 3.376 \times 10^{-8} \text{ M}$

$pH = -\log[H_3O^+] = -\log(3.376 \times 10^{-8}) = 7.472$

fraction dissociated $= \dfrac{[H_2O]_{diss}}{[H_2O]_{initial}} = \dfrac{3.376 \times 10^{-8} \text{ M}}{55.48 \text{ M}} = 6.09 \times 10^{-10}$

% dissociation $= \dfrac{[H_2O]_{diss}}{[H_2O]_{initial}} \times 100\% = \dfrac{3.376 \times 10^{-8} \text{ M}}{55.48 \text{ M}} \times 100\% = 6.09 \times 10^{-8} \%$

15.110 For $C_{10}H_{14}N_2H^+$, $K_{a1} = \dfrac{K_w}{K_{b1} \text{ for } C_{10}H_{14}N_2} = \dfrac{1.0 \times 10^{-14}}{1.0 \times 10^{-6}} = 1.0 \times 10^{-8}$

For $C_{10}H_{14}N_2H_2^{2+}$, $K_{a2} = \dfrac{K_w}{K_{b2} \text{ for } C_{10}H_{14}N_2H^+} = \dfrac{1.0 \times 10^{-14}}{1.3 \times 10^{-11}} = 7.7 \times 10^{-4}$

15.112 For the dissociation of the first proton, the following equilibrium must be considered:

	$H_3PO_4(aq)$	$+ H_2O(l)$	$\rightleftharpoons$	$H_3O^+(aq)$	$+ H_2PO_4^-(aq)$
initial (M)	0.10			~0	0
change (M)	−x			+x	+x
equil (M)	0.10 − x			x	x

$$K_{a1} = \frac{[H_3O^+][H_2PO_4^-]}{[H_3PO_4]} = 7.5 \times 10^{-3} = \frac{x^2}{0.10 - x}$$

$x^2 + (7.5 \times 10^{-3})x - (7.5 \times 10^{-4}) = 0$

Solve for x using the quadratic formula.

$$x = \frac{-(7.5 \times 10^{-3}) \pm \sqrt{(7.5 \times 10^{-3})^2 - (4)(-7.5 \times 10^{-4})}}{2(1)} = \frac{(-7.5 \times 10^{-3}) \pm 0.055}{2}$$

$x = 0.024$ and -0.031;   Of the two solutions for x, only the positive value of x has physical meaning, because x is the $[H_3O^+]$.

$x = 0.024$ M $= [H_2PO_4^-] = [H_3O^+]$

For the dissociation of the second proton, the following equilibrium must be considered:

$$H_2PO_4^-(aq) + H_2O(l) \rightleftharpoons H_3O^+(aq) + HPO_4^{2-}(aq)$$

initial (M)	0.024	0.024	0
change (M)	$-y$	$+y$	$+y$
equil (M)	$0.024 - y$	$0.024 + y$	y

$$K_{a2} = \frac{[H_3O^+][HPO_4^{2-}]}{[H_2PO_4^-]} = 6.2 \times 10^{-8} = \frac{(0.024 + y)(y)}{0.024 - y} \approx \frac{(0.024)(y)}{0.024} = y$$

$y = 6.2 \times 10^{-8}$ M $= [HPO_4^{2-}]$

For the dissociation of the third proton, the following equilibrium must be considered:

$$HPO_4^{2-}(aq) + H_2O(l) \rightleftharpoons H_3O^+(aq) + PO_4^{3-}(aq)$$

initial (M)	6.2×10^{-8}	0.024	0
change (M)	$-z$	$+z$	$+z$
equil (M)	$(6.2 \times 10^{-8}) - z$	$0.024 + z$	z

$$K_{a3} = \frac{[H_3O^+][PO_4^{3-}]}{[HPO_4^{2-}]} = 4.8 \times 10^{-13} = \frac{(0.024 + z)(z)}{(6.2 \times 10^{-8}) - z} \approx \frac{(0.024)(z)}{6.2 \times 10^{-8}}$$

$z = 1.2 \times 10^{-18}$ M $= [PO_4^{3-}]$

$[H_3PO_4] = 0.10 - x = 0.076$ M; $[H_2PO_4^-] = [H_3O^+] = 0.024$ M

$[HPO_4^{2-}] = 6.2 \times 10^{-8}$ M; $[PO_4^{3-}] = 1.2 \times 10^{-18}$ M

$$[OH^-] = \frac{K_w}{[H_3O^+]} = \frac{1.0 \times 10^{-14}}{0.024} = 4.2 \times 10^{-13} \text{ M}$$

$pH = -\log[H_3O^+] = -\log(0.024) = 1.62$

15.114   % dissociation $= \dfrac{[HA]_{diss}}{[HA]_{initial}}$

For a weak acid, $[HA]_{diss} = [H_3O^+] = [A^-]$

$$K_a = \frac{[H_3O^+][A^-]}{[HA]} = \frac{[H_3O^+]^2}{[HA]}; \qquad [H_3O^+] = \sqrt{K_a[HA]}$$

$$\text{% dissociation} = \frac{[HA]_{diss}}{[HA]} = \frac{[H_3O^+]}{[HA]} = \frac{\sqrt{K_a[HA]}}{[HA]} = \sqrt{\frac{K_a}{[HA]}}$$

15.116 (a) NH_4F;   For NH_4^+, $K_a = 5.6 \times 10^{-10}$ and for F^-, $K_b = 2.9 \times 10^{-11}$
Because $K_a > K_b$, the salt solution is acidic.
(b) $NH_4(CH_3CO_2)$;   For NH_4^+, $K_a = 5.6 \times 10^{-10}$ and for $CH_3CO_2^-$, $K_b = 5.6 \times 10^{-10}$
Because $K_a = K_b$, the salt solution is neutral.
(c) $(NH_4)_2SO_3$;   For NH_4^+, $K_a = 5.6 \times 10^{-10}$ and for SO_3^{2-}, $K_b = 1.6 \times 10^{-7}$
Because $K_b > K_a$, the salt solution is basic.

15.118 Both reactions occur together.
Let $x = [H_3O^+]$ from CH_3COOH and $y = [H_3O^+]$ from C_6H_5COOH
The following two equilibria must be considered:

$$CH_3COOH(aq) + H_2O(l) \rightleftharpoons H_3O^+(aq) + CH_3CO_2^-(aq)$$

initial (M)	0.10	y	0
change (M)	$-x$	$+x$	$+x$
equil (M)	$0.10 - x$	$x + y$	x

$$C_6H_5COOH(aq) + H_2O(l) \rightleftharpoons H_3O^+(aq) + C_6H_5CO_2^-(aq)$$

initial (M)	0.10	x	0
change (M)	$-y$	$+y$	$+y$
equil (M)	$0.10 - y$	$x + y$	y

$$K_a(\text{for } CH_3COOH) = \frac{[H_3O^+][CH_3CO_2^-]}{[CH_3COOH]} = 1.8 \times 10^{-5} = \frac{(x+y)(x)}{0.10-x} \approx \frac{(x+y)(x)}{0.10}$$

$1.8 \times 10^{-6} = (x+y)(x)$

$$K_a(\text{for } C_6H_5COOH) = \frac{[H_3O^+][C_6H_5CO_2^-]}{[C_6H_5COOH]} = 6.5 \times 10^{-5} = \frac{(x+y)(y)}{0.10-y} \approx \frac{(x+y)(y)}{0.10}$$

$6.5 \times 10^{-6} = (x+y)(y)$

$1.8 \times 10^{-6} = (x+y)(x)$
$6.5 \times 10^{-6} = (x+y)(y)$

These two equations must be solved simultaneously for x and y. Divide the first equation by the second.

$$\frac{x}{y} = \frac{1.8 \times 10^{-6}}{6.5 \times 10^{-6}}; \quad x = 0.277y$$

$6.5 \times 10^{-6} = (x+y)(y)$; substitute $x = 0.277y$ into this equation and solve for y.
$6.5 \times 10^{-6} = (0.277y + y)(y) = 1.277y^2$
$y = 0.002\ 256$
$x = 0.277y = (0.277)(0.002\ 256) = 0.000\ 624\ 9$
$[H_3O^+] = (x + y) = (0.000\ 624\ 9 + 0.002\ 256) = 0.002\ 881$ M
$pH = -\log[H_3O^+] = -\log(0.002\ 881) = 2.54$

15.120 For 1.0×10^{-10} M HCl, pH = 7.00, and the principal source of H_3O^+ is the dissociation of H_2O.

For 1.0×10^{-7} M HCl,

$$2\ H_2O(l) \rightleftharpoons H_3O^+(aq) \quad + \quad OH^-(aq)$$

initial (M)	1.0×10^{-7}	~0
change (M)	+x	+x
equil (M)	$(1.0 \times 10^{-7}) + x$	x

$K_w = 1.0 \times 10^{-14} = [H_3O^+][OH^-] = [(1.0 \times 10^{-7}) + x](x)$

$x^2 + (1.0 \times 10^{-7})x - (1.0 \times 10^{-14}) = 0$

Solve for x using the quadratic formula.

$$x = \frac{-(1.0 \times 10^{-7}) \pm \sqrt{(1.0 \times 10^{-7})^2 - (4)(-1.0 \times 10^{-14})}}{2(1)} = \frac{(-1.0 \times 10^{-7}) \pm (2.236 \times 10^{-7})}{2}$$

$x = 6.18 \times 10^{-8}$ and -1.62×10^{-7}

Of the two solutions for x, only the positive value of x has physical meaning, because x is the $[OH^-]$.

$[H_3O^+] = (1.0 \times 10^{-7}) + x = (1.0 \times 10^{-7}) + (6.18 \times 10^{-8}) = 1.618 \times 10^{-7}$ M

pH $= -\log[H_3O^+] = -\log(1.618 \times 10^{-7}) = 6.79$

15.122 Na_3PO_4, 163.94 amu

$$3.28\ g\ Na_3PO_4 \times \frac{1\ mol\ Na_3PO_4}{163.94\ g\ Na_3PO_4} = 0.0200\ mol = 20.0\ mmol\ Na_3PO_4$$

$300.0\ mL \times 0.180\ mmol/mL = 54.0\ mmol\ HCl$

$$H_3O^+(aq) + PO_4^{3-}(aq) \rightleftharpoons HPO_4^{2-}(aq) + H_2O(l)$$

before (mmol)	54.0	20.0	0
change (mmol)	−20.0	−20.0	+20.0
after (mmol)	34.0	0	20.0

$$H_3O^+(aq) + HPO_4^{2-}(aq) \rightleftharpoons H_2PO_4^-(aq) + H_2O(l)$$

before (mmol)	34.0	20.0	0
change (mmol)	−20.0	−20.0	+20.0
after (mmol)	14.0	0	20.0

$$H_3O^+(aq) + H_2PO_4^-(aq) \rightleftharpoons H_3PO_4(aq) + H_2O(l)$$

before (mmol)	14.0	20.0	0
change (mmol)	−14.0	−14.0	+14.0
after (mmol)	0	6.0	14.0

$$[H_3PO_4] = \frac{14.0\ mmol}{300.0\ mL} = 0.047\ M; \quad [H_2PO_4^-] = \frac{6.0\ mmol}{300.0\ mL} = 0.020\ M$$

$$H_3PO_4(aq) + H_2O(l) \rightleftharpoons H_3O^+(aq) + H_2PO_4^-(aq)$$

initial (M)	0.047	~0	0.020
change (M)	−x	+x	+x
equil (M)	0.047 − x	x	0.020 + x

$$K_a = \frac{[H_3O^+][H_2PO_4^-]}{[H_3PO_4]} = 7.5 \times 10^{-3} = \frac{x(0.020 + x)}{(0.047 - x)}$$

$x^2 + 0.0275x - (3.525 \times 10^{-4}) = 0$

Solve for x using the quadratic formula.

$$x = \frac{-(0.0275) \pm \sqrt{(0.0275)^2 - (4)(-3.525 \times 10^{-4})}}{2(1)} = \frac{-0.0275 \pm 0.0465}{2}$$

$x = 0.009\ 52$ and -0.0370

Of the two solutions for x, only the positive value of x has physical meaning, because x is the $[H_3O^+]$.

$pH = -\log[H_3O^+] = -\log(0.009\ 52) = 2.02$

16   Applications of Aqueous Equilibria

16.1   (a) $HNO_2(aq) + OH^-(aq) \rightleftharpoons NO_2^-(aq) + H_2O(l)$; NO_2^- (basic anion), pH > 7.00

      (b) $H_3O^+(aq) + NH_3(aq) \rightleftharpoons NH_4^+(aq) + H_2O(l)$; NH_4^+ (acidic cation), pH < 7.00

      (c) $OH^-(aq) + H_3O^+(aq) \rightleftharpoons 2\ H_2O(l)$;   pH = 7.00

16.2   (a) $HF(aq) + OH^-(aq) \rightleftharpoons H_2O(l) + F^-(aq)$

$$K_n = \frac{K_a}{K_w} = \frac{3.5 \times 10^{-4}}{1.0 \times 10^{-14}} = 3.5 \times 10^{10}$$

      (b) $H_3O^+(aq) + OH^-(aq) \rightleftharpoons 2\ H_2O(l)$

$$K_n = \frac{1}{K_w} = \frac{1}{1.0 \times 10^{-14}} = 1.0 \times 10^{14}$$

      (c) $HF(aq) + NH_3(aq) \rightleftharpoons NH_4^+(aq) + F^-(aq)$

$$K_n = \frac{K_a K_b}{K_w} = \frac{(3.5 \times 10^{-4})(1.8 \times 10^{-5})}{1.0 \times 10^{-14}} = 6.3 \times 10^5$$

The tendency to proceed to completion is determined by the magnitude of K_n. The larger the value of K_n, the further does the reaction proceed to completion.
The tendency to proceed to completion is: reaction (c) < reaction (a) < reaction (b)

16.3

	$HCN(aq)$	+	$H_2O(l)$	$\rightleftharpoons$	$H_3O^+(aq)$	+	$CN^-(aq)$
initial (M)	0.025				~0		0.010
change (M)	−x				+x		+x
equil (M)	0.025 − x				x		0.010 + x

$$K_a = \frac{[H_3O^+][CN^-]}{[HCN]} = 4.9 \times 10^{-10} = \frac{x(0.010 + x)}{0.025 - x} \approx \frac{x(0.010)}{0.025}$$

Solve for x. $x = 1.23 \times 10^{-9}\ M = 1.2 \times 10^{-9}\ M = [H_3O^+]$

$pH = -\log[H_3O^+] = -\log(1.23 \times 10^{-9}) = 8.91$

$$[OH^-] = \frac{K_w}{[H_3O^+]} = \frac{1.0 \times 10^{-14}}{1.23 \times 10^{-9}} = 8.2 \times 10^{-6}\ M$$

$[Na^+] = [CN^-] = 0.010\ M$; $[HCN] = 0.025\ M$

$$\%\ \text{dissociation} = \frac{[HCN]_{diss}}{[HCN]_{initial}} \times 100\% = \frac{1.23 \times 10^{-9}\ M}{0.025\ M} \times 100\% = 4.9 \times 10^{-6}\ \%$$

16.4   From $NH_4Cl(s)$, $[NH_4^+]_{initial} = \dfrac{0.10\ mol}{0.500\ L} = 0.20\ M$

$$NH_3(aq) + H_2O(l) \rightleftharpoons NH_4^+(aq) + OH^-(aq)$$

initial (M)	0.40	0.20	~0
change (M)	–x	+x	+x
equil (M)	0.40 – x	0.20 + x	x

$$K_b = \frac{[NH_4^+][OH^-]}{[NH_3]} = 1.8 \times 10^{-5} = \frac{(0.20 + x)(x)}{(0.40 - x)} \approx \frac{(0.20)(x)}{(0.40)}$$

Solve for x. $x = [OH^-] = 3.6 \times 10^{-5}$ M

$$[H_3O^+] = \frac{K_w}{[OH^-]} = \frac{1.0 \times 10^{-14}}{3.6 \times 10^{-5}} = 2.8 \times 10^{-10} \text{ M}$$

$$pH = -\log[H_3O^+] = -\log(2.8 \times 10^{-10}) = 9.55$$

16.5
$$HF(aq) + H_2O(l) \rightleftharpoons H_3O^+(aq) + F^-(aq)$$

initial (M)	0.25	~0	0.50
change (M)	–x	+x	+x
equil (M)	0.25 – x	x	0.50 + x

$$K_a = \frac{[H_3O^+][F^-]}{[HF]} = 3.5 \times 10^{-4} = \frac{x(0.50 + x)}{0.25 - x} \approx \frac{x(0.50)}{0.25}$$

Solve for x. $x = 1.75 \times 10^{-4}$ M $= [H_3O^+]$

For the buffer, $pH = -\log[H_3O^+] = -\log(1.75 \times 10^{-4}) = 3.76$

(a) mol HF = 0.025 mol; mol F$^-$ = 0.050 mol; vol = 0.100 L

$$\overset{100\%}{F^-(aq) + H_3O^+(aq) \rightarrow HF(aq) + H_2O(l)}$$

before (mol)	0.050	0.002	0.025
change (mol)	–0.002	–0.002	+0.002
after (mol)	0.048	0	0.027

$$[H_3O^+] = K_a \frac{[HF]}{[F^-]} = (3.5 \times 10^{-4})\left(\frac{0.27}{0.48}\right) = 1.97 \times 10^{-4} \text{ M}$$

$$pH = -\log[H_3O^+] = -\log(1.97 \times 10^{-4}) = 3.71$$

(b) mol HF = 0.025 mol; mol F$^-$ = 0.050 mol; vol = 0.100 L

$$\overset{100\%}{HF(aq) + OH^-(aq) \rightarrow F^-(aq) + H_2O(l)}$$

before (mol)	0.025	0.004	0.050
change (mol)	–0.004	–0.004	+0.004
after (mol)	0.021	0	0.054

$$[H_3O^+] = K_a \frac{[HF]}{[F^-]} = (3.5 \times 10^{-4})\left(\frac{0.21}{0.54}\right) = 1.36 \times 10^{-4} \text{ M}$$

$$pH = -\log[H_3O^+] = -\log(1.36 \times 10^{-4}) = 3.87$$

16.6
$$HF(aq) + H_2O(l) \rightleftharpoons H_3O^+(aq) + F^-(aq)$$

initial (M)	0.050	~0	0.100
change (M)	–x	+x	+x
equil (M)	0.050 – x	x	0.100 + x

$$K_a = \frac{[H_3O^+][F^-]}{[HF]} = 3.5 \times 10^{-4} = \frac{x(0.100 + x)}{0.050 - x} \approx \frac{x(0.100)}{0.050}$$

Solve for x. x = $[H_3O^+]$ = 1.75×10^{-4} M

pH = $-\log[H_3O^+]$ = $-\log(1.75 \times 10^{-4})$ = 3.76

mol HF = 0.050 mol/L x 0.100 L = 0.0050 mol HF

mol F$^-$ = 0.100 mol/L x 0.100 L = 0.0100 mol F$^-$

mol HNO_3 = mol H_3O^+ = 0.002 mol

Neutralization reaction:

	F$^-$(aq)	+	H$_3$O$^+$(aq)	$\xrightarrow{100\%}$	HF(aq)	+	H$_2$O(l)
before reaction (mol)	0.0100		0.002		0.0050		
change (mol)	−0.002		−0.002		+0.002		
after reaction (mol)	0.008		0		0.007		

$$[HF] = \frac{0.007\ mol}{0.100\ L} = 0.07\ M; \qquad [F^-] = \frac{0.008\ mol}{0.100\ L} = 0.08\ M$$

$$[H_3O^+] = K_a \frac{[HF]}{[F^-]} = (3.5 \times 10^{-4})\frac{(0.07)}{(0.08)} = 3 \times 10^{-4}\ M$$

pH = $-\log[H_3O^+]$ = $-\log(3 \times 10^{-4})$ = 3.5

This solution has less buffering capacity than the solution in Problem 16.5 because it contains less HF and F$^-$ per 100 mL. Note that the change in pH is greater than that in Problem 16.5.

16.7   When equal volumes of two solutions are mixed together, the concentration of each solution is cut in half.

$$pH = pK_a + \log\frac{[base]}{[acid]} = pK_a + \log\frac{[CO_3^{2-}]}{[HCO_3^-]}$$

For HCO$_3^-$, K_a = 5.6×10^{-11}, pK$_a$ = $-\log K_a$ = $-\log(5.6 \times 10^{-11})$ = 10.25

$$pH = 10.25 + \log\left(\frac{0.050}{0.10}\right) = 10.25 - 0.30 = 9.95$$

16.8 $$pH = pK_a + \log\frac{[base]}{[acid]} = pK_a + \log\frac{[CO_3^{2-}]}{[HCO_3^-]}$$

For HCO$_3^-$, K_a = 5.6×10^{-11}, pK$_a$ = $-\log K_a$ = $-\log(5.6 \times 10^{-11})$ = 10.25

$$10.40 = 10.25 + \log\frac{[CO_3^{2-}]}{[HCO_3^-]}; \qquad \log\frac{[CO_3^{2-}]}{[HCO_3^-]} = 10.40 - 10.25 = 0.15$$

$$\frac{[CO_3^{2-}]}{[HCO_3^-]} = 10^{0.15} = 1.4$$

To obtain a buffer solution with pH 10.40, make the Na$_2$CO$_3$ concentration 1.4 times the concentration of NaHCO$_3$.

16.9   Look for an acid with pK_a near the required pH of 7.50.
$K_a = 10^{-pH} = 10^{-7.50} = 3.2 \times 10^{-8}$
Suggested buffer system: HOCl ($K_a = 3.5 \times 10^{-8}$) and NaOCl.

16.10   (a) serine is 66% dissociated at $pH = 9.15 + \log\left(\dfrac{66}{34}\right) = 9.44$

(b) serine is 5% dissociated at $pH = 9.15 + \log\left(\dfrac{5}{95}\right) = 7.87$

16.11   (a) mol HCl = mol H_3O^+ = 0.100 mol/L x 0.0400 L = 0.004 00 mol
mol NaOH = mol OH^- = 0.100 mol/L x 0.0350 L = 0.003 50 mol

Neutralization reaction:	$H_3O^+(aq)$	+	$OH^-(aq)$	$\rightarrow$	$2 H_2O(l)$
before reaction (mol)	0.004 00		0.003 50		
change (mol)	−0.003 50		−0.003 50		
after reaction (mol)	0.000 50		0		

$[H_3O^+] = \dfrac{0.000\ 50\ mol}{(0.0400\ L\ +\ 0.0350\ L)} = 6.7 \times 10^{-3}$ M

$pH = -\log[H_3O^+] = -\log(6.7 \times 10^{-3}) = 2.17$
(b) mol HCl = mol H_3O^+ = 0.100 mol/L x 0.0400 L = 0.004 00 mol
mol NaOH = mol OH^- = 0.100 mol/L x 0.0450 L = 0.004 50 mol

Neutralization reaction:	$H_3O^+(aq)$	+	$OH^-(aq)$	$\rightarrow$	$2 H_2O(l)$
before reaction (mol)	0.004 00		0.004 50		
change (mol)	−0.004 00		−0.004 00		
after reaction (mol)	0		0.000 50		

$[OH^-] = \dfrac{0.000\ 50\ mol}{(0.0400\ L\ +\ 0.0450\ L)} = 5.9 \times 10^{-3}$ M

$[H_3O^+] = \dfrac{K_w}{[OH^-]} = \dfrac{1.0 \times 10^{-14}}{5.9 \times 10^{-3}} = 1.7 \times 10^{-12}$ M

$pH = -\log[H_3O^+] = -\log(1.7 \times 10^{-12}) = 11.77$
The results obtained here are consistent with the pH data in Table 16.1

16.12   (a) mol NaOH = mol OH^- = 0.100 mol/L x 0.0400 L = 0.004 00 mol
mol HCl = mol H_3O^+ = 0.0500 mol/L x 0.0600 L = 0.003 00 mol

Neutralization reaction:	$H_3O^+(aq)$	+	$OH^-(aq)$	$\rightarrow$	$2 H_2O(l)$
before reaction (mol)	0.003 00		0.004 00		
change (mol)	−0.003 00		−0.003 00		
after reaction (mol)	0		0.001 00		

$[OH^-] = \dfrac{0.001\ 00\ mol}{(0.0400\ L\ +\ 0.0600\ L)} = 1.0 \times 10^{-2}$ M

$[H_3O^+] = \dfrac{K_w}{[OH^-]} = \dfrac{1.0 \times 10^{-14}}{1.0 \times 10^{-2}} = 1.0 \times 10^{-12}$ M

$pH = -\log[H_3O^+] = -\log(1.0 \times 10^{-12}) = 12.00$

(b) mol NaOH = mol OH⁻ = 0.100 mol/L x 0.0400 L = 0.004 00 mol
mol HCl = mol H_3O^+ = 0.0500 mol/L x 0.0802 L = 0.004 01 mol

Neutralization reaction:	$H_3O^+(aq)$ +	$OH^-(aq)$	→	$2 H_2O(l)$
before reaction (mol)	0.004 01	0.004 00		
change (mol)	−0.004 00	−0.004 00		
after reaction (mol)	0.000 01	0		

$[H_3O^+] = \dfrac{0.000\ 01\ mol}{(0.0400\ L\ +\ 0.0802\ L)} = 8.3 \times 10^{-5}\ M$

pH = $-\log[H_3O^+]$ = $-\log(8.3 \times 10^{-5})$ = 4.08

(c) mol NaOH = mol OH⁻ = 0.100 mol/L x 0.0400 L = 0.004 00 mol
mol HCl = mol H_3O^+ = 0.0500 mol/L x 0.1000 L = 0.005 00 mol

Neutralization reaction:	$H_3O^+(aq)$ +	$OH^-(aq)$	→	$2 H_2O(l)$
before reaction (mol)	0.005 00	0.004 00		
change (mol)	−0.004 00	−0.004 00		
after reaction (mol)	0.001 00	0		

$[H_3O^+] = \dfrac{0.001\ 00\ mol}{(0.0400\ L\ +\ 0.1000\ L)} = 7.1 \times 10^{-3}\ M$

pH = $-\log[H_3O^+]$ = $-\log(7.1 \times 10^{-3})$ = 2.15

16.13   mol NaOH required = $\left(\dfrac{0.016\ mol\ HOCl}{L}\right)(0.100\ L)\left(\dfrac{1\ mol\ NaOH}{1\ mol\ HOCl}\right)$ = 0.0016 mol

vol NaOH required = $(0.0016\ mol)\left(\dfrac{1\ L}{0.0400\ mol}\right)$ = 0.040 L = 40 mL

40 mL of 0.0400 M NaOH are required to reach the equivalence point.
(a) mmol HOCl = 0.016 mmol/mL x 100.0 mL = 1.6 mmol
mmol NaOH = mmol OH⁻ = 0.0400 mmol/mL x 10.0 mL = 0.400 mmol

Neutralization reaction:	$HOCl(aq)$ +	$OH^-(aq)$	→	$OCl^-(aq)$ +	$H_2O(l)$
before reaction (mmol)	1.6	0.400		0	
change (mmol)	−0.400	−0.400		+0.400	
after reaction (mmol)	1.2	0		0.400	

$[HOCl] = \dfrac{1.2\ mmol}{(100.0\ mL\ +\ 10.0\ mL)} = 1.09 \times 10^{-2}\ M$

$[OCl^-] = \dfrac{0.400\ mmol}{(100.0\ mL\ +\ 10.0\ mL)} = 3.64 \times 10^{-3}\ M$

	$HOCl(aq)$ +	$H_2O(l)$	⇌	$H_3O^+(aq)$ +	$OCl^-(aq)$
initial (M)	0.0109			~0	0.003 64
change (M)	−x			+x	+x
equil (M)	0.0109 − x			x	0.003 64 + x

$K_a = \dfrac{[H_3O^+][OCl^-]}{[HOCl]} = 3.5 \times 10^{-8} = \dfrac{x(0.003\ 64 + x)}{0.0109 - x} \approx \dfrac{x(0.003\ 64)}{0.0109}$

Solve for x. x = $[H_3O^+]$ = 1.05×10^{-7} M
pH = $-\log[H_3O^+]$ = $-\log(1.05 \times 10^{-7})$ = 6.98

(b) Halfway to the equivalence point, $[OCl^-] = [HOCl]$
$pH = pK_a = -\log K_a = -\log(3.5 \times 10^{-8}) = 7.46$
(c) At the equivalence point the solution contains the salt, NaOCl.
mol NaOCl = initial mol HOCl = 0.0016 mol = 1.6 mmol

$$[OCl^-] = \frac{1.6 \text{ mmol}}{(100.0 \text{ mL} + 40.0 \text{ mL})} = 1.1 \times 10^{-2} \text{ M}$$

$$\text{For } OCl^-, \ K_b = \frac{K_w}{K_a \text{ for HOCl}} = \frac{1.0 \times 10^{-14}}{3.5 \times 10^{-8}} = 2.9 \times 10^{-7}$$

$$OCl^-(aq) + H_2O(l) \ \rightleftharpoons \ HOCl(aq) + OH^-(aq)$$

initial (M)	0.011	0	~0
change (M)	$-x$	$+x$	$+x$
equil (M)	$0.011 - x$	x	x

$$K_b = \frac{[HOCl][OH^-]}{[OCl^-]} = 2.9 \times 10^{-7} = \frac{x^2}{0.011 - x} \approx \frac{x^2}{0.011}$$

Solve for x. $x = [OH^-] = 5.65 \times 10^{-5} \text{ M}$

$$[H_3O^+] = \frac{K_w}{[OH^-]} = \frac{1.0 \times 10^{-14}}{5.65 \times 10^{-5}} = 1.77 \times 10^{-10} = 1.8 \times 10^{-10} \text{ M}$$

$pH = -\log[H_3O^+] = -\log(1.77 \times 10^{-10}) = 9.75$

16.14   From Problem 16.13, pH = 9.75 at the equivalence point.
Use thymolphthalein (pH 9.4 – 10.6). Bromthymol blue is unacceptable because it changes color halfway to the equivalence point.

16.15   (a) mol NaOH required to reach first equivalence point

$$= \left(\frac{0.0800 \text{ mol } H_2SO_3}{L} \right) (0.0400 \text{ L}) \left(\frac{1 \text{ mol NaOH}}{1 \text{ mol } H_2SO_3} \right) = 0.003 \ 20 \text{ mol}$$

vol NaOH required to reach first equivalence point

$$= (0.003 \ 20 \text{ mol}) \left(\frac{1 \text{ L}}{0.160 \text{ mol}} \right) = 0.020 \text{ L} = 20.0 \text{ mL}$$

20.0 mL is enough NaOH solution to reach the first equivalence point for the titration of the diprotic acid, H_2SO_3.
For H_2SO_3,
$K_{a1} = 1.5 \times 10^{-2}, \ pK_{a1} = -\log K_{a1} = -\log(1.5 \times 10^{-2}) = 1.82$

$K_{a2} = 6.3 \times 10^{-8}, \ pK_{a2} = -\log K_{a2} = -\log(6.3 \times 10^{-8}) = 7.20$

At the first equivalence point, $pH = \dfrac{pK_{a1} + pK_{a2}}{2} = \dfrac{1.82 + 7.20}{2} = 4.51$

(b) mol NaOH required to reach second equivalence point

$$= \left(\frac{0.0800 \text{ mol } H_2SO_3}{L} \right) (0.0400 \text{ L}) \left(\frac{2 \text{ mol NaOH}}{1 \text{ mol } H_2SO_3} \right) = 0.006 \ 40 \text{ mol}$$

vol NaOH required to reach second equivalence point

$$= (0.006\ 40\ \text{mol})\left(\frac{1\ \text{L}}{0.160\ \text{mol}}\right) = 0.040\ \text{L} = 40.0\ \text{mL}$$

30.0 mL is enough NaOH solution to reach halfway to the second equivalent point. Halfway to the second equivalence point

$$pH = pK_{a2} = -\log K_{a2} = -\log(6.3 \times 10^{-8}) = 7.20$$

(c) mmol $HSO_3^- = 0.0800$ mmol/mL × 40.0 mL = 3.2 mmol

volume NaOH added after first equivalence point = 35.0 mL – 20.0 mL = 15.0 mL

mmol NaOH = mmol $OH^- = 0.160$ mmol/L × 15.0 mL = 2.4 mmol

Neutralization reaction:	$HSO_3^-(aq)$	+	$OH^-(aq)$	$\rightleftarrows$	$SO_3^{2-}(aq)$	+	$H_2O(l)$
before reaction (mmol)	3.2		2.4		0		
change (mmol)	–2.4		–2.4		+2.4		
after reaction (mmol)	0.8		0		2.4		

$$[HSO_3^-] = \frac{0.8\ \text{mmol}}{(40.0\ \text{mL} + 35.0\ \text{mL})} = 0.0107\ M$$

$$[SO_3^{2-}] = \frac{2.4\ \text{mmol}}{(40.0\ \text{mL} + 35.0\ \text{mL})} = 0.032\ M$$

	$HSO_3^-(aq)$	+	$H_2O(l)$	$\rightleftarrows$	$H_3O^+(aq)$	+	$SO_3^{2-}(aq)$
initial (M)	0.0107				~0		0.032
change (M)	–x				+x		+x
equil (M)	0.0107 – x				x		0.032 + x

$$K_a = \frac{[H_3O^+][SO_3^{2-}]}{[HSO_3^-]} = 6.3 \times 10^{-8} = \frac{x(0.032 + x)}{0.0107 - x} \approx \frac{x(0.032)}{0.0107}$$

Solve for x. $x = [H_3O^+] = 2.1 \times 10^{-8}\ M$

$$pH = -\log[H_3O^+] = -\log(2.1 \times 10^{-8}) = 7.7$$

16.16 Let H_2A^+ = phenylalanine cation

(a) mol NaOH required to reach first equivalence point

$$= \left(\frac{0.0250\ \text{mol}\ H_2A^+}{L}\right)(0.0400\ L)\left(\frac{1\ \text{mol NaOH}}{1\ \text{mol}\ H_2A^+}\right) = 0.001\ 00\ \text{mol}$$

vol NaOH required to reach first equivalence point

$$= (0.001\ 00\ \text{mol})\left(\frac{1\ \text{L}}{0.100\ \text{mol}}\right) = 0.010\ \text{L} = 10.0\ \text{mL}$$

10.0 mL is enough NaOH solution to reach the first equivalence point for the titration of the diprotic acid, H_2A^+.

For H_2A^+,

$$K_{a1} = 1.5 \times 10^{-2},\ pK_{a1} = -\log K_{a1} = -\log(1.5 \times 10^{-2}) = 1.82$$

$$K_{a2} = 7.4 \times 10^{-10},\ pK_{a2} = -\log K_{a2} = -\log(7.4 \times 10^{-10}) = 9.13$$

At the first equivalence point, $pH = \dfrac{pK_{a1} + pK_{a2}}{2} = \dfrac{1.82 + 9.13}{2} = 5.48$

(b) mol NaOH required to reach second equivalence point

$$= \left(\frac{0.0250 \text{ mol H}_2\text{A}^+}{\text{L}} \right)(0.0400 \text{ L})\left(\frac{2 \text{ mol NaOH}}{1 \text{ mol H}_2\text{A}^+} \right) = 0.002\ 00 \text{ mol}$$

vol NaOH required to reach second equivalence point

$$= (0.002\ 00 \text{ mol})\left(\frac{1 \text{ L}}{0.100 \text{ mol}} \right) = 0.020 \text{ L} = 20.0 \text{ mL}$$

15.0 mL is enough NaOH solution to reach halfway to the second equivalent point.
Halfway to the second equivalence point

$$\text{pH} = \text{p}K_{a2} = -\log K_{a2} = -\log(7.4 \times 10^{-10}) = 9.13$$

(c) 20.0 mL is enough NaOH to reach the second equivalence point.
At the second equivalence point

$$\text{mmol A}^- = (0.0250 \text{ mmol/mL})(40.0 \text{ mL}) = 1.00 \text{ mmol A}^-$$

$$\text{solution volume} = 40.0 \text{ mL} + 20.0 \text{ mL} = 60.0 \text{ mL}$$

$$[\text{A}^-] = \frac{1.00 \text{ mmol}}{60.0 \text{ mL}} = 0.0167 \text{ M}$$

	$\text{A}^-(aq)$	+	$\text{H}_2\text{O}(l)$	$\rightleftharpoons$	$\text{HA}(aq)$	+	$\text{OH}^-(aq)$
initial (M)	0.0167				0		~0
change (M)	$-x$				$+x$		$+x$
equil (M)	$0.0167 - x$				x		x

$$K_b = \frac{K_w}{K_a \text{ for HA}} = \frac{K_w}{K_{a2}} = \frac{1.0 \times 10^{-14}}{4.75 \times 10^{-4}} = 1.35 \times 10^{-5}$$

$$K_b = \frac{[\text{HA}][\text{OH}^-]}{[\text{A}^-]} = 1.35 \times 10^{-5} = \frac{x^2}{0.0167 - x}$$

$$x^2 + (1.35 \times 10^{-5})x - (2.254 \times 10^{-7}) = 0$$

Use the quadratic formula to solve for x.

$$x = \frac{-(1.35 \times 10^{-5}) \pm \sqrt{(1.35 \times 10^{-5})^2 - 4(-2.254 \times 10^{-7})}}{2(1)} = \frac{(-1.35 \times 10^{-5}) \pm (9.50 \times 10^{-4})}{2}$$

$$x = 4.68 \times 10^{-4} \text{ and } -4.82 \times 10^{-4}$$

Of the two solutions for x, only the positive value has physical meaning because x is the $[\text{OH}^-]$.

$$x = [\text{OH}^-] = 4.68 \times 10^{-4} \text{ M}$$

$$[\text{H}_3\text{O}^+] = \frac{K_w}{[\text{OH}^-]} = \frac{1.0 \times 10^{-14}}{4.68 \times 10^{-4}} = 2.14 \times 10^{-11} \text{ M}$$

$$\text{pH} = -\log[\text{H}_3\text{O}^+] = -\log(2.14 \times 10^{-11}) = 10.67$$

16.17 (a) $K_{sp} = [\text{Ag}^+][\text{Cl}^-]$ (b) $K_{sp} = [\text{Pb}^{2+}][\text{I}^-]^2$ (c) $K_{sp} = [\text{Ca}^{2+}]^3[\text{PO}_4^{3-}]^2$

16.18 $K_{sp} = [\text{Ca}^{2+}]^3[\text{PO}_4^{3-}]^2 = (2.01 \times 10^{-8})^3(1.6 \times 10^{-5})^2 = 2.1 \times 10^{-33}$

16.19 $[\text{Ba}^{2+}] = [\text{SO}_4^{2-}] = 1.05 \times 10^{-5} \text{ M};$ $K_{sp} = [\text{Ba}^{2+}][\text{SO}_4^{2-}] = (1.05 \times 10^{-5})^2 = 1.10 \times 10^{-10}$

16.20 (a)
$$AgCl(s) \rightleftharpoons Ag^+(aq) + Cl^-(aq)$$

equil (M)   x   x

$K_{sp} = [Ag^+][Cl^-] = 1.8 \times 10^{-10} = (x)(x)$

molar solubility $= x = \sqrt{K_{sp}} = 1.3 \times 10^{-5}$ mol/L

AgCl, 143.32 amu

$$\text{solubility} = \frac{\left(1.3 \times 10^{-5} \text{ mol} \times \frac{143.32 \text{ g}}{1 \text{ mol}}\right)}{1 \text{ L}} = 0.0019 \text{ g/L}$$

(b)
$$Ag_2CrO_4(s) \rightleftharpoons 2 Ag^+(aq) + CrO_4^{2-}(aq)$$

equil (M)   2x   x

$K_{sp} = [Ag^+]^2[CrO_4^{2-}] = 1.1 \times 10^{-12} = (2x)^2(x) = 4x^3$

molar solubility $= x = \sqrt[3]{\dfrac{1.1 \times 10^{-12}}{4}} = 6.5 \times 10^{-5}$ mol/L

Ag_2CrO_4, 331.73 amu

$$\text{solubility} = \frac{\left(6.5 \times 10^{-5} \text{ mol} \times \frac{331.73 \text{ g}}{1 \text{ mol}}\right)}{1 \text{ L}} = 0.022 \text{ g/L}$$

Ag_2CrO_4 has both the higher molar and gram solubility, despite its smaller value of K_{sp}.

16.21 $[Mg^{2+}]_0$ is from 0.10 M $MgCl_2$.

$$MgF_2(s) \rightleftharpoons Mg^{2+}(aq) + 2 F^-(aq)$$

initial (M)   0.10   0

change (M)   +x   +2x

equil (M)   0.10 + x   2x

$K_{sp} = 7.4 \times 10^{-11} = [Mg^{2+}][F^-]^2 = (0.10 + x)(2x)^2 \approx (0.10)(4x^2)$

$x = 1.4 \times 10^{-5}$,   molar solubility $= x = 1.4 \times 10^{-5}$ M

16.22 Compounds that contain basic anions are more soluble in acidic solution than in pure water. AgCN, $Al(OH)_3$, and ZnS all contain basic anions.

16.23 $[Cu^{2+}] = (5.0 \times 10^{-3} \text{ mol})/(0.500 \text{ L}) = 0.010$ M

$$Cu^{2+}(aq) + 4 NH_3(aq) \rightleftharpoons Cu(NH_3)_4^{2+}(aq)$$

before reaction (M)   0.010   0.40   0

assume 100 % reaction (M)   −0.010   −4(0.010)   +0.010

after reaction (M)   0   0.36   0.010

assume small back reaction (M)   +x   +4x   −x

equil (M)   x   0.36 + 4x   0.010 − x

$$K_f = \frac{[Cu(NH_3)_4^{2+}]}{[Cu^{2+}][NH_3]^4} = 1.1 \times 10^{13} = \frac{(0.010 - x)}{(x)(0.36 + 4x)^4} \approx \frac{0.010}{x(0.36)^4}$$

Solve for x. $x = [Cu^{2+}] = 5.4 \times 10^{-14}$ M

16.24

$$AgBr(s) \rightleftharpoons Ag^+(aq) + Br^-(aq) \qquad\qquad K_{sp} = 5.4 \times 10^{-13}$$
$$\underline{Ag^+(aq) + 2\,S_2O_3^{2-} \rightarrow Ag(S_2O_3)_2^{3-}(aq)} \qquad\qquad K_f = 2.9 \times 10^{13}$$

dissolution $\quad AgBr(s) + 2\,S_2O_3^{2-}(aq) \rightleftharpoons Ag(S_2O_3)_2^{3-}(aq) + Br^-(aq)$
reaction

$K = (K_{sp})(K_f) = (5.4 \times 10^{-13})(2.9 \times 10^{13}) = 16$

$$AgBr(s) + 2\,S_2O_3^{2-}(aq) \rightleftharpoons Ag(S_2O_3)_2^{3-}(aq) + Br^-(aq)$$

initial (M)	0.10	0	0
change (M)	$-2x$	x	x
equil (M)	$0.10 - 2x$	x	x

$$K = \frac{[Ag(S_2O_3)_2^{3-}][Br^-]}{[S_2O_3^{2-}]^2} = 16 = \frac{x^2}{(0.10 - 2x)^2}$$

Take the square root of both sides and solve for x.

$$\sqrt{16} = \sqrt{\frac{x^2}{(0.10 - 2x)^2}}; \qquad 4.0 = \frac{x}{0.10 - 2x}; \qquad x = \text{molar solubility} = 0.044 \text{ mol/L}$$

16.25   On mixing equal volumes of two solutions, the concentrations of both solutions are cut in half.
For $BaCO_3$, $K_{sp} = 2.6 \times 10^{-9}$
(a) $IP = [Ba^{2+}][CO_3^{2-}] = (1.5 \times 10^{-3})(1.0 \times 10^{-3}) = 1.5 \times 10^{-6}$
    $IP > K_{sp}$; a precipitate of $BaCO_3$ will form.
(b) $IP = [Ba^{2+}][CO_3^{2-}] = (5.0 \times 10^{-6})(2.0 \times 10^{-5}) = 1.0 \times 10^{-10}$
    $IP < K_{sp}$; no precipitate will form.

16.26 $pH = pK_a + \log\dfrac{[\text{base}]}{[\text{acid}]} = pK_a + \log\dfrac{[NH_3]}{[NH_4^+]}$

For NH_4^+, $K_a = 5.6 \times 10^{-10}$, $pK_a = -\log K_a = -\log(5.6 \times 10^{-10}) = 9.25$

$pH = 9.25 + \log\dfrac{(0.20)}{(0.20)} = 9.25;\quad [H_3O^+] = 10^{-pH} = 10^{-9.25} = 5.6 \times 10^{-10}$ M

$[OH^-] = \dfrac{K_w}{[H_3O^+]} = \dfrac{1.0 \times 10^{-14}}{5.6 \times 10^{-10}} = 1.8 \times 10^{-5}$ M

$[Fe^{2+}] = [Mn^{2+}] = \dfrac{(25 \text{ mL})(1.0 \times 10^{-3} \text{ M})}{250 \text{ mL}} = 1.0 \times 10^{-4}$ M

For $Mn(OH)_2$, $K_{sp} = 2.1 \times 10^{-13}$
    $IP = [Mn^{2+}][OH^-]^2 = (1.0 \times 10^{-4})(1.8 \times 10^{-5})^2 = 3.2 \times 10^{-14}$
    $IP < K_{sp}$; no precipitate will form.

For $Fe(OH)_2$, $K_{sp} = 4.9 \times 10^{-17}$
    $IP = [Fe^{2+}][OH^-]^2 = (1.0 \times 10^{-4})(1.8 \times 10^{-5})^2 = 3.2 \times 10^{-14}$
    $IP > K_{sp}$; a precipitate of $Fe(OH)_2$ will form.

16.27 $MS(s) + 2 H_3O^+(aq) \rightleftharpoons M^{2+}(aq) + H_2S(aq) + 2 H_2O(l)$

$K_{spa} = \dfrac{[M^{2+}][H_2S]}{[H_3O^+]^2}$

For ZnS, $K_{spa} = 3 \times 10^{-2}$;   for CdS, $K_{spa} = 8 \times 10^{-7}$

$[Cd^{2+}] = [Zn^{2+}] = 0.005$ M

Because the two cation concentrations are equal, Q_c is the same for both.

$Q_c = \dfrac{[M^{2+}]_t[H_2S]_t}{[H_3O^+]_t^2} = \dfrac{(0.005)(0.10)}{(0.3)^2} = 6 \times 10^{-3}$

$Q_c > K_{spa}$ for CdS; CdS will precipitate. $Q_c < K_{spa}$ for ZnS; Zn^{2+} will remain in solution.

Understanding Key Concepts

16.28   (a) (1) and (3). Both pictures show equal concentrations of HA and A^-.
         (b) (3). It contains a higher concentration of HA and A^-.

16.29   (a) (2) has the highest pH, $[A^-] > [HA]$
              (3) has the lowest pH, $[HA] > [A^-]$

(b)   (c)

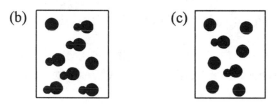

16.30   (4); only A^- and water should be present

16.31   (a) (1) corresponds to (iii); (2) to (i); (3) to (iv); and (4) to (ii)
         (b) Solution (3) has the highest pH; solution (2) has the lowest pH.

16.32   (2) is supersaturated; (3) is unsaturated; (4) is unsaturated

16.33   (a) The bottom curve represents the titration of a strong acid; the top curve represents
              the titration of a weak acid.
         (b) pH = 7 for titration of the strong acid; pH = 10 for titration of the weak acid.
         (c) $pK_a \sim 6.3$

16.34 (a)

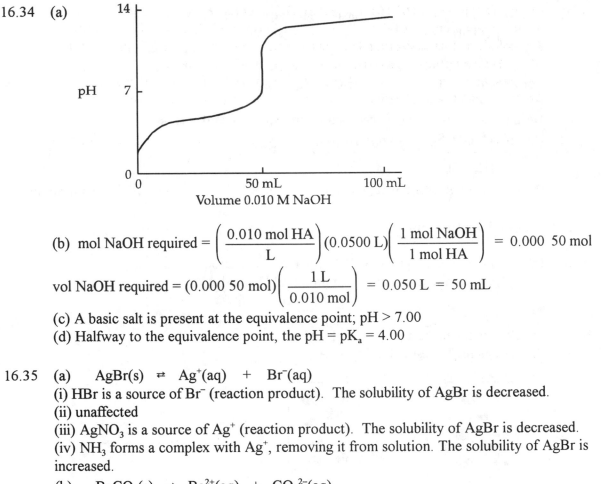

(b) mol NaOH required = $\left(\dfrac{0.010 \text{ mol HA}}{\text{L}}\right)(0.0500 \text{ L})\left(\dfrac{1 \text{ mol NaOH}}{1 \text{ mol HA}}\right)$ = 0.000 50 mol

vol NaOH required = $(0.000\ 50 \text{ mol})\left(\dfrac{1 \text{ L}}{0.010 \text{ mol}}\right)$ = 0.050 L = 50 mL

(c) A basic salt is present at the equivalence point; pH > 7.00
(d) Halfway to the equivalence point, the pH = pK_a = 4.00

16.35 (a) $AgBr(s) \rightleftharpoons Ag^+(aq) + Br^-(aq)$
(i) HBr is a source of Br^- (reaction product). The solubility of AgBr is decreased.
(ii) unaffected
(iii) $AgNO_3$ is a source of Ag^+ (reaction product). The solubility of AgBr is decreased.
(iv) NH_3 forms a complex with Ag^+, removing it from solution. The solubility of AgBr is increased.
(b) $BaCO_3(s) \rightleftharpoons Ba^{2+}(aq) + CO_3^{2-}(aq)$
(i) unaffected
(ii) HNO_3 reacts with CO_3^{2-}, removing it from the solution. The solubility of $BaCO_3$ is increased.
(iii) $Ba(NO_3)_2$ is a source of Ba^{2+} (reaction product). The solubility of $BaCO_3$ is decreased.
(iv) Na_2CO_3 is a source of CO_3^{2-} (reaction product). The solubility of $BaCO_3$ is decreased.

Additional Problems
Neutralization Reactions

16.36 (a) $HI(aq) + NaOH(aq) \rightarrow H_2O(l) + NaI(aq)$
net ionic equation: $H_3O^+(aq) + OH^-(aq) \rightarrow 2\ H_2O(l)$
The solution at neutralization contains a neutral salt (NaI); pH = 7.00.
(b) $2\ HOCl(aq) + Ba(OH)_2(aq) \rightarrow 2\ H_2O(l) + Ba(OCl)_2(aq)$
net ionic equation: $HOCl(aq) + OH^-(aq) \rightarrow H_2O(l) + OCl^-(aq)$
The solution at neutralization contains a basic anion (OCl^-); pH > 7.00

(c) $HNO_3(aq) + C_6H_5NH_2(aq) \rightarrow C_6H_5NH_3NO_3(aq)$
net ionic equation: $H_3O^+(aq) + C_6H_5NH_2(aq) \rightarrow H_2O(l) + C_6H_5NH_3^+(aq)$
The solution at neutralization contains an acidic cation ($C_6H_5NH_3^+$); pH < 7.00.
(d) $C_6H_5COOH(aq) + KOH(aq) \rightarrow H_2O(l) + C_6H_5CO_2K(aq)$
net ionic equation: $C_6H_5COOH(aq) + OH^-(aq) \rightarrow H_2O(l) + C_6H_5CO_2^-(aq)$
The solution at neutralization contains a basic anion ($C_6H_5COO^-$); pH > 7.00.

16.38   (a) Strong acid – strong base reaction $K_n = \dfrac{1}{K_w} = \dfrac{1}{1.0 \times 10^{-14}} = 1.0 \times 10^{14}$

     (b) Weak acid – strong base reaction $K_n = \dfrac{K_a}{K_w} = \dfrac{3.5 \times 10^{-8}}{1.0 \times 10^{-14}} = 3.5 \times 10^6$

     (c) Strong acid – weak base reaction $K_n = \dfrac{K_b}{K_w} = \dfrac{4.3 \times 10^{-10}}{1.0 \times 10^{-14}} = 4.3 \times 10^4$

     (d) Weak acid – strong base reaction $K_n = \dfrac{K_a}{K_w} = \dfrac{6.5 \times 10^{-5}}{1.0 \times 10^{-14}} = 6.5 \times 10^9$

     (c) < (b) < (d) < (a)

16.40   (a) After mixing, the solution contains the basic salt, NaF; pH > 7.00
     (b) After mixing, the solution contains the neutral salt, NaCl; pH = 7.00
     Solution (a) has the higher pH.

16.42   Weak acid – weak base reaction $K_n = \dfrac{K_a K_b}{K_w} = \dfrac{(1.3 \times 10^{-10})(1.8 \times 10^{-9})}{1.0 \times 10^{-14}} = 2.3 \times 10^{-5}$

K_n is small so the neutralization reaction does not proceed very far to completion.

The Common–Ion Effect

16.44 $HNO_2(aq) + H_2O(l) \rightleftharpoons H_3O^+(aq) + NO_2^-(aq)$
     (a) $NaNO_2$ is a source of NO_2^- (reaction product). The equilibrium shifts towards reactants, and the percent dissociation of HNO_2 decreases.
     (c) HCl is a source of H_3O^+ (reaction product). The equilibrium shifts towards reactants, and the percent dissociation of HNO_2 decreases.
     (d) $Ba(NO_2)_2$ is a source of NO_2^- (reaction product). The equilibrium shifts towards reactants, and the percent dissociation of HNO_2 decreases.

16.46   (a) $HF(aq) + H_2O(l) \rightleftharpoons H_3O^+(aq) + F^-(aq)$
     LiF is a source of F^- (reaction product). The equilibrium shifts toward reactants, and the $[H_3O^+]$ decreases. The pH increases.
     (b) Because HI is a strong acid, addition of KI, a neutral salt, does not change the pH.
     (c) $NH_3(aq) + H_2O(l) \rightleftharpoons NH_4^+(aq) + OH^-(aq)$
     NH_4Cl is a source of NH_4^+ (reaction product). The equilibrium shifts toward reactants, and the $[OH^-]$ decreases. The pH decreases.

16.48   For 0.25 M HF and 0.10 M NaF

$$HF(aq) \; + \; H_2O(l) \; \rightleftharpoons \; H_3O^+(aq) \; + \; F^-(aq)$$

initial (M)	0.25	~0	0.10
change (M)	–x	+x	+x
equil (M)	0.25 – x	x	0.10 + x

$$K_a = \frac{[H_3O^+][F^-]}{[HF]} = 3.5 \times 10^{-4} = \frac{x(0.10 + x)}{0.25 - x} \approx \frac{x(0.10)}{0.25}$$

Solve for x. $x = [H_3O^+] = 8.8 \times 10^{-4}$ M

$pH = -\log[H_3O^+] = -\log(8.8 \times 10^{-4}) = 3.06$

16.50   For 0.10 M HN_3:

$$HN_3(aq) \; + \; H_2O(l) \; \rightleftharpoons \; H_3O^+(aq) \; + \; N_3^-(aq)$$

initial (M)	0.10	~0	0
change (M)	–x	+x	+x
equil (M)	0.10 – x	x	x

$$K_a = \frac{[H_3O^+][N_3^-]}{[HN_3]} = 1.9 \times 10^{-5} = \frac{x^2}{0.10 - x} \approx \frac{x^2}{0.10}$$

Solve for x. $x = 1.4 \times 10^{-3}$ M

$$\% \text{ dissociation} = \frac{[HN_3]_{diss}}{[HN_3]_{initial}} \times 100\% = \frac{1.4 \times 10^{-3} \text{ M}}{0.10 \text{ M}} \times 100\% = 1.4\%$$

For 0.10 M HN_3 in 0.10 M HCl:

$$HN_3(aq) \; + \; H_2O(l) \; \rightleftharpoons \; H_3O^+(aq) \; + \; N_3^-(aq)$$

initial (M)	0.10	0.10	0
change (M)	–x	+x	+x
equil (M)	0.10 – x	0.10 + x	x

$$K_a = \frac{[H_3O^+][N_3^-]}{[HN_3]} = 1.9 \times 10^{-5} = \frac{(0.10 + x)(x)}{0.10 - x} \approx \frac{(0.10)(x)}{0.10} = x$$

Solve for x. $x = 1.9 \times 10^{-5}$ M

$$\% \text{ dissociation} = \frac{[HN_3]_{diss}}{[HN_3]_{initial}} \times 100\% = \frac{1.9 \times 10^{-5} \text{ M}}{0.10 \text{ M}} \times 100\% = 0.019\%$$

The % dissociation is less because of the common ion (H_3O^+) effect.

Buffer Solutions

16.52   Solutions (a), (c) and (d) are buffer solutions. Neutralization reactions for (c) and (d) result in solutions with equal concentrations of HF and F^-.

16.54   Both solutions buffer at the same pH because in both cases the $[NO_2^-]/[HNO_2] = 1$. Solution (a), however, has a higher concentration of both HNO_2 and NO_2^-, and therefore it has the greater buffer capacity.

16.56   When blood absorbs acid, the equilibrium shifts to the left, decreasing the pH, but not by much because the $[HCO_3^-]/[H_2CO_3]$ ratio remains nearly constant. When blood absorbs base, the equilibrium shifts to the right, increasing the pH, but not by much because the $[HCO_3^-]/[H_2CO_3]$ ratio remains nearly constant.

16.58 $pH = pK_a + \log \dfrac{[base]}{[acid]} = pK_a + \log \dfrac{[CN^-]}{[HCN]}$

For HCN, $K_a = 4.9 \times 10^{-10}$, $pK_a = -\log K_a = -\log(4.9 \times 10^{-10}) = 9.31$

$pH = 9.31 + \log\left(\dfrac{0.12}{0.20}\right) = 9.09$

The pH of a buffer solution will not change on dilution because the acid and base concentrations will change by the same amount and their ratio will remain the same.

16.60 $pH = pK_a + \log \dfrac{[base]}{[acid]} = pK_a + \log \dfrac{[NH_3]}{[NH_4^+]}$

For NH_4^+, $K_a = 5.6 \times 10^{-10}$, $pK_a = -\log K_a = -\log(5.6 \times 10^{-10}) = 9.25$

For the buffer: $pH = 9.25 + \log \dfrac{(0.200)}{(0.200)} = 9.25$

(a) add 0.0050 mol NaOH, $[OH^-] = 0.0050$ mol/0.500 L = 0.010 M

$$NH_4^+(aq) + OH^-(aq) \rightleftharpoons NH_3(aq) + H_2O(l)$$

before reaction (M)	0.200	0.010	0.200
change (M)	−0.010	−0.010	+0.010
after reaction (M)	0.200 − 0.010		0.200 + 0.010

$pH = 9.25 + \log \dfrac{[NH_3]}{[NH_4^+]} = 9.25 + \log \dfrac{(0.200 + 0.010)}{(0.200 - 0.010)} = 9.29$

(b) add 0.020 mol HCl, $[H_3O^+] = 0.020$ mol/0.500 L = 0.040 M

$$NH_3(aq) + H_3O^+(aq) \rightleftharpoons NH_4^+(aq) + H_2O(l)$$

before reaction (M)	0.200	0.040	0.200
change (M)	−0.040	−0.040	+0.040
after reaction (M)	0.200 − 0.040	0	0.200 + 0.040

$pH = 9.25 + \log \dfrac{[NH_3]}{[NH_4^+]} = 9.25 + \log \dfrac{(0.200 - 0.040)}{(0.200 + 0.040)} = 9.07$

16.62

	Acid	K_a	$pK_a = -\log K_a$
(a)	H_3BO_3	5.8×10^{-10}	9.24
(b)	HCOOH	1.8×10^{-4}	3.74
(c)	HOCl	3.5×10^{-8}	7.46

The stronger the acid (the larger the K_a), the smaller is the pK_a.

16.64 $pH = pK_a + \log \dfrac{[base]}{[acid]} = pK_a + \log \dfrac{[HCO_2^-]}{[HCOOH]}$

For HCOOH, $K_a = 1.8 \times 10^{-4}$; $pK_a = -\log K_a = -\log(1.8 \times 10^{-4}) = 3.74$

$pH = 3.74 + \log \dfrac{(0.50)}{(0.25)} = 4.04$

16.66 $pH = pK_a + \log \dfrac{[base]}{[acid]} = pK_a + \log \dfrac{[NH_3]}{[NH_4^+]}$

For NH_4^+, $K_a = 5.6 \times 10^{-10}$; $pK_a = -\log K_a = -\log(5.6 \times 10^{-10}) = 9.25$

$9.80 = 9.25 + \log \dfrac{[NH_3]}{[NH_4^+]}$; $0.550 = \log \dfrac{[NH_3]}{[NH_4^+]}$

$\dfrac{[NH_3]}{[NH_4^+]} = 10^{0.55} = 3.5$

The volume of the 1.0 M NH_3 solution should be 3.5 times the volume of the 1.0 M NH_4Cl solution so that the mixture will buffer at pH 9.80.

16.68 H_3PO_4, $K_{a1} = 7.5 \times 10^{-3}$; $pK_{a1} = -\log K_{a1} = 2.12$

$H_2PO_4^-$, $K_{a2} = 6.2 \times 10^{-8}$; $pK_{a2} = -\log K_{a2} = 7.21$

HPO_4^{2-}, $K_{a3} = 4.8 \times 10^{-13}$; $pK_{a3} = -\log K_{a3} = 12.32$

The buffer system of choice for pH 7.00 is (b) $H_2PO_4^- - HPO_4^{2-}$ because the pK_a for $H_2PO_4^-$ (7.21) is closest to 7.00.

pH Titration Curves

16.70   (a) $(0.060\ L)(0.150\ mol/L)(1000\ mmol/mol) = 9.00$ mmol HNO_3

(b) vol NaOH = $(9.00\ mmol\ HNO_3)\left(\dfrac{1\ mmol\ NaOH}{1\ mmol\ HNO_3}\right)\left(\dfrac{1\ mL\ NaOH}{0.450\ mmol\ NaOH}\right) = 20.0$ mL NaOH

(c) At the equivalence point the solution contains the neutral salt $NaNO_3$. The pH is 7.00.

(d)

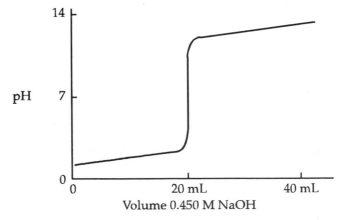

16.72 mmol $OH^- = (20.0 \text{ mL})(0.150 \text{ mmol/mL}) = 3.00$ mmol

mmol acid present = mmol OH^- added = 3.00 mmol

$$[\text{acid}] = \frac{3.00 \text{ mmol}}{60.0 \text{ mL}} = 0.0500 \text{ M}$$

16.74 $HBr(aq) + NaOH(aq) \rightarrow Na^+(aq) + Br^-(aq) + H_2O(l)$

(a) $[H_3O^+] = 0.120$ M; pH = $-\log[H_3O^+] = -\log (0.120) = 0.92$

(b) $(50.0 \text{ mL})(0.120 \text{ mmol/mL}) = 6.00$ mmol HBr

$(20.0 \text{ mL})(0.240 \text{ mmol/mL}) = 4.80$ mmol NaOH

6.00 mmol HBr – 4.80 mmol NaOH = 1.20 mmol HBr after neutralization

$$[H_3O^+] = \frac{1.20 \text{ mmol}}{(50.0 \text{ mL} + 20.0 \text{ mL})} = 0.0171 \text{ M}$$

pH = $-\log[H_3O^+] = -\log(0.0171) = 1.77$

(c) $(24.9 \text{ mL})(0.240 \text{ mmol/mL}) = 5.98$ mmol NaOH

6.00 mmol HBr – 5.98 mmol NaOH = 0.02 mmol HBr after neutralization

$$[H_3O^+] = \frac{0.02 \text{ mmol}}{(50.0 \text{ mL} + 24.9 \text{ mL})} = 3 \times 10^{-4} \text{ M}$$

pH = $-\log[H_3O^+] = -\log(3 \times 10^{-4}) = 3.5$

(d) The titration reaches the equivalence point when 25.0 mL of 0.240 M NaOH is added. At the equivalence point the solution contains the neutral salt NaBr. The pH is 7.00.

(e) $(25.1 \text{ mL})(0.240 \text{ mmol/mL}) = 6.024$ mmol NaOH

6.024 mmol NaOH – 6.00 mmol HBr = 0.024 mmol NaOH after neutralization

$$[OH^-] = \frac{0.024 \text{ mmol}}{(50.0 \text{ mL} + 25.1 \text{ mL})} = 3.2 \times 10^{-4} \text{ M}$$

$$[H_3O^+] = \frac{K_w}{[OH^-]} = \frac{1.0 \times 10^{-14}}{3.2 \times 10^{-4}} = 3.1 \times 10^{-11} \text{ M}$$

pH = $-\log[H_3O^+] = -\log(3.1 \times 10^{-11}) = 10.5$

(f) $(40.0 \text{ mL})(0.240 \text{ mmol/mL}) = 9.60$ mmol NaOH

9.60 mmol NaOH – 6.00 mmol HBr = 3.60 mmol NaOH after neutralization

$$[OH^-] = \frac{3.60 \text{ mmol}}{(50.0 \text{ mL} + 40.0 \text{ mL})} = 0.040 \text{ M}$$

$$[H_3O^+] = \frac{K_w}{[OH^-]} = \frac{1.0 \times 10^{-14}}{0.040} = 2.5 \times 10^{-13} \text{ M}$$

pH = $-\log[H_3O^+] = -\log(2.5 \times 10^{-13}) = 12.60$

16.76 mmol HF = (40.0 mL)(0.250 mmol/mL) = 10.0 mmol

mmol NaOH required = mmol HF = 10.0 mmol

$$\text{mL NaOH required} = (10.0 \text{ mmol})\left(\frac{1.00 \text{ mL}}{0.200 \text{ mmol}}\right) = 50.0 \text{ mL}$$

50.0 mL of 0.200 M NaOH is required to reach the equivalence point.

For HF, $K_a = 3.5 \times 10^{-4}$; $pK_a = -\log K_a = -\log(3.5 \times 10^{-4}) = 3.46$

(a) mmol HF = 10.0 mmol

mmol NaOH = (0.200 mmol/mL)(10.0 mL) = 2.00 mmol

Neutralization reaction:	$HF(aq)$	+	$OH^-(aq)$	$\rightarrow$	$F^-(aq)$	+	$H_2O(l)$
before reaction (mmol)	10.0		2.00		0		
change (mmol)	−2.00		−2.00		+2.00		
after reaction (mmol)	8.0		0		2.00		

$$[HF] = \frac{8.0 \text{ mmol}}{(40.0 \text{ mL} + 10.0 \text{ mL})} = 0.16 \text{ M}; \quad [F^-] = \frac{2.00 \text{ mmol}}{(40.0 \text{ mL} + 10.0 \text{ mL})} = 0.0400 \text{ M}$$

	$HF(aq)$	+	$H_2O(l)$	$\rightleftharpoons$	$H_3O^+(aq)$	+	$F^-(aq)$
initial (M)	0.16				~0		0.0400
change (M)	−x				+x		+x
equil (M)	0.16 − x				x		0.0400 + x

$$K_a = \frac{[H_3O^+][F^-]}{[HF]} = 3.5 \times 10^{-4} = \frac{x(0.0400 + x)}{0.16 - x} \approx \frac{x(0.0400)}{0.16}$$

Solve for x. $x = [H_3O^+] = 1.4 \times 10^{-3}$ M

$pH = -\log[H_3O^+] = -\log(1.4 \times 10^{-3}) = 2.85$

(b) Halfway to the equivalence point,

$pH = pK_a = -\log K_a = -\log(3.5 \times 10^{-4}) = 3.46$

(c) At the equivalence point only the salt NaF is in solution.

$$[F^-] = \frac{10.0 \text{ mmol}}{(40.0 \text{ mL} + 50.0 \text{ mL})} = 0.111 \text{ M}$$

	$F^-(aq)$	+	$H_2O(l)$	$\rightleftharpoons$	$HF(aq)$	+	$OH^-(aq)$
initial (M)	0.111				0		~0
change (M)	−x				+x		+x
equil (M)	0.111 − x				x		x

$$\text{For } F^-, \ K_b = \frac{K_w}{K_a \text{ for HF}} = \frac{1.0 \times 10^{-14}}{3.5 \times 10^{-4}} = 2.9 \times 10^{-11}$$

$$K_b = \frac{[HF][OH^-]}{[F^-]} = 2.9 \times 10^{-11} = \frac{x^2}{0.111 - x} \approx \frac{x^2}{0.111}$$

Solve for x. $x = [OH^-] = 1.8 \times 10^{-6}$ M

$$[H_3O^+] = \frac{K_w}{[OH^-]} = \frac{1.0 \times 10^{-14}}{1.8 \times 10^{-6}} = 5.6 \times 10^{-9} \text{ M}$$

$pH = -\log[H_3O^+] = -\log(5.6 \times 10^{-9}) = 8.25$

(d) mmol HF = 10.0 mmol

mol NaOH = (0.200 mmol/mL)(80.0 mL) = 16.0 mmol

Neutralization reaction: $\quad$ HF(aq) + OH$^-$(aq) $\rightarrow$ F$^-$(aq) + H$_2$O(l)

before reaction (mmol)	10.0	16.0	0
change (mmol)	−10.0	−10.0	+10.0
after reaction (mmol)	0	6.0	10.0

After the equivalence point, the pH of the solution is determined by the [OH$^-$].

$$[\text{OH}^-] = \frac{6.0 \text{ mmol}}{(40.0 \text{ mL} + 80.0 \text{ mL})} = 5.0 \times 10^{-2} \text{ M}$$

$$[\text{H}_3\text{O}^+] = \frac{\text{K}_\text{w}}{[\text{OH}^-]} = \frac{1.0 \times 10^{-14}}{5.0 \times 10^{-2}} = 2.0 \times 10^{-13} \text{ M}$$

$$\text{pH} = -\log[\text{H}_3\text{O}^+] = -\log(2.0 \times 10^{-13}) = 12.70$$

16.78 $\quad$ For H$_2$A$^+$, K$_{a1}$ = 4.6 x 10^{-3} and K$_{a2}$ = 2.0 x 10^{-10}

(a) (10.0 mL)(0.100 mmol/mL) = 1.00 mmol NaOH added = 1.00 mmol HA produced.

(50.0 mL)(0.100 mmol/mL) = 5.00 mmol H$_2$A$^+$

5.00 mmol H$_2$A$^+$ − 1.00 mmol NaOH = 4.00 mmol H$_2$A$^+$ after neutralization

$$[\text{H}_2\text{A}^+] = \frac{4.00 \text{ mmol}}{(50.0 \text{ mL} + 10.0 \text{ mL})} = 6.67 \times 10^{-2} \text{ M}$$

$$[\text{HA}] = \frac{1.00 \text{ mmol}}{(50.0 \text{ mL} + 10.0 \text{ mL})} = 1.67 \times 10^{-2} \text{ M}$$

$$\text{pH} = \text{pK}_{a1} + \log\frac{[\text{HA}]}{[\text{H}_2\text{A}^+]} = -\log(4.6 \times 10^{-3}) + \log\left(\frac{1.67 \times 10^{-2}}{6.67 \times 10^{-2}}\right) = 1.74$$

(b) Halfway to the first equivalence point, pH = pK$_{a1}$ = 2.34

(c) At the first equivalence point, pH = $\dfrac{\text{pK}_{a1} + \text{pK}_{a2}}{2}$ = 6.02

(d) Halfway between the first and second equivalence points, pH = pK$_{a2}$ = 9.70

(e) At the second equivalence point only the basic salt, NaA, is in solution.

$$\text{K}_\text{h} = \frac{\text{K}_\text{w}}{\text{K}_\text{a for HA}} = \frac{\text{K}_\text{w}}{\text{K}_{a2}} = \frac{1.0 \times 10^{-14}}{2.0 \times 10^{-10}} = 5.0 \times 10^{-5}$$

mmol A$^-$ = (50.0 mL)(0.100 mmol/L) = 5.00 mmol

$$[\text{A}^-] = \frac{5.0 \text{ mmol}}{(50.0 \text{ mL} + 100.0 \text{ mL})} = 3.3 \times 10^{-2} \text{ M}$$

	A$^-$(aq) + H$_2$O(l) $\rightleftharpoons$	HA(aq) +	OH$^-$(aq)
initial (M)	0.033	0	~0
change (M)	−x	+x	+x
equil (M)	0.033 − x	x	x

$$\text{K}_\text{b} = \frac{[\text{HA}][\text{OH}^-]}{[\text{A}^-]} = 5.0 \times 10^{-5} = \frac{(x)(x)}{0.033 - x} \approx \frac{x^2}{0.033}$$

Solve for x.

$$x = [\text{OH}^-] = \sqrt{(5.0 \times 10^{-5})(0.033)} = 1.3 \times 10^{-3} \text{ M}$$

$$[H_3O^+] = \frac{K_w}{[OH^-]} = \frac{1.0 \times 10^{-14}}{1.3 \times 10^{-3}} = 7.7 \times 10^{-12} \text{ M}$$

$$pH = -\log[H_3O^+] = -\log(7.7 \times 10^{-12}) = 11.11$$

16.80 When equal volumes of acid and base react, all concentrations are cut in half.
(a) At the equivalence point only the salt $NaNO_2$ is in solution.
$[NO_2^-] = 0.050$ M

$$\text{For } NO_2^-, K_b = \frac{K_w}{K_a \text{ for } HNO_2} = \frac{1.0 \times 10^{-14}}{4.5 \times 10^{-4}} = 2.2 \times 10^{-11}$$

	$NO_2^-(aq)$	$+ H_2O(l)$	$\rightleftharpoons$	$HNO_2(aq)$	$+ OH^-(aq)$
Initial (M)	0.050			0	~0
change (M)	−x			+x	+x
equil (M)	0.050 − x			x	x

$$K_b = \frac{[HNO_2][OH^-]}{[NO_2^-]} = 2.2 \times 10^{-11} = \frac{(x)(x)}{0.050 - x} \approx \frac{x^2}{0.050}$$

Solve for x. $x = [OH^-] = 1.1 \times 10^{-6}$ M

$$[H_3O^+] = \frac{K_w}{[OH^-]} = \frac{1.0 \times 10^{-14}}{1.1 \times 10^{-6}} = 9.1 \times 10^{-9} \text{ M}$$

$$pH = -\log[H_3O^+] = -\log(9.1 \times 10^{-9}) = 8.04$$

Cresol red, phenol red, or m-nitrophenol would be suitable indicators. (see Figure 15.4)
(b) The pH is 7.00 at the equivalence point for the titration of a strong acid (HI) with a strong base (NaOH).
Alizarin, bromthymol blue, or phenol red would be suitable indicators. (Any indicator that changes color in the pH range 4 − 10 is satisfactory for a strong acid − strong base titration.)
(c) At the equivalence point only the salt CH_3NH_3Cl is in solution.
$[CH_3NH_3^+] = 0.050$ M

$$\text{For } CH_3NH_3^+, K_a = \frac{K_w}{K_b \text{ for } CH_3NH_2} = \frac{1.0 \times 10^{-14}}{3.7 \times 10^{-4}} = 2.7 \times 10^{-11}$$

	$CH_3NH_3^+(aq)$	$+ H_2O(l)$	$\rightleftharpoons$	$H_3O^+(aq)$	$+ CH_3NH_2(aq)$
initial (M)	0.050			~0	0
change (M)	−x			+x	+x
equil (M)	0.050 − x			x	x

$$K_a = \frac{[H_3O^+][CH_3NH_2]}{[CH_3NH_3^+]} = 2.7 \times 10^{-11} = \frac{(x)(x)}{0.050 - x} \approx \frac{x^2}{0.050}$$

Solve for x. $x = [H_3O^+] = 1.2 \times 10^{-6}$ M
$$pH = -\log[H_3O^+] = -\log(1.2 \times 10^{-6}) = 5.92$$
Eriochrome black T and bromcresol purple would be suitable indicators.

Chapter 16 – Applications of Aqueous Equilibria

Solubility Equilibria

16.82 (a) $Ag_2CO_3(s) \rightleftharpoons 2\ Ag^+(aq) + CO_3^{2-}(aq)$ $K_{sp} = [Ag^+]^2[CO_3^{2-}]$

      (b) $PbCrO_4(s) \rightleftharpoons Pb^{2+}(aq) + CrO_4^{2-}(aq)$ $K_{sp} = [Pb^{2+}][CrO_4^{2-}]$

      (c) $Al(OH)_3(s) \rightleftharpoons Al^{3+}(aq) + 3\ OH^-(aq)$ $K_{sp} = [Al^{3+}][OH^-]^3$

      (d) $Hg_2Cl_2(s) \rightleftharpoons Hg_2^{2+}(aq) + 2\ Cl^-(aq)$ $K_{sp} = [Hg_2^{2+}][Cl^-]^2$

16.84 (a) $K_{sp} = [Pb^{2+}][I^-]^2 = (5.0 \times 10^{-3})(1.3 \times 10^{-3})^2 = 8.4 \times 10^{-9}$

      (b) $[I^-] = \sqrt{\dfrac{K_{sp}}{[Pb^{2+}]}} = \sqrt{\dfrac{(8.4 \times 10^{-9})}{(2.5 \times 10^{-4})}} = 5.8 \times 10^{-3}\ M$

      (c) $[Pb^{2+}] = \dfrac{K_{sp}}{[I^-]^2} = \dfrac{(8.4 \times 10^{-9})}{(2.5 \times 10^{-4})^2} = 0.13\ M$

16.86 $Ag_2CO_3(s) \rightleftharpoons 2\ Ag^+(aq) + CO_3^{2-}(aq)$

      equil (M) $2x$ x

      $[Ag^+] = 2x = 2.56 \times 10^{-4}\ M;$ $[CO_3^{2-}] = x = (2.56 \times 10^{-4}\ M)/2 = 1.28 \times 10^{-4}\ M$

      $K_{sp} = [Ag^+]^2[CO_3^{2-}] = (2.56 \times 10^{-4})^2(1.28 \times 10^{-4}) = 8.39 \times 10^{-12}$

16.88 (a) $CuCO_3(s) \rightleftharpoons Cu^{2+}(aq) + CO_3^{2-}(aq)$

         equil (M) x x

      $K_{sp} = [Cu^{2+}][CO_3^{2-}] = 2.5 \times 10^{-10} = (x)(x)$

      molar solubility $= x = \sqrt{2.5 \times 10^{-10}} = 1.6 \times 10^{-5}\ M$

      (b) $Ag_2SO_4(s) \rightleftharpoons 2\ Ag^+(aq) + SO_4^{2-}(aq)$

         equil (M) $2x$ x

      $K_{sp} = [Ag^+]^2[SO_4^{2-}] = 1.2 \times 10^{-5} = (2x)^2 x = 4x^3$

      molar solubility $= x = \sqrt[3]{\dfrac{1.2 \times 10^{-5}}{4}} = 1.4 \times 10^{-2}\ M$

      (c) $Cr(OH)_3(s) \rightleftharpoons Cr^{3+}(aq) + 3\ OH^-(aq)$

         equil (M) x $3x$

      $K_{sp} = [Cr^{3+}][OH^-]^3 = 6.7 \times 10^{-31} = (x)(3x)^3 = 27x^4$

      molar solubility $= x = \sqrt[4]{\dfrac{6.7 \times 10^{-31}}{27}} = 1.3 \times 10^{-8}\ M$

Factors That Affect Solubility

16.90 $Ag_2CO_3(s) \rightleftharpoons 2\ Ag^+(aq) + CO_3^{2-}(aq)$

      (a) $AgNO_3$, source of Ag^+; equilibrium shifts left

      (b) HNO_3, source of H_3O^+, removes CO_3^{2-}; equilibrium shifts right

      (c) Na_2CO_3, source of CO_3^{2-}; equilibrium shifts left

      (d) NH_3, forms $Ag(NH_3)_2^+$; removes Ag^+; equilibrium shifts right

16.92 (a)
$$AgBr(s) \rightleftharpoons Ag^+(aq) + Br^-(aq)$$

equil (M)   x   x

$K_{sp} = [Ag^+][Br^-] = 5.4 \times 10^{-13} = (x)(x)$

molar solubility $= x = \sqrt{5.4 \times 10^{-13}} = 7.3 \times 10^{-7}$ M

(b) $[Br^-] = 0.050$ M

$$AgBr(s) \rightleftharpoons Ag^+(aq) + Br^-(aq)$$

initial (M)   0   0.050

equil (M)   x   0.050 + x

$K_{sp} = [Ag^+][Br^-] = 5.4 \times 10^{-13} = x(0.050 + x) \approx x(0.050)$

molar solubility $= x = \dfrac{5.4 \times 10^{-13}}{0.050} = 1.1 \times 10^{-11}$ M

16.94 (a) pH = 12.00; $[H_3O^+] = 10^{-pH} = 10^{-12.00} = 1.0 \times 10^{-12}$ M

$[OH^-] = \dfrac{K_w}{[H_3O^+]} = \dfrac{1.0 \times 10^{-14}}{1.0 \times 10^{-12}} = 0.010$ M

$$Mg(OH)_2(s) \rightleftharpoons Mg^{2+}(aq) + 2\,OH^-(aq)$$

equil (M)   x   0.010 (fixed by buffer)

$K_{sp} = [Mg^{2+}][OH^-]^2 = 5.6 \times 10^{-12} = x(0.010)^2$

molar solubility $= x = \dfrac{5.6 \times 10^{-12}}{(0.010)^2} = 5.6 \times 10^{-8}$ M

(b) pH = 9.00; $[H_3O^+] = 10^{-pH} = 10^{-9.00} = 1.0 \times 10^{-9}$ M

$[OH^-] = \dfrac{K_w}{[H_3O^+]} = \dfrac{1.0 \times 10^{-14}}{1.0 \times 10^{-9}} = 1.0 \times 10^{-5}$ M

$$Mg(OH)_2(s) \rightleftharpoons Mg^{2+}(aq) + 2\,OH^-(aq)$$

equil (M)   x 1.0×10^{-5} (fixed by buffer)

$K_{sp} = [Mg^{2+}][OH^-]^2 = 5.6 \times 10^{-12} = x(1.0 \times 10^{-5})^2$

molar solubility $= x = \dfrac{5.6 \times 10^{-12}}{(1.0 \times 10^{-5})^2} = 0.056$ M

16.96 (b), (c), and (d) are more soluble in acidic solution.

(a) $AgBr(s) \rightleftharpoons Ag^+(aq) + Br^-(aq)$

(b) $CaCO_3(s) + H_3O^+(aq) \rightleftharpoons Ca^{2+}(aq) + HCO_3^-(aq) + H_2O(l)$

(c) $Ni(OH)_2(s) + 2\,H_3O^+(aq) \rightleftharpoons Ni^{2+}(aq) + 4\,H_2O(l)$

(d) $Ca_3(PO_4)_2(s) + 2\,H_3O^+(aq) \rightleftharpoons 3\,Ca^{2+}(aq) + 2\,HPO_4^{2-}(aq) + 2\,H_2O(l)$

16.98 On mixing equal volumes of two solutions, the concentrations of both solutions are cut in half.

$$Ag^+(aq) \quad + \quad 2\ CN^-(aq) \quad \rightleftharpoons \quad Ag(CN)_2^-(aq)$$

before reaction (M)	0.0010	0.10	0
assume 100% reaction	−0.0010	−2(0.0010)	0.0010
after reaction (M)	0	0.098	0.0010
assume small back rxn	+x	+2x	−x
equil (M)	x	0.098 + 2x	0.0010 − x

$$K_f = 1 \times 10^{21} = \frac{[Ag(CN)_2^-]}{[Ag^+][CN^-]^2} = \frac{(0.0010 - x)}{x(0.098 + 2x)^2} \approx \frac{0.0010}{x(0.098)^2}$$

Solve for x. $x = [Ag^+] = 1 \times 10^{-22}$ M

16.100 (a)
$$AgI(s) \rightleftharpoons Ag^+(aq) + I^-(aq) \qquad\qquad K_{sp} = 8.5 \times 10^{-17}$$
$$\underline{Ag^+(aq) + 2\ CN^-(aq) \rightarrow Ag(CN)_2^-(aq)} \qquad K_f = 1 \times 10^{21}$$

dissolution $\quad AgI(s) + 2\ CN^-(aq) \rightleftharpoons Ag(CN)_2^-(aq) + I^-(aq)$
reaction

$K = (K_{sp})(K_f) = (8.5 \times 10^{-17})(1 \times 10^{21}) = 8 \times 10^4$

(b)
$$Al(OH)_3(s) \rightleftharpoons Al^{3+}(aq) + 3\ OH^-(aq) \qquad K_{sp} = 1.9 \times 10^{-33}$$
$$\underline{Al^{3+}(aq) + 4\ OH^-(aq) \rightarrow Al(OH)_4^-(aq)} \qquad K_f = 2.1 \times 10^{34}$$

dissolution $\quad Al(OH)_3(s) + OH^-(aq) \rightleftharpoons Al(OH)_4^-(aq)$
reaction

$K = (K_{sp})(K_f) = (1.9 \times 10^{-33})(2.1 \times 10^{34}) = 40$

(c)
$$Zn(OH)_2(s) \rightleftharpoons Zn^{2+}(aq) + 2\ OH^-(aq) \qquad K_{sp} = 4.1 \times 10^{-17}$$
$$\underline{Zn^{2+}(aq) + 4\ NH_3(aq) \rightarrow Zn(NH_3)_4^{2+}(aq)} \qquad K_f = 2.9 \times 10^9$$

dissolution $\quad Zn(OH)_2(s) + 4\ NH_3(aq) \rightleftharpoons Zn(NH_3)_4^{2+} + 2\ OH^-(aq)$
reaction

$K = (K_{sp})(K_f) = (4.1 \times 10^{-17})(2.9 \times 10^9) = 1.2 \times 10^{-7}$

16.102 (a)
$$AgI(s) \rightleftharpoons Ag^+(aq) + I^-(aq)$$

equil (M)	x	x

$K_{sp} = [Ag^+][I^-] = 8.5 \times 10^{-17} = (x)(x)$

molar solubility $= x = \sqrt{8.5 \times 10^{-17}} = 9.2 \times 10^{-9}$ M

(b)
$$AgI(s) + 2\ CN^-(aq) \rightleftharpoons Ag(CN)_2^-(aq) + I^-(aq)$$

initial (M)	0.10	0	0
change (M)	−2x	+x	+x
equil (M)	0.10 − 2x	x	x

$K = (K_{sp})(K_f) = (8.5 \times 10^{-17})(1 \times 10^{21}) = 8.5 \times 10^4$

$$K = 8.5 \times 10^4 = \frac{[Ag(CN)_2^-][I^-]}{[CN^-]^2} = \frac{x^2}{(0.10 - 2x)^2}$$

Take the square root of both sides and solve for x.

molar solubility $= x = 0.05$ M

Precipitation; Qualitative Analysis

16.104 For $BaSO_4$, $K_{sp} = 1.1 \times 10^{-10}$

Total volume = 300 mL + 100 mL = 400 mL

$$[Ba^{2+}] = \frac{(4.0 \times 10^{-3}\ M)(100\ mL)}{(400\ mL)} = 1.0 \times 10^{-3}\ M$$

$$[SO_4^{2-}] = \frac{(6.0 \times 10^{-4}\ M)(300\ mL)}{(400\ mL)} = 4.5 \times 10^{-4}\ M$$

IP = $[Ba^{2+}]_t[SO_4^{2-}]_t = (1.0 \times 10^{-3})(4.5 \times 10^{-4}) = 4.5 \times 10^{-7}$

IP > K_{sp}; $BaSO_4(s)$ will precipitate.

16.106 $BaSO_4$, $K_{sp} = 1.1 \times 10^{-10}$; $Fe(OH)_3$, $K_{sp} = 2.6 \times 10^{-39}$

Total volume = 80 mL + 20 mL = 100 mL

$$[Ba^{2+}] = \frac{(1.0 \times 10^{-5}\ M)(80\ mL)}{(100\ mL)} = 8.0 \times 10^{-6}\ M$$

$[OH^-] = 2[Ba^{2+}] = 2(8.0 \times 10^{-6}) = 1.6 \times 10^{-5}\ M$

$$[Fe^{3+}] = \frac{2(1.0 \times 10^{-5}\ M)(20\ mL)}{(100\ mL)} = 4.0 \times 10^{-6}\ M$$

$$[SO_4^{2-}] = \frac{3(1.0 \times 10^{-5}\ M)(20\ mL)}{(100\ mL)} = 6.0 \times 10^{-6}\ M$$

For $BaSO_4$, IP = $[Ba^{2+}]_t[SO_4^{2-}]_t = (8.0 \times 10^{-6})(6.0 \times 10^{-6}) = 4.8 \times 10^{-11}$

IP < K_{sp}; $BaSO_4$ will not precipitate.

For $Fe(OH)_3$, IP = $[Fe^{3+}]_t[OH^-]_t^3 = (4.0 \times 10^{-6})(1.6 \times 10^{-5})^3 = 1.6 \times 10^{-20}$

IP > K_{sp}; $Fe(OH)_3(s)$ will precipitate.

16.108 pH = 10.80; $[H_3O^+] = 10^{-pH} = 10^{-10.80} = 1.6 \times 10^{-11}\ M$

$$[OH^-] = \frac{K_w}{[H_3O^+]} = \frac{1.0 \times 10^{-14}}{1.6 \times 10^{-11}} = 6.2 \times 10^{-4}\ M$$

For $Mg(OH)_2$, $K_{sp} = 5.6 \times 10^{-12}$

IP = $[Mg^{2+}]_t[OH^-]_t^2 = (2.5 \times 10^{-4})(6.2 \times 10^{-4})^2 = 9.6 \times 10^{-11}$

IP > K_{sp}; $Mg(OH)_2(s)$ will precipitate

16.110 $K_{spa} = \dfrac{[M^{2+}][H_2S]}{[H_3O^+]^2}$; FeS, $K_{spa} = 6 \times 10^2$; SnS, $K_{spa} = 1 \times 10^{-5}$

Fe^{2+} and Sn^{2+} can be separated by bubbling H_2S through an acidic solution containing the two cations because their K_{spa} values are so different.

For FeS and SnS, $Q_c = \dfrac{(0.01)(0.10)}{(0.3)^2} = 1.1 \times 10^{-2}$

For FeS, $Q_c < K_{spa}$, and no FeS will precipitate.

For SnS, $Q_c > K_{spa}$, and SnS will precipitate.

16.112 (a) add Cl^- to precipitate $AgCl$
     (b) add CO_3^{2-} to precipitate $CaCO_3$
     (c) add H_2S to precipitate MnS
     (d) add NH_3 and NH_4Cl to precipitate $Cr(OH)_3$
     (Need buffer to control $[OH^-]$; excess OH^- produces the soluble $Cr(OH)_4^-$.)

General Problems

16.114 Prepare aqueous solutions of the three salts. Add a solution of $(NH_4)_2HPO_4$. If a white precipitate forms, the solution contains Mg^{2+}. Perform flame test on the other two solutions. A yellow flame test indicates Na^+. A violet flame test indicates K^+.

16.116 (a), solution contains H_2CO_3 and HCO_3^-
     (b), solution contains HCO_3^- and CO_3^{2-}
     (d), solution contains HCO_3^- and CO_3^{2-}

16.118 For NH_4^+, $K_a = \dfrac{K_w}{K_b \text{ for } NH_3} = \dfrac{1.0 \times 10^{-14}}{1.8 \times 10^{-5}} = 5.6 \times 10^{-10}$

$pK_a = -\log K_a = -\log(5.6 \times 10^{-10}) = 9.25$

$pH = pK_a + \log \dfrac{[NH_3]}{[NH_4^+]}$; $9.40 = 9.25 + \log \dfrac{[NH_3]}{[NH_4^+]}$

$\log \dfrac{[NH_3]}{[NH_4^+]} = 9.40 - 9.25 = 0.15$; $\dfrac{[NH_3]}{[NH_4^+]} = 10^{0.15} = 1.41$

Because the volume is the same for both NH_3 and NH_4^+, $\dfrac{\text{mol } NH_3}{\text{mol } NH_4^+} = 1.41$.

$\text{mol } NH_3 = (0.20 \text{ mol/L})(0.250 \text{ L}) = 0.050 \text{ mol } NH_3$

$\text{mol } NH_4^+ = \dfrac{\text{mol } NH_3}{1.41} = \dfrac{0.050}{1.41} = 0.035 \text{ mol } NH_4^+$

$\text{vol } NH_4^+ = (0.035 \text{ mol})\left(\dfrac{1 \text{ L}}{3.0 \text{ mol}}\right) = 0.012 \text{ L} = 12 \text{ mL}$

12 mL of 3.0 M NH_4Cl must be added to 250 mL of 0.20 M NH_3 to obtain a buffer solution having a pH = 9.40.

16.120 $pH = 10.35$; $[H_3O^+] = 10^{-pH} = 10^{-10.35} = 4.5 \times 10^{-11}$ M

$[OH^-] = \dfrac{K_w}{[H_3O^+]} = \dfrac{1.0 \times 10^{-14}}{4.5 \times 10^{-11}} = 2.2 \times 10^{-4}$ M

$[Mg^{2+}] = \dfrac{[OH^-]}{2} = \dfrac{2.2 \times 10^{-4}}{2} = 1.1 \times 10^{-4}$ M

$K_{sp} = [Mg^{2+}][OH^-]^2 = (1.1 \times 10^{-4})(2.2 \times 10^{-4})^2 = 5.3 \times 10^{-12}$

16.122 NaOH, 40.0 amu; $20 \text{ g} \times \dfrac{1 \text{ mol}}{40.0 \text{ g}} = 0.50 \text{ mol NaOH}$

$(0.500 \text{ L})(1.5 \text{ mol/L}) = 0.75 \text{ mol NH}_4\text{Cl}$

	$\text{NH}_4^+(aq)$	$+$	$\text{OH}^-(aq)$	$\rightleftharpoons$	$\text{NH}_3(aq)$	$+$	$\text{H}_2\text{O}(l)$
before reaction (mol)	0.75		0.50		0		
change (mol)	−0.50		−0.50		+0.50		
after reaction (mol)	0.25		0		0.50		

This reaction produces a buffer solution.

$[\text{NH}_4^+] = 0.25 \text{ mol}/0.500 \text{ L} = 0.50 \text{ M};\quad [\text{NH}_3] = 0.50 \text{ mol}/0.500 \text{ L} = 1.0 \text{ M}$

$$\text{pH} = \text{p}K_a + \log\dfrac{[\text{base}]}{[\text{acid}]} = \text{p}K_a + \log\dfrac{[\text{NH}_3]}{[\text{NH}_4^+]}$$

For NH_4^+, $K_a = \dfrac{K_w}{K_b \text{ for NH}_3} = \dfrac{1.0 \times 10^{-14}}{1.8 \times 10^{-5}} = 5.6 \times 10^{-10}$; $\text{p}K_a = -\log K_a = 9.25$

$$\text{pH} = 9.25 + \log\left(\dfrac{1.0}{0.5}\right) = 9.55$$

16.124 For NH_4^+, $K_a = \dfrac{K_w}{K_b \text{ for NH}_3} = \dfrac{1.0 \times 10^{-14}}{1.8 \times 10^{-5}} = 5.6 \times 10^{-10}$; $\text{p}K_a = -\log K_a = 9.25$

$$\text{pH} = \text{p}K_a + \log\dfrac{[\text{NH}_3]}{[\text{NH}_4^+]} = 9.25 + \log\dfrac{(0.50)}{(0.30)} = 9.47$$

$[\text{H}_3\text{O}^+] = 10^{-\text{pH}} = 10^{-9.47} = 3.4 \times 10^{-10} \text{ M}$

For MnS, $K_{spa} = \dfrac{[\text{Mn}^{2+}][\text{H}_2\text{S}]}{[\text{H}_3\text{O}^+]^2} = 3 \times 10^{10}$

molar solubility $= [\text{Mn}^{2+}] = \dfrac{K_{spa}[\text{H}_3\text{O}^+]^2}{[\text{H}_2\text{S}]} = \dfrac{(3 \times 10^{10})(3.4 \times 10^{-10})^2}{(0.10)} = 3.5 \times 10^{-8} \text{ M}$

MnS, 87.00 amu; solubility $= (3.5 \times 10^{-8} \text{ mol/L})(87.00 \text{ g/mol}) = 3 \times 10^{-6} \text{ g/L}$

16.126 (a) $\text{HA}^-(aq) + \text{H}_2\text{O}(l) \rightleftharpoons \text{H}_3\text{O}^+(aq) + \text{A}^{2-}(aq) \qquad K_{a2} = 10^{-10}$

$\text{HA}^-(aq) + \text{H}_2\text{O}(l) \rightleftharpoons \text{H}_2\text{A}(aq) + \text{OH}^-(aq) \qquad K_b = \dfrac{K_w}{K_{a1}} = 10^{-10}$

$2 \text{ HA}^-(aq) \rightleftharpoons \text{H}_2\text{A}(aq) + \text{A}^{2-}(aq) \qquad K = \dfrac{K_{a2}}{K_{a1}} = 10^{-6}$

$2 \text{ H}_2\text{O}(l) \rightleftharpoons \text{H}_3\text{O}^+(aq) + \text{OH}^-(aq) \qquad K_w = 1.0 \times 10^{-14}$

The principal reaction of the four is the one with the largest K, and that is the third reaction.

(b) $K_{a1} = \dfrac{[\text{H}_3\text{O}^+][\text{HA}^-]}{[\text{H}_2\text{A}]}$ and $K_{a2} = \dfrac{[\text{H}_3\text{O}^+][\text{A}^{2-}]}{[\text{HA}^-]}$

$$[H_3O^+] = \frac{K_{a1}[H_2A]}{[HA^-]} \quad \text{and} \quad [H_3O^+] = \frac{K_{a2}[HA^-]}{[A^{2-}]}$$

$$\frac{K_{a1}[H_2A]}{[HA^-]} \times \frac{K_{a2}[HA^-]}{[A^{2-}]} = [H_3O^+]^2; \qquad \frac{K_{a1}K_{a2}[H_2A]}{[A^{2-}]} = [H_3O^+]^2$$

Because the principal reaction is $2\,HA^-(aq) \rightleftharpoons H_2A(aq) + A^{2-}(aq)$, $[H_2A] = [A^{2-}]$.

$$K_{a1}K_{a2} = [H_3O^+]^2$$
$$\log K_{a1} + \log K_{a2} = 2\log[H_3O^+]$$
$$\frac{\log K_{a1} + \log K_{a2}}{2} = \log[H_3O^+]; \qquad \frac{-\log K_{a1} + (-\log K_{a2})}{2} = -\log[H_3O^+]$$

$$\frac{pK_{a1} + pK_{a2}}{2} = pH$$

16.128 (a) The first equivalence point is reached when all the H_3O^+ from the HCl and the H_3O^+ form the first ionization of H_3PO_4 is consumed.

At the first equivalence point $pH = \dfrac{pK_{a1} + pK_{a2}}{2} = 4.66$

$[H_3O^+] = 10^{-pH} = 10^{(-4.66)} = 2.2 \times 10^{-5}$ M

(88.0 mL)(0.100 mmol/mL) = 8.80 mmol NaOH are used to get to the first equivalence point

(b) mmol ($HCl + H_3PO_4$) = mmol NaOH = 8.8 mmol

mmol H_3PO_4 = (126.4 mL − 88.0 mL)(0.100 mmol/mL) = 3.84 mmol

mmol HCl = (8.8 − 3.84) = 4.96 mmol

$$[HCl] = \frac{4.96 \text{ mmol}}{40.0 \text{ mL}} = 0.124 \text{ M}; \qquad [H_3PO_4] = \frac{3.84 \text{ mmol}}{40.0 \text{ mL}} = 0.0960 \text{ M}$$

(c) 100% of the HCl is neutralized at the first equivalence point.

(d)

	$H_3PO_4(aq)$ + $H_2O(l)$	$\rightleftharpoons$	$H_3O^+(aq)$ +	$H_2PO_4^-(aq)$
initial (M)	0.0960		0.124	0
change (M)	−x		+x	+x
equil (M)	0.0960 − x		x	x

$$K_{a1} = \frac{[H_3O^+][H_2PO_4^-]}{[H_3PO_4]} = 7.5 \times 10^{-3} = \frac{(0.124 + x)(x)}{0.0960 - x}$$

$x^2 + 0.132x - (7.2 \times 10^{-4}) = 0$

Use the quadratic formula to solve for x.

$$x = \frac{-(0.132) \pm \sqrt{(0.132)^2 - 4(1)(-7.2 \times 10^{-4})}}{2(1)} = \frac{-0.132 \pm 0.142}{2}$$

x = −0.137 and 0.005

Of the two solutions for x, only the positive value of x has physical meaning because the other root would give a negative $[H_3O^+]$.

$[H_3O^+] = 0.124 + x = 0.124 + 0.005 = 0.129$ M

$pH = -\log[H_3O^+] = -\log(0.129) = 0.89$

(e)

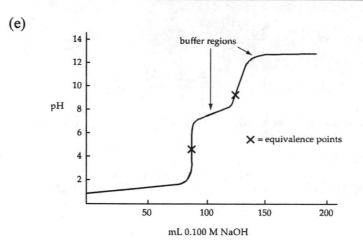

(f) Bromcresol green or methyl orange are suitable indicators for the first equivalence point. Thymolphthalein is a suitable indicator for the second equivalence point.

17 Thermodynamics: Entropy, Free Energy, and Equilibrium

17.1   (a) spontaneous;   (b), (c), and (d) nonspontaneous

17.2   (a) $H_2O(g) \rightarrow H_2O(l)$
A liquid is more ordered than a gas. Therefore, ΔS is negative.
(b) $I_2(g) \rightarrow 2\,I(g)$
ΔS is positive because the reaction increases the number of gaseous particles from 1 mol to 2 mol.
(c) $CaCO_3(s) \rightarrow CaO(s) + CO_2(g)$
ΔS is positive because the reaction increases the number of gaseous molecules.
(d) $Ag^+(aq) + Br^-(aq) \rightarrow AgBr(s)$
A solid is more ordered than +1 and −1 charged ions in an aqueous solution. Therefore, ΔS is negative.

17.3   (a) shuffled deck (more disorder)   (b) disordered crystal
(c) 1 mole N_2 at STP (larger volume, more disorder)
(d) 1 mole N_2 at 273 K and 0.25 atm (larger volume, more disorder)

17.4 $CaCO_3(s) \rightarrow CaO(s) + CO_2(g)$
$\Delta S^\circ = [S^\circ(CaO) + S^\circ(CO_2)] - S^\circ(CaCO_3)$
$\Delta S^\circ = [(1\ mol)(39.7\ J/(K \cdot mol)) + (1\ mol)(213.6\ J/(K \cdot mol))]$
$$- (1\ mol)(92.9\ J/(K \cdot mol)) = +160.4\ J/K$$

17.5   From Problem 17.4, $\Delta S_{sys} = \Delta S^\circ = 160.4\ J/K$
$CaCO_3(s) \rightarrow CaO(s) + CO_2(g)$
$\Delta H^\circ = [\Delta H^\circ_f(CaO) + \Delta H^\circ_f(CO_2)] - \Delta H^\circ_f(CaCO_3)$
$\Delta H^\circ = [(1\ mol)(-635.1\ kJ/mol) + (1\ mol)(-393.5\ kJ/mol)]$
$$- (1\ mol)(-1206.9\ kJ/mol) = +178.3\ kJ$$

$$\Delta S_{surr} = \frac{-\Delta H^\circ}{T} = \frac{-178,300\ J}{298\ K} = -598\ J/K$$

$\Delta S_{total} = \Delta S_{sys} + \Delta S_{surr} = 160.4\ J/K + (-598\ J/K) = -438\ J/K$
Because ΔS_{total} is negative, the reaction is not spontaneous under standard-state conditions at 25°C.

17.6   (a) $\Delta G = \Delta H - T\Delta S = 57.1\ kJ - (298\ K)(0.1758\ kJ/K) = +4.7\ kJ$
Because $\Delta G > 0$, the reaction is nonspontaneous at 25°C (298 K)
(b) Set $\Delta G = 0$ and solve for T.
$$0 = \Delta H - T\Delta S; \qquad T = \frac{\Delta H}{\Delta S} = \frac{57.1\ kJ}{0.1758\ kJ/K} = 325\ K = 52°C$$

17.7   (a) $\Delta G = \Delta H - T\Delta S = 58.5 \text{ kJ/mol} - (598 \text{ K})[0.0929 \text{ kJ/(K} \cdot \text{mol)}] = +2.9 \text{ kJ/mol}$
Because $\Delta G > 0$, Hg does not boil at 325°C and 1 atm.
(b) The boiling point (phase change) is associated with an equilibrium. Set $\Delta G = 0$ and solve for T, the boiling point.

$$0 = \Delta H_{vap} - T\Delta S_{vap}, \qquad T_{bp} = \frac{\Delta H_{vap}}{\Delta S_{vap}} = \frac{58.5 \text{ kJ/mol}}{0.0929 \text{ kJ/(K} \cdot \text{mol)}} = 630 \text{ K} = 357°C$$

17.8   From Problems 17.4 and 17.5: $\Delta H° = 178.3 \text{ kJ}$ and $\Delta S° = 160.4 \text{ J/K} = 0.1604 \text{ kJ/K}$
(a) $\Delta G° = \Delta H° - T\Delta S° = 178.3 \text{ kJ} - (298 \text{ K})(0.1604 \text{ kJ/K}) = +130.5 \text{ kJ}$
(b) Because $\Delta G > 0$, the reaction is nonspontaneous at 25°C (298 K).
(c) Set $\Delta G = 0$ and solve for T, the temperature above which the reaction becomes spontaneous.

$$0 = \Delta H - T\Delta S; \qquad T = \frac{\Delta H}{\Delta S} = \frac{178.3 \text{ kJ}}{0.1604 \text{ kJ/K}} = 1112 \text{ K} = 839°C$$

17.9   (a) $CaC_2(s) + 2 H_2O(l) \rightarrow C_2H_2(g) + Ca(OH)_2(s)$
$\Delta G° = [\Delta G°_f(C_2H_2) + \Delta G°_f(Ca(OH)_2)] - [\Delta G°_f(CaC_2) + 2 \Delta G°_f(H_2O)]$
$\Delta G° = [(1 \text{ mol})(209.2 \text{ kJ/mol}) + (1 \text{ mol})(-898.6 \text{ kJ/mol})]$
            $- [(1 \text{ mol})(-64.8 \text{ kJ/mol}) + (2 \text{ mol})(-237.2 \text{ kJ/mol})] = -150.2 \text{ kJ}$
This reaction can be used for the synthesis of C_2H_2 because $\Delta G < 0$.
(b) It is not possible to synthesize acetylene from solid graphite and gaseous H_2 at 25°C and 1 atm because $\Delta G°_f(C_2H_2) > 0$.

17.10 $C(s) + 2 H_2(g) \rightarrow C_2H_4(g)$

$$Q_p = \frac{P_{C_2H_4}}{(P_{H_2})^2} = \frac{(0.10)}{(100)^2} = 1.0 \times 10^{-5}$$

$\Delta G = \Delta G° + 2.303RT \log Q_p$
$\Delta G = 68.1 \text{ kJ} + (2.303)(8.314 \times 10^{-3} \text{ kJ/K})(298 \text{ K})\log(1.0 \times 10^{-5}) = +39.6 \text{ kJ}$
Because $\Delta G > 0$, the reaction is spontaneous in the reverse direction.

17.11   From Problem 17.8, $\Delta G° = +130.5 \text{ kJ}$
$\Delta G° = -2.303RT \log K$

$$\log K = \frac{-\Delta G°}{2.303RT} = \frac{-130.5 \text{ kJ}}{(2.303)(8.314 \times 10^{-3} \text{ kJ/K})(298 \text{ K})} = -22.9$$

$K = K_p = 10^{-22.9} = 1 \times 10^{-23}$

17.12 $H_2O(l) \rightleftharpoons H_2O(g)$
$K_p = P_{H_2O};$ K_p is equal to the vapor pressure for H_2O.
$\Delta G° = \Delta G°_f(H_2O(g)) - \Delta G°_f(H_2O(l))$
$\Delta G° = (1 \text{ mol})(-228.6 \text{ kJ/mol}) - (1 \text{ mol})(-237.2 \text{ kJ/mol}) = +8.6 \text{ kJ}$
$\Delta G° = -2.303RT \log K$

$$\log K = \frac{-\Delta G^\circ}{2.303 RT} = \frac{-8.6 \text{ kJ}}{(2.303)(8.314 \times 10^{-3} \text{ kJ/K})(298 \text{ K})} = -1.5$$

$$K = K_p = P_{H_2O} = 10^{-1.5} = 0.03 \text{ atm}$$

17.13 $\Delta G^\circ = -2.303 RT \log K = -(2.303)(8.314 \times 10^{-3} \text{ kJ/K})(298 \text{ K})\log(1.0 \times 10^{-14}) = +80 \text{ kJ}$

Understanding Key Concepts

17.14   (a)

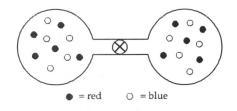

$\bullet$ = red $\circ$ = blue

(b) $\Delta H = 0$ (no heat is gained or lost in the mixing of ideal gases)
$\Delta S = +$ (the mixture of the two gases is more disordered)
$\Delta G = -$ (the mixing of the two gases is spontaneous)
(c) For an isolated system, $\Delta S_{surr} = 0$ and $\Delta S_{sys} = \Delta S_{Total} > 0$ for the spontaneous process.
(d) $\Delta G = +$ and the process is nonspontaneous.

17.15 $\Delta H = +$ (heat is absorbed during sublimation)
$\Delta S = +$ (gas is more disordered than solid)
$\Delta G = -$ (the reaction is spontaneous)

17.16 $\Delta H = -$ (heat is lost during condensation)
$\Delta S = -$ (liquid is more ordered than vapor)
$\Delta G = -$ (the reaction is spontaneous)

17.17 $\Delta H = 0$ (system is an ideal gas at constant temperature)
$\Delta S = -$ (there is more order in the smaller volume)
$\Delta G = +$ (compression of a gas is not spontaneous)

17.18   (a) $2 \text{ A}_2 + \text{ B}_2 \rightarrow 2 \text{ A}_2\text{B}$
(b) $\Delta H = -$ (because ΔS is negative, ΔH must also be negative in order for ΔG to be
         negative)
$\Delta S = -$ (the reaction becomes more ordered in going from reactants (3 molecules) to
         products (2 molecules))
$\Delta G = -$ (the reaction is spontaneous)

17.19   (a) For <u>initial state 1</u>, $Q_p < K_p$
     (more reactant (A_2) than product (A) compared to the equilibrium state)
     For <u>initial state 2</u>, $Q_p > K_p$
     (more product (A) than reactant (A_2) compared to the equilibrium state)

(b) $\Delta H = +$ (reaction involves bond breaking - endothermic)

$\Delta S = +$ (equilibrium state is more disordered than initial state 1)

$\Delta G = -$ (reaction spontaneously proceeds toward equilibrium)

(c) $\Delta H = -$ (reaction involves bond making - exothermic)

$\Delta S = -$ (equilibrium state is more ordered than initial state 2)

$\Delta G = -$ (reaction spontaneously proceeds toward equilibrium)

(d) State 1 lies to the left of the minimum in Figure 17.10. State 2 lies to the right of the minimum.

17.20   (a) $\Delta H^{\circ} = +$ (reaction involves bond breaking - endothermic)

$\Delta S^{\circ} = +$ (2 A's are more disordered than A_2)

(b) ΔS° is for the complete conversion of 1 mole of A_2 in its standard state to 2 moles of A in its standard state.

(c) There is not enough information to say anything about the sign of ΔG°. ΔG° decreases (becomes less positive or more negative) as the temperature increases.

(d) K_p increases as the temperature increases. As the temperature increases there will be more A and less A_2.

(e) $\Delta G = 0$ at equilibrium.

17.21

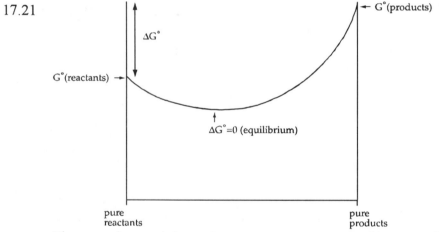

The curve has a minimum between pure reactants and pure products because the free energy decreases when pure reactants (or pure products) react to form the equilibrium mixture. The minimum is on the left side of the diagram because a positive value of ΔG° corresponds to a value of K less than 1.

Additional Problems
Spontaneous Processes

17.22   A spontaneous process is one that proceeds on its own without any external influence.

For example: $H_2O(s) \rightarrow H_2O(l)$ at 25°C

A nonspontaneous process takes place only in the presence of some continuous external influence.

For example: $2 NaCl(s) \rightarrow 2 Na(s) + Cl_2(g)$

17.24 (a) and (d) nonspontaneous; (b) and (c) spontaneous

17.26 (b) and (d) spontaneous (because of the large positive K_p's)

Entropy

17.28 Molecular randomness or disorder is called entropy. For the following reaction, the entropy (disorder) increases: $H_2O(s) \rightarrow H_2O(l)$ at 25°C.

17.30 (a) + (solid → gas)   (b) – (liquid → solid)
       (c) – (aqueous ions → solid)   (d) + (CO_2(aq) → CO_2(g))

17.32 (a) – (liquid → solid)
       (b) – (decrease in number of O_2 molecules)
       (c) + (gas is more disordered in larger volume)
       (d) – (aqueous ions → solid)

17.34 $S = k \ln W$, $k = 1.38 \times 10^{-23}$ J/K
       (a) $S = k \ln (1) = 0$ (b) $S = k \ln (1) = 0$
       (c) $S = (1.38 \times 10^{-23}$ J/K$) \ln (2^6) = 5.74 \times 10^{-23}$ J/K

17.36 $S = k \ln W$, $k = 1.38 \times 10^{-23}$ J/K
       (a) $S = (1.38 \times 10^{-23}$ J/K$) \ln (4^{12}) = 2.30 \times 10^{-22}$ J/K
       (b) $S = (1.38 \times 10^{-23}$ J/K$) \ln (4^{120}) = 2.30 \times 10^{-21}$ J/K
       (c) $S = (1.38 \times 10^{-23}$ J/K$) \ln (4^{6.02 \times 10^{23}}) = 11.5$ J/K
       If all C–D bonds point in the same direction, $S = 0$.

17.38 (a) H_2 at 25°C in 50 L (larger volume)
       (b) O_2 at 25°C, 1 atm (larger volume)
       (c) H_2 at 100°C, 1 atm (larger volume and higher T)
       (d) CO_2 at 100°C, 0.1 atm (larger volume and higher T)

Standard Molar Entropies and Standard Entropies of Reaction

17.40 The standard molar entropy of a substance is the entropy of 1 mol of the pure substance at 1 atm pressure and 25°C.
       $\Delta S° = S°(\text{products}) - S°(\text{reactants})$

17.42 (a) C_2H_6(g); more atoms/molecule   (b) CO_2(g); more atoms/molecule
       (c) I_2(g); gas is more disordered than the solid
       (d) CH_3OH(g); gas is more disordered than the liquid.

17.44 (a) $2 H_2O_2$(l) → $2 H_2O$(l) + O_2(g)
       $\Delta S° = [2 S°(H_2O(l)) + S°(O_2)] - 2 S°(H_2O_2)$

$\Delta S° = [(2 \text{ mol})(69.9 \text{ J/(K} \cdot \text{mol)}) + (1 \text{ mol})(205.0 \text{ J/(K} \cdot \text{mol)})]$
$\quad - (2 \text{ mol})(110 \text{ J/(K} \cdot \text{mol)}) = +125 \text{ J/K}$ (+, because moles of gas increase)

(b) $2 \text{ Na(s)} + \text{Cl}_2\text{(g)} \rightarrow 2 \text{ NaCl(s)}$
$\Delta S° = 2 \text{ S°(NaCl)} - [2 \text{ S°(Na)} + \text{S°(Cl}_2)]$
$\Delta S° = (2 \text{ mol})(72.1 \text{ J/(K} \cdot \text{mol)})$
$\quad - [(2 \text{ mol})(51.2 \text{ J/(K} \cdot \text{mol)}) + (1 \text{ mol})(223.0 \text{ J/(K} \cdot \text{mol)})]$
$\Delta S° = -181.2 \text{ J/K}$ (–, because moles of gas decrease)

(c) $2 \text{ O}_3\text{(g)} \rightarrow 3 \text{ O}_2\text{(g)}$
$\Delta S° = 3 \text{ S°(O}_2) - 2 \text{ S°(O}_3)$
$\Delta S° = (3 \text{ mol})(205.0 \text{ J/(K} \cdot \text{mol)}) - (2 \text{ mol})(238.8 \text{ J/(K} \cdot \text{mol)})$
$\Delta S° = +137.4 \text{ J/K}$ (+, because moles of gas increase)

(d) $4 \text{ Al(s)} + 3 \text{ O}_2\text{(g)} \rightarrow 2 \text{ Al}_2\text{O}_3\text{(s)}$
$\Delta S° = 2 \text{ S°(Al}_2\text{O}_3) - [4 \text{ S°(Al)} + 3 \text{ S°(O}_2)]$
$\Delta S° = (2 \text{ mol})(50.9 \text{ J/(K} \cdot \text{mol)})$
$\quad - [(4 \text{ mol})(28.3 \text{ J/(K} \cdot \text{mol)}) + (3 \text{ mol})(205.0 \text{ J/(K} \cdot \text{mol)})]$
$\Delta S° = -626.4 \text{ J/K}$ (–, because moles of gas decrease)

Entropy and the Second Law of Thermodynamics

17.46 In any spontaneous process, the total entropy of a system and its surroundings always increases.

17.48 $\Delta S_{surr} = \dfrac{-\Delta H}{T}$;   The temperature (T) is always positive.

(a) For an exothermic reaction, ΔH is negative and ΔS_{surr} is positive.
(b) For an endothermic reaction, ΔH is positive and ΔS_{surr} is negative.

17.50 $\text{N}_2\text{(g)} + 2 \text{ O}_2\text{(g)} \rightarrow \text{N}_2\text{O}_4\text{(g)}$
$\Delta H° = \Delta H°_f(\text{N}_2\text{O}_4) = 9.16 \text{ kJ}$
$\Delta S_{sys} = \Delta S° = \text{S°(N}_2\text{O}_4) - [\text{S°(N}_2) + 2 \text{ S°(O}_2)]$
$\Delta S_{sys} = (1 \text{ mol})(304.2 \text{ J/(K} \cdot \text{mol)})$
$\quad - [(1 \text{ mol})(191.5 \text{ J/(K} \cdot \text{mol)}) + (2 \text{ mol})(205.0 \text{ J/(K} \cdot \text{mol)})] = -297.3 \text{ J/K}$
$\Delta S_{surr} = \dfrac{-\Delta H°}{T} = \dfrac{-9.16 \text{ kJ}}{298 \text{ K}} = -0.0307 \text{ kJ/K} = -30.7 \text{ J/K}$
$\Delta S_{total} = \Delta S_{sys} + \Delta S_{surr} = -297.3 \text{ J/K} + (-30.7 \text{ J/K}) = -328.0 \text{ J/K}$
Because $\Delta S_{total} < 0$, the reaction is nonspontaneous.

17.52 (a) $\Delta S_{surr} = \dfrac{-\Delta H_{vap}}{T} = \dfrac{-30,700 \text{ J/mol}}{343 \text{ K}} = -89.5 \text{ J/(K} \cdot \text{mol)}$

$\Delta S_{total} = \Delta S_{vap} + \Delta S_{surr} = 87.0 \text{ J/(K} \cdot \text{mol)} + (-89.5 \text{ J/(K} \cdot \text{mol)}) = -2.5 \text{ J/(K} \cdot \text{mol)}$

(b) $\Delta S_{surr} = \dfrac{-\Delta H_{vap}}{T} = \dfrac{-30,700 \text{ J/mol}}{353 \text{ K}} = -87.0 \text{ J/(K} \cdot \text{mol)}$

$\Delta S_{total} = \Delta S_{vap} + \Delta S_{surr} = 87.0 \text{ J/(K} \cdot \text{mol)} + (-87.0 \text{ J/(K} \cdot \text{mol)}) = 0$

(c) $\Delta S_{surr} = \dfrac{-\Delta H_{vap}}{T} = \dfrac{-30,700 \text{ J/mol}}{363 \text{ K}} = -84.6 \text{ J/(K} \cdot \text{mol)}$

$\Delta S_{total} = \Delta S_{vap} + \Delta S_{surr} = 87.0 \text{ J/(K} \cdot \text{mol)} + (-84.6 \text{ J/(K} \cdot \text{mol)}) = +2.4 \text{ J/(K} \cdot \text{mol)}$
Benzene does not boil at 70°C (343 K) because ΔS_{total} is negative.
The normal boiling point for benzene is 80°C (353 K), where $\Delta S_{total} = 0$.

Free Energy

17.54 The free energy (G) is that part of the system's energy that is still ordered and therefore free (available) to cause spontaneous change.
$G = H - TS$ or $\Delta G = \Delta H - T\Delta S$ (at constant T)

17.56

$\underline{\Delta H}$	$\underline{\Delta S}$	$\underline{\Delta G = \Delta H - T\Delta S}$	$\underline{\text{Reaction Spontaneity}}$
–	+	–	Spontaneous at all temperatures
–	–	– or +	Spontaneous at low temperatures where $|\Delta H| > |T\Delta S|$
			Nonspontaneous at high temperatures where $|\Delta H| < |T\Delta S|$
+	–	+	Nonspontaneous at all temperatures
+	+	– or +	Spontaneous at high temperatures where $T\Delta S > \Delta H$
			Nonspontaneous at low temperature where $T\Delta S < \Delta H$

17.58 $\Delta H_{vap} = 30.7 \text{ kJ/mol}$
$\Delta S_{vap} = 87.0 \text{ J/(K} \cdot \text{mol)} = 87.0 \times 10^{-3} \text{ kJ/(K} \cdot \text{mol)}$
$\Delta G_{vap} = \Delta H_{vap} - T\Delta S_{vap}$
(a) $\Delta G_{vap} = 30.7 \text{ kJ/mol} - (343 \text{ K})(87.0 \times 10^{-3} \text{ kJ/(K} \cdot \text{mol)}) = +0.9 \text{ kJ/mol}$
At 70°C (343 K), benzene does not boil because ΔG_{vap} is positive.
(b) $\Delta G_{vap} = 30.7 \text{ kJ/mol} - (353 \text{ K})(87.0 \times 10^{-3} \text{ kJ/(K} \cdot \text{mol)}) = 0$
80°C (353 K) is the boiling point for benzene because $\Delta G_{vap} = 0$
(c) $\Delta G_{vap} = 30.7 \text{ kJ/mol} - (363 \text{ K})(87.0 \times 10^{-3} \text{ kJ/(K} \cdot \text{mol)}) = -0.9 \text{ kJ/mol}$
At 90°C (363 K), benzene boils because ΔG_{vap} is negative.

17.60 At the melting point (phase change), $\Delta G_{fusion} = 0$
$\Delta G_{fusion} = \Delta H_{fusion} - T\Delta S_{fusion}$

$0 = \Delta H_{fusion} - T\Delta S_{fusion}, \quad T = \dfrac{\Delta H_{fusion}}{\Delta S_{fusion}} = \dfrac{17.3 \text{ kJ/mol}}{43.8 \times 10^{-3} \text{ kJ/(K} \cdot \text{mol)}} = 395 \text{ K} = 122°C$

Standard Free-Energy Changes and Standard Free Energies of Formation

17.62 (a) $\Delta G°$ is the change in free energy that occurs when reactants in their standard states are converted to products in their standard states.
(b) $\Delta G°_f$ is the free-energy change for formation of one mole of a substance in its standard state from the most stable form of the constituent elements in their standard states.

17.64   (a) $N_2(g) + 2 O_2(g) \rightarrow 2 NO_2(g)$
$\Delta H° = 2 \Delta H°_f(NO_2) = (2 \text{ mol})(33.2 \text{ kJ/mol}) = 66.4 \text{ kJ}$
$\Delta S° = 2 S°(NO_2) - [S°(N_2) + 2 S°(O_2)]$
$\Delta S° = (2 \text{ mol})(240.0 \text{ J/(K} \cdot \text{mol)})$
                    $- [(1 \text{ mol})(191.5 \text{ J/(K} \cdot \text{mol)}) + (2 \text{ mol})(205.0 \text{ J/(K} \cdot \text{mol)})]$
$\Delta S° = -121.5 \text{ J/K} = -121.5 \times 10^{-3} \text{ kJ/K}$
$\Delta G° = \Delta H° - T\Delta S° = 66.4 \text{ kJ} - (298 \text{ K})(-121.5 \times 10^{-3} \text{ kJ/K}) = +102.6 \text{ kJ}$
Because $\Delta G°$ is positive, the reaction is nonspontaneous under standard-state conditions at 25°C.
(b) $2 KClO_3(s) \rightarrow 2 KCl(s) + 3 O_2(g)$
$\Delta H° = 2 \Delta H°_f(KCl) - 2 \Delta H°_f(KClO_3)$
$\Delta H° = (2 \text{ mol})(-436.7 \text{ kJ/mol}) - (2 \text{ mol})(-397.7 \text{ kJ/mol}) = -78.0 \text{ kJ}$
$\Delta S° = [2 S°(KCl) + 3 S°(O_2)] - 2 S°(KClO_3)$
$\Delta S° = [(2 \text{ mol})(82.6 \text{ J/(K} \cdot \text{mol)}) + (3 \text{ mol})(205.0 \text{ J/(K} \cdot \text{mol)})]$
                    $- (2 \text{ mol})(143 \text{ J/(K} \cdot \text{mol)})$
$\Delta S° = 494.2 \text{ J/(K} \cdot \text{mol}) = 494.2 \times 10^{-3} \text{ kJ/(K} \cdot \text{mol)}$
$\Delta G° = \Delta H° - T\Delta S° = -78.0 \text{ kJ} - (298 \text{ K})(494.2 \times 10^{-3} \text{ kJ/(K} \cdot \text{mol)}) = -225.3 \text{ kJ}$
Because $\Delta G°$ is negative, the reaction is spontaneous under standard-state conditions at 25°C.
(c) $CH_3CH_2OH(l) + O_2(g) \rightarrow CH_3COOH(l) + H_2O(l)$
$\Delta H° = [\Delta H°_f(CH_3COOH) + \Delta H°_f(H_2O)] - \Delta H°_f(CH_3CH_2OH)$
$\Delta H° = [(1 \text{ mol})(-484.5 \text{ kJ/mol}) + (1 \text{ mol})(-285.8 \text{ kJ/mol})]$
                    $- (1 \text{ mol})(-277.7 \text{ kJ/mol}) = -492.6 \text{ kJ}$
$\Delta S° = [S°(CH_3COOH) + S°(H_2O)] - [S°(CH_3CH_2OH) + S°(O_2)]$
$\Delta S° = [(1 \text{ mol})(160 \text{ J/(K} \cdot \text{mol)}) + (1 \text{ mol})(69.9 \text{ J/(K} \cdot \text{mol)})]$
                    $- [(1 \text{ mol})(161 \text{ J/(K} \cdot \text{mol)}) + (1 \text{ mol})(205.0 \text{ J/(K} \cdot \text{mol)})]$
$\Delta S° = -136.1 \text{ J/(K} \cdot \text{mol}) = -136.1 \times 10^{-3} \text{ kJ/(K} \cdot \text{mol)}$
$\Delta G° = \Delta H° - T\Delta S° = -492.6 \text{ kJ} - (298 \text{ K})(-136.1 \times 10^{-3} \text{ kJ/(K} \cdot \text{mol)}) = -452.0 \text{ kJ}$
Because $\Delta G°$ is negative, the reaction is spontaneous under standard-state conditions at 25°C.

17.66   (a) $N_2(g) + 2 O_2(g) \rightarrow 2 NO_2(g)$
$\Delta G° = 2 \Delta G°_f(NO_2) = (2 \text{ mol})(51.3 \text{ kJ/mol}) = +102.6 \text{ kJ}$
(b) $2 KClO_3(s) \rightarrow 2 KCl(s) + 3 O_2(g)$
$\Delta G° = 2 \Delta G°_f(KCl) - 2 \Delta G°_f(KClO_3)$
$\Delta G° = (2 \text{ mol})(-409.2 \text{ kJ/mol}) - (2 \text{ mol})(-296.3 \text{ kJ/mol}) = -225.8 \text{ kJ}$
(c) $CH_3CH_2OH(l) + O_2(g) \rightarrow CH_3COOH(l) + H_2O(l)$
$\Delta G° = [\Delta G°_f(CH_3COOH) + \Delta G°_f(H_2O)] - \Delta G°_f(CH_3CH_2OH)$
$\Delta G° = [(1 \text{ mol})(-390 \text{ kJ/mol}) + (1 \text{ mol})(-237.2 \text{ kJ/mol})]$
                    $- (1 \text{ mol})(-174.9 \text{ kJ/mol}) = -452 \text{ kJ}$

17.68 A compound is thermodynamically stable with respect to its constituent elements at 25°C if ΔG°_f is negative.

	ΔG°_f (kJ/mol)	Stable
(a) $BaCO_3(s)$	−1138	yes
(b) $HBr(g)$	−53.4	yes
(c) $N_2O(g)$	+104.2	no
(d) $C_2H_4(g)$	+68.1	no

17.70 $CH_2=CH_2(g) + H_2O(l) \rightarrow CH_3CH_2OH(l)$
$\Delta H^\circ = \Delta H^\circ_f(CH_2CH_2OH) - [\Delta H^\circ_f(CH_2=CH_2) + \Delta H^\circ_f(H_2O)]$
$\Delta H^\circ = (1\ mol)(-277.7\ kJ/mol) - [(1\ mol)(52.3\ kJ/mol) + (1\ mol)(-285.8\ kJ/mol)]$
$\Delta H^\circ = -44.2\ kJ$
$\Delta S^\circ = S^\circ(CH_3CH_2OH) - [S^\circ(CH_2=CH_2) + S^\circ(H_2O)]$
$\Delta S^\circ = (1\ mol)(161\ J/(K \cdot mol)) - [(1\ mol)(219.5\ J/(K \cdot mol))$
$\qquad\qquad + (1\ mol)(69.9\ J/(K \cdot mol))]$
$\Delta S^\circ = -128\ J/(K \cdot mol) = -128 \times 10^{-3}\ kJ/(K \cdot mol)$
$\Delta G^\circ = \Delta H^\circ - T\Delta S^\circ = -44.2\ kJ - (298\ K)(-128 \times 10^{-3}\ kJ/K) = -6.1\ kJ$
Because ΔG° is negative, the reaction is spontaneous under standard-state conditions at 25°C.
The reaction becomes nonspontaneous at high temperatures because ΔS° is negative.
To find the crossover temperature, set $\Delta G = 0$ and solve for T.
$$T = \frac{\Delta H^\circ}{\Delta S^\circ} = \frac{-44,200\ J}{-128\ J/K} = 345\ K = 72°C$$
The reaction becomes nonspontaneous at 72°C.

17.72 $3\ C_2H_2(g) \rightarrow C_6H_6(l)$
$\Delta G^\circ = \Delta G^\circ_f(C_6H_6) - 3\ \Delta G^\circ_f(C_2H_2)$
$\Delta G^\circ = (1\ mol)(124.5\ kJ/mol) - (3\ mol)(209.2\ kJ/mol) = -503.1\ kJ$
Because ΔG° is negative, the reaction is possible. Look for a catalyst.
Because ΔG°_f for benzene is positive (+124.5 kJ/mol), the synthesis of benzene from graphite and gaseous H_2 at 25°C and 1 atm pressure is not possible.

Free Energy, Composition, and Chemical Equilibrium

17.74 $\Delta G = \Delta G^\circ + RT \ln Q$

17.76 $\Delta G = \Delta G^\circ + 2.303RT \log \left[\dfrac{(P_{SO_3})^2}{(P_{SO_2})^2(P_{O_2})} \right]$

(a) $\Delta G = (-141.8\ kJ) + (2303)(8.314 \times 10^{-3}\ kJ/K)(298\ K)\log\left[\dfrac{(1.0)^2}{(100)^2(100)}\right] = -176.0\ kJ$

(b) $\Delta G = (-141.8\ kJ) + (2.303)(8.314 \times 10^{-3}\ kJ/K)(298\ K)\log\left[\dfrac{(10)^2}{(2.0)^2(1.0)}\right] = -133.8\ kJ$

(c) $Q = 1$, $\log Q = 0$, $\Delta G = \Delta G^\circ = -141.8\ kJ$

17.78 $\Delta G^\circ = -RT \ln K = -2.303RT \log K$
(a) If $K > 1$, ΔG° is negative.   (b) If $K = 1$, $\Delta G^\circ = 0$.
(c) If $K < 1$, ΔG° is positive.

17.80 $\Delta G^\circ = -141.8$ kJ
$\Delta G^\circ = -2.303RT \log K_p$

$$\log K_p = \frac{-\Delta G^\circ}{2.303\,RT} = \frac{-(-141.8 \text{ kJ})}{(2.303)(8.314 \times 10^{-3} \text{ kJ/K})(298 \text{ K})} = 24.85$$

$K_p = 10^{24.85} = 7.1 \times 10^{24}$

17.82 $C_2H_5OH(l) \rightleftharpoons C_2H_5OH(g)$
$\Delta G^\circ = \Delta G^\circ_f(C_2H_5OH(g)) - \Delta G^\circ_f(C_2H_5OH(l))$
$\Delta G^\circ = (1 \text{ mol})(-168.6 \text{ kJ/mol}) - (1 \text{ mol})(-174.9 \text{ kJ/mol}) = +6.3$ kJ

$\Delta G^\circ = -2.303RT \log K$

$$\log K = \frac{-\Delta G^\circ}{2.303\,RT} = \frac{-(6.3 \text{ kJ})}{(2.303)(8.314 \times 10^{-3} \text{ kJ/K})(298 \text{ K})} = -1.10$$

$K = 10^{-1.10} = 0.079$; $K = K_p = P_{C_2H_5OH} = 0.079$ atm

17.84 $2\,CH_2\text{=}CH_2(g) + O_2(g) \rightarrow 2\,C_2H_4O(g)$
$\Delta G^\circ = 2\,\Delta G^\circ_f(C_2H_4O) - 2\,\Delta G^\circ_f(CH_2\text{=}CH_2)$
$\Delta G^\circ = (2 \text{ mol})(-13.1 \text{ kJ/mol}) - (2 \text{ mol})(68.1 \text{ kJ/mol}) = -162.4$ kJ
$\Delta G^\circ = -2.303RT \log K$

$$\log K = \frac{-\Delta G^\circ}{2.303\,RT} = \frac{-(-162.4 \text{ kJ})}{(2.303)(8.314 \times 10^{-3} \text{ kJ/K})(298 \text{ K})} = 28.46$$

$K = K_p = 10^{28.46} = 2.9 \times 10^{28}$

General Problems

17.86   The kinetic parameters [(a), (b), and (h)] are affected by a catalyst. The thermodynamic
and equilibrium parameters [(c), (d), (e), (f), and (g)] are not affected by a catalyst.

17.88   (a) Spontaneous does not mean fast, just possible.
(b) For a spontaneous reaction $\Delta S_{total} > 0$. ΔS_{sys} can be positive or negative.
(c) An endothermic reaction can be spontaneous if $\Delta S_{sys} > 0$.
(d) This statement is true because the sign of ΔG changes when the direction of a
reaction is reversed.

17.90

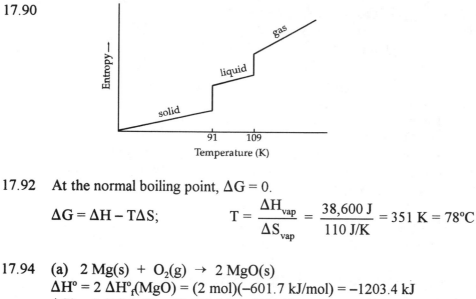

17.92 At the normal boiling point, $\Delta G = 0$.

$$\Delta G = \Delta H - T\Delta S; \qquad T = \frac{\Delta H_{vap}}{\Delta S_{vap}} = \frac{38,600\ J}{110\ J/K} = 351\ K = 78°C$$

17.94 (a) $2\ Mg(s) + O_2(g) \rightarrow 2\ MgO(s)$
$\Delta H° = 2\ \Delta H°_f(MgO) = (2\ mol)(-601.7\ kJ/mol) = -1203.4\ kJ$
$\Delta S° = 2\ S°(MgO) - [2\ S°(Mg) + S°(O_2)]$
$\Delta S° = (2\ mol)(26.9\ J/(K \cdot mol))$
$\qquad\qquad - [(2\ mol)(32.7\ J/(K \cdot mol)) + (1\ mol)(205.0\ J/(K \cdot mol))]$
$\Delta S° = -216.6\ J/K = -216.6 \times 10^{-3}\ kJ/K$
$\Delta G° = \Delta H° - T\Delta S° = -1203.4\ kJ - (298\ K)(-216.6 \times 10^{-3}\ kJ/K) = -1138.8\ kJ$
Because $\Delta G°$ is negative, the reaction is spontaneous at 25°C. $\Delta G°$ becomes less negative as the temperature is raised.
(b) $MgCO_3(s) \rightarrow MgO(s) + CO_2(g)$
$\Delta H° = [\Delta H°_f(MgO) + \Delta H°_f(CO_2)] - \Delta H°_f(MgCO_3)$
$\Delta H° = [(1\ mol)(-601.1\ kJ/mol) + (1\ mol)(-393.5\ kJ/mol)]$
$\qquad\qquad - (1\ mol)(-1096\ kJ/mol) = +101\ kJ$
$\Delta S° = [S°(MgO) + S°(CO_2)] - S°(MgCO_3)$
$\Delta S° = [(1\ mol)(26.9\ J/(K \cdot mol)) + (1\ mol)(213.6\ J/(K \cdot mol))]$
$\qquad\qquad - (1\ mol)(65.7\ J/(K \cdot mol))$
$\Delta S° = 174.8\ J/K = 174.8 \times 10^{-3}\ kJ/K$
$\Delta G° = \Delta H° - T\Delta S° = 101\ kJ - (298\ K)(174.8 \times 10^{-3}\ kJ/K) = +49\ kJ$
Because $\Delta G°$ is positive, the reaction is not spontaneous at 25°C. $\Delta G°$ becomes less positive as the temperature is raised.
(c) $Fe_2O_3(s) + 2\ Al(s) \rightarrow Al_2O_3(s) + 2\ Fe(s)$
$\Delta H° = \Delta H°_f(Al_2O_3) - \Delta H°_f(Fe_2O_3)$
$\Delta H° = (1\ mol)(-1676\ kJ/mol) - (1\ mol)(-824.2\ kJ/mol) = -852\ kJ$
$\Delta S° = [S°(Al_2O_3) + 2\ S°(Fe)] - [S°(Fe_2O_3) + 2\ S°(Al)]$
$\Delta S° = [(1\ mol)(50.9\ J/(K \cdot mol)) + (2\ mol)(27.3\ J/(K \cdot mol))]$
$\qquad - [(1\ mol)(87.4\ J/(K \cdot mol)) + (2\ mol)(28.3\ J/(K \cdot mol))]$
$\Delta S° = -38.5\ J/K = -38.5 \times 10^{-3}\ kJ/K$
$\Delta G° = \Delta H° - T\Delta S° = -852\ kJ - (298\ K)(-38.5 \times 10^{-3}\ kJ/K) = -840\ kJ$
Because $\Delta G°$ is negative, the reaction is spontaneous at 25°C. $\Delta G°$ becomes less negative as the temperature is raised.

(d) 2 NaHCO₃(s) → Na₂CO₃(s) + CO₂(g) + H₂O(g)

$\Delta H° = [\Delta H°_f(Na_2CO_3) + \Delta H°_f(CO_2) + \Delta H°_f(H_2O)] - 2\,\Delta H°_f(NaHCO_3)$

$\Delta H° = [(1\ mol)(-1130.7\ kJ/mol) + (1\ mol)(-393.5\ kJ/mol)$
$\qquad + (1\ mol)(-241.8\ kJ/mol)] - (2\ mol)(-950.8\ kJ/mol) = +135.6\ kJ$

$\Delta S° = [S°(Na_2CO_3) + S°(CO_2) + S°(H_2O)] - 2\,S°(NaHCO_3)$

$\Delta S° = [(1\ mol)(135.0\ J/(K\cdot mol)) + (1\ mol)(213.6\ J/(K\cdot mol))$
$\qquad + (1\ mol)(188.7\ J/(K\cdot mol))] - (2\ mol)(102\ J/(K\cdot mol))$

$\Delta S° = +333\ J/K = +333 \times 10^{-3}\ kJ/K$

$\Delta G° = \Delta H° - T\Delta S° = +135.6\ kJ - (298\ K)(+333 \times 10^{-3}\ kJ/K) = +36.4\ kJ$

Because $\Delta G°$ is positive, the reaction is not spontaneous at 25°C. $\Delta G°$ becomes less positive as the temperature is raised.

17.96 For C₆H₆(s) → C₆H₆(l), both ΔH and ΔS are positive.
(a) ΔH and ΔS are positive, and because 0°C is below the melting point, ΔG is also positive.
(b) ΔH and ΔS are positive, and because 15°C is above the melting point, the reaction takes place and ΔG is negative.

17.98 (a) $\Delta G = \Delta H - T\Delta S$; $\Delta H = \Delta G + T\Delta S$
Because ΔS is positive and ΔG is positive (the reaction is nonspontaneous), ΔH must be positive. Therefore, the reaction is endothermic.
(b) $\Delta G = \Delta H - T\Delta S$
If ΔG is set to 0, the minimum value for ΔH can be calculated.
$0 = \Delta H - T\Delta S$; $\Delta H = T\Delta S = (323\ K)(104 \times 10^{-3}\ kJ/K) = 33.6\ kJ$

17.100 (a) 6 C(s) + 3 H₂(g) → C₆H₆(l)
$\Delta S°_f = S°(C_6H_6) - [6\,S°(C) + 3\,S°(H_2)]$
$\Delta S°_f = (1\ mol)(172.8\ J/(K\cdot mol))$
$\qquad - [(6\ mol)(5.7\ J/(K\cdot mol)) + (3\ mol)(130.6\ J/(K\cdot mol))]$
$\Delta S°_f = -253\ J/K = -253\ J/(K\cdot mol)$
$\Delta G°_f = \Delta H°_f - T\Delta S°_f$
$\Delta S°_f = \dfrac{\Delta H°_f - \Delta G°_f}{T} = \dfrac{49.0\ kJ/mol - 124.5\ kJ/mol}{298\ K} = -0.253\ kJ/(K\cdot mol)$
$\Delta S°_f = -253\ J/(K\cdot mol)$
Both calculations lead to the same value of $\Delta S°_f$.
(b) Ca(s) + S(s) + 2 O₂(g) → CaSO₄(s)
$\Delta S°_f = S°(CaSO_4) - [S°(Ca) + S°(S) + 2\,S°(O_2)]$
$\Delta S°_f = (1\ mol)(107\ J/(K\cdot mol))$
$- [(1\ mol)(41.4\ J/(K\cdot mol)) + (1\ mol)(31.8\ J/(K\cdot mol)) + (2\ mol)(205.0\ J/(K\cdot mol))]$
$\Delta S°_f = -376\ J/K = -376\ J/(K\cdot mol)$
$\Delta G°_f = \Delta H°_f - T\Delta S°_f$
$\Delta S°_f = \dfrac{\Delta H°_f - \Delta G°_f}{T} = \dfrac{-1434.1\ kJ/mol - (-1321.9\ kJ/mol)}{298\ K} = -0.376\ kJ/(K\cdot mol)$
$\Delta S°_f = -376\ J/(K\cdot mol)$
Both calculations lead to the same value of $\Delta S°_f$.

(c) $2 C(s) + 3 H_2(g) + 1/2 O_2(g) \rightarrow C_2H_5OH(l)$

$\Delta S^\circ_f = S^\circ(C_2H_5OH) - [S^\circ(C + S^\circ(H_2) + 1/2 S^\circ(O_2)]$

$\Delta S^\circ_f = (1 \text{ mol})(161 \text{ J/(K} \cdot \text{mol}))$

$- [(2 \text{ mol})(5.7 \text{ J/(K} \cdot \text{mol})) + (3 \text{ mol})(130.6 \text{ J/(K} \cdot \text{mol})) + (0.5 \text{ mol})(205.0 \text{ J/(K} \cdot \text{mol}))]$

$\Delta S^\circ_f = -345 \text{ J/K} = -345 \text{ J/(K} \cdot \text{mol})$

$\Delta G^\circ_f = \Delta H^\circ_f - T\Delta S^\circ_f$

$\Delta S^\circ_f = \dfrac{\Delta H^\circ_f - \Delta G^\circ_f}{T} = \dfrac{-277.7 \text{ kJ/mol} - (-174.9 \text{ kJ/mol})}{298 \text{ K}} = -0.345 \text{ kJ/(K} \cdot \text{mol})$

$\Delta S^\circ_f = -345 \text{ J/(K} \cdot \text{mol})$

Both calculations lead to the same value of ΔS°_f.

17.102 $\Delta G^\circ = -2.303RT \log K_b$

At 20 °C: $\Delta G^\circ = -(2.303)(8.314 \times 10^{-3} \text{ kJ/K})(293 \text{ K})\log(1.710 \times 10^{-5}) = +26.74 \text{ kJ}$

At 50 °C: $\Delta G^\circ = -(2.303)(8.314 \times 10^{-3} \text{ kJ/K})(323 \text{ K})\log(1.892 \times 10^{-5}) = +29.21 \text{ kJ}$

$\Delta G^\circ = \Delta H^\circ - T\Delta S^\circ$

$26.74 = \Delta H^\circ - 293\Delta S^\circ$

$29.21 = \Delta H^\circ - 323\Delta S^\circ$ Solve these two equations simultaneously for ΔH° and ΔS°.

$26.74 + 293\Delta S^\circ = \Delta H^\circ$

$29.21 + 323\Delta S^\circ = \Delta H^\circ$ Set these two equations equal to each other.

$26.74 + 293\Delta S^\circ = 29.21 + 323\Delta S^\circ$

$26.74 - 29.21 = 323\Delta S^\circ - 293\ \Delta S^\circ$

$-2.47 = 30\Delta S^\circ$

$\Delta S^\circ = -2.47/30 = -0.0823 = -0.0823 \text{ kJ/K} = -82.3 \text{ J/K}$

$26.74 + 293\Delta S^\circ = 26.74 + 293(-0.0823) = \Delta H^\circ = +2.62 \text{ kJ}$

17.104 $Br_2(l) \rightleftharpoons Br_2(g)$

$\Delta S^\circ = S^\circ(Br_2(g)) - S^\circ(Br_2(l))$

$\Delta S^\circ = (1 \text{ mol})(245.4 \text{ J/K}) - (1 \text{ mol})(152.2 \text{ J/K}) = 93.2 \text{ J/K} = 93.2 \times 10^{-3} \text{ kJ/K}$

$T_{bp} = \dfrac{\Delta H_{vap}}{\Delta S_{vap}} \approx \dfrac{\Delta H^\circ}{\Delta S^\circ}$

$\Delta H^\circ = T_{bp}\ \Delta S^\circ = (332 \text{ K})(93.2 \times 10^{-3} \text{ kJ/K}) = 30.9 \text{ kJ}$

$K_p = P_{Br_2} = \left(227 \text{ mm Hg} \times \dfrac{1 \text{ atm}}{760 \text{ mm Hg}} \right) = 0.299 \text{ atm}$

$\Delta G^\circ = -2.303RT \log K_p$ and $\Delta G^\circ = \Delta H^\circ - T\Delta S^\circ$ (set equations equal to each other)

$\Delta H^\circ - T\Delta S^\circ = -2.303RT \log K_p$ (rearrange)

$\log K_p = \dfrac{-\Delta H^\circ}{2.303\,R}\dfrac{1}{T} + \dfrac{\Delta S^\circ}{2.303\,R}$ (solve for T)

$$T = \frac{\left(\dfrac{-\Delta H^\circ}{2.303\,R}\right)}{\left(\log K_p - \dfrac{\Delta S^\circ}{2.303\,R}\right)} = \frac{\left(\dfrac{-30.9\ \text{kJ}}{(2.303)(8.314 \times 10^{-3}\ \text{kJ})}\right)}{\left(\log(0.299) - \dfrac{93.2 \times 10^{-3}\ \text{kJ}}{(2.303)(8.314 \times 10^{-3}\ \text{kJ})}\right)} = 299\ \text{K} = 26°\text{C}$$

$Br_2(l)$ has a vapor pressure of 227 mm Hg at 26°C.

17.106 $N_2(g) + 3\,H_2(g) \rightleftharpoons 2\,NH_3(g)$

$\Delta H^\circ = 2\,\Delta H^\circ_f(NH_3) - [\Delta H^\circ_f(N_2) + 3\,\Delta H^\circ_f(H_2)] = (2\ \text{mol})(-46.1\ \text{kJ}) - [0] = -92.2\ \text{kJ}$

$\Delta S^\circ = 2\,S^\circ(NH_3) - [S^\circ(N_2) + 3\,S^\circ(H_2)]$

$\Delta S^\circ = (2\ \text{mol})(192.3\ \text{J/(K} \cdot \text{mol)})$
$\qquad\quad - [(1\ \text{mol})(191.5\ \text{J/(K} \cdot \text{mol)}) + (3\ \text{mol})(130.6\ \text{J/(K} \cdot \text{mol)})] = -198.7\ \text{J/K}$

$\Delta G^\circ = \Delta H^\circ - T\Delta S^\circ = -92.2\ \text{kJ} - (673\ \text{K})(-198.7 \times 10^{-3}\ \text{kJ/K}) = 41.5\ \text{kJ}$

$\Delta G^\circ = -2.303\,RT \log K_p$

$$\log K_p = \frac{-\Delta G^\circ}{2.303\,RT} = \frac{-(41.5\ \text{kJ})}{(2.303)(8.314 \times 10^{-3}\ \text{kJ/K})(673\ \text{K})} = -3.22$$

$K_p = 10^{-3.22} = 6.0 \times 10^{-4}$

Because $K_p = K_c(RT)^{\Delta n}$, $K_c = K_p(RT)^{-\Delta n}$

$K_c = K_p(RT)^2 = (6.0 \times 10^{-4})[(0.082\,06)(673)]^2 = 1.83$

N_2, 28.01 amu; H_2, 2.016 amu

Initial concentrations:

$$[N_2] = \frac{(14.0\ \text{g})\left(\dfrac{1\ \text{mol}}{28.01\ \text{g}}\right)}{5.00\ \text{L}} = 0.100\ \text{M} \quad \text{and} \quad [H_2] = \frac{(3.024\ \text{g})\left(\dfrac{1\ \text{mol}}{2.016\ \text{g}}\right)}{5.00\ \text{L}} = 0.300\ \text{M}$$

	$N_2(g)$	$+$	$3\,H_2(g)$	$\rightleftharpoons$	$2\,NH_3(g)$
initial (M)	0.100		0.300		0
change (M)	$-x$		$-3x$		$+2x$
equil (M)	$0.100 - x$		$0.300 - 3x$		$2x$

$$K_c = \frac{[NH_3]^2}{[N_2][H_2]^3} = \frac{(2x)^2}{(0.100-x)(0.300-3x)^3} = \frac{4x^2}{27(0.100-x)^4} = 1.83$$

$$\left(\frac{x}{(0.100-x)^2}\right)^2 = \frac{(27)(1.83)}{4} = 12.35; \qquad \frac{x}{(0.100-x)^2} = \sqrt{12.35} = 3.515$$

$3.515x^2 - 1.703x + 0.03515 = 0$

Use the quadratic formula to solve for x.

$$x = \frac{-(-1.703) \pm \sqrt{(-1.703)^2 - (4)(3.515)(0.03515)}}{2(3.515)} = \frac{1.703 \pm 1.551}{7.030}$$

$x = 0.463$ and 0.0216

Of the two solutions for x, only 0.0216 has physical meaning because $x = 0.463$ would lead to negative concentrations N_2 and H_2.

$[N_2] = 0.100 - x = 0.100 - 0.0216 = 0.078\ \text{M}$

$[H_2] = 0.300 - 3x = 0.300 - 3(0.0216) = 0.235\ \text{M}$

$[NH_3] = 2x = 2(0.0216) = 0.043\ \text{M}$

17.108 $Pb(s) + PbO_2(s) + 2 H^+(aq) + 2 HSO_4^-(aq) \rightarrow 2 PbSO_4(s) + 2 H_2O(l)$

(a) $\Delta G° = [2 \Delta G°_f(PbSO_4) + 2 \Delta G°_f(H_2O)] - [\Delta G°_f(PbO_2) + 2 \Delta G°_f(HSO_4^-)]$

$\Delta G° = (2 \text{ mol})(-813.2 \text{ kJ/mol}) + (2 \text{ mol})(-237.2 \text{ kJ/mol})]$

$\quad - [(1 \text{ mol})(-217.4 \text{ kJ/mol}) + (2 \text{ mol})(-756.0 \text{ kJ/mol})] = -371.4 \text{ kJ}$

(b) $°C = 5/9(°F - 32) = 5/9(10 - 32) = -12.2°C; \quad -12.2°C = 261 \text{ K}$

$\Delta H° = [2 \Delta H°_f(PbSO_4) + 2 \Delta H°_f(H_2O)] - [\Delta H°_f(PbO_2) + 2 \Delta H°_f(HSO_4^-)]$

$\Delta H° = [(2 \text{ mol})(-919.9 \text{ kJ/mol}) + (2 \text{ mol})(-285.8 \text{ kJ/mol})]$

$\quad - [(1 \text{ mol})(-277 \text{ kJ/mol}) + (2 \text{ mol})(-887.3 \text{ kJ/mol})] = -359.8 \text{ kJ}$

$\Delta S° = [2 S°(PbSO_4) + 2 S°(H_2O)] - [S°(Pb) + S°(PbO_2) + 2 S°(H^+) + 2 S°(HSO_4^-)]$

$\Delta S° = [(2 \text{ mol})(148.6 \text{ J/(K} \cdot \text{mol)}) + (2 \text{ mol})(69.9 \text{ J/(K} \cdot \text{mol)})]$

$\quad - [(1 \text{ mol})(64.8 \text{ J/(K} \cdot \text{mol)}) + (1 \text{ mol})(68.6 \text{ J/(K} \cdot \text{mol)})$

$\quad + (2 \text{ mol})(132 \text{ J/(K} \cdot \text{mol)})] = 39.6 \text{ J/K} = 39.6 \times 10^{-3} \text{ kJ/K}$

$\Delta G° = \Delta H° - T\Delta S° = -359.8 \text{ kJ} - (261 \text{ K})(39.6 \times 10^{-3} \text{ kJ/K}) = -370.1 \text{ kJ at } 261 \text{ K}$

$$HSO_4^-(aq) + H_2O(l) \rightleftharpoons H_3O^+(aq) + SO_4^{2-}(aq)$$

initial (M)	0.100	0.100	0
change (M)	−x	+x	+x
equil (M)	0.100 − x	0.100 + x	x

$$K_{a2} = \frac{[H_3O^+][SO_4^{2-}]}{[HSO_4^-]} = 1.2 \times 10^{-2} = \frac{(0.100 + x)x}{0.100 - x}$$

$x^2 + 0.112x - (1.2 \times 10^{-3}) = 0$

Use the quadratic formula to solve for x.

$$x = \frac{-(0.112) \pm \sqrt{(0.112)^2 - (4)(1)(-1.2 \times 10^{-3})}}{2(1)} = \frac{-0.112 \pm 0.132}{2}$$

$x = -0.122 \text{ and } 0.010$

Of the two solutions for x, only 0.010 has physical meaning because x = −0.122 would lead to a negative concentration of H_3O^+.

$[H^+] = 0.100 + x = 0.100 + 0.010 = 0.110 \text{ M}$

$[HSO_4^-] = 0.100 - x = 0.100 - 0.010 = 0.090 \text{ M}$

$$\Delta G = \Delta G° + 2.303 \text{ RT} \log \frac{1}{[H^+]^2[HSO_4^-]^2}$$

$$\Delta G = (-370.1 \text{ kJ}) + (2.303)(8.314 \times 10^{-3} \text{ kJ/K})(261 \text{ K}) \log \frac{1}{(0.110)^2(0.090)^2} = -350.1 \text{ kJ}$$

18

Electrochemistry

18.1 $2\,Ag^+(aq) + Ni(s) \rightarrow 2\,Ag(s) + Ni^{2+}(aq)$

There is a Ni anode in an aqueous solution of Ni^{2+}, and a Ag cathode in an aqueous solution of Ag^+. A salt bridge connects the anode and cathode compartment. The electrodes are connected through an external circuit.

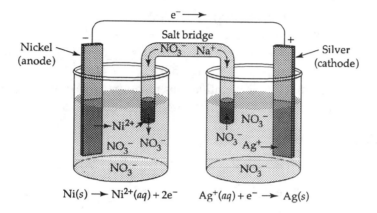

18.2 $Fe(s)\,|\,Fe^{2+}(aq)\,\|\,Sn^{2+}(aq)\,|\,Sn(s)$

18.3 $Pb(s)\ +\ Br_2(l)\ \rightarrow\ Pb^{2+}(aq)\ +\ 2\,Br^-(aq)$
There is a Pb anode in an aqueous solution of Pb^{2+}. The cathode is a Pt wire that dips into a pool of liquid Br_2 and an aqueous solution that is saturated with Br_2. A salt bridge connects the anode and cathode compartment. The electrodes are connected through an external circuit.

18.4 $Al(s)\ +\ Cr^{3+}(aq)\ \rightarrow\ Al^{3+}(aq)\ +\ Cr(s)$

$$\Delta G^\circ = -nFE^\circ = -(3\text{ mol }e^-)\left(\frac{96{,}500\text{ C}}{1\text{ mol }e^-}\right)(0.92\text{ V})\left(\frac{1\text{ J}}{1\text{ C}\cdot\text{V}}\right) = -266{,}340\text{ J} = -270\text{ kJ}$$

18.5 oxidation: $Al(s)\ \rightarrow\ Al^{3+}(aq)\ +\ 3\ e^-$ $E^\circ = 1.66$ V
reduction: $\underline{Cr^{3+}(aq)\ +\ 3\ e^-\ \rightarrow\ Cr(s)}$ $\underline{E^\circ = ?}$
overall $Al(s) + Cr^{3+}(aq) \rightarrow Al^{3+}(aq) + Cr(s)$ $E^\circ = 0.92$ V
The standard reduction potential for the Cr^{3+}/Cr half cell is: $E^\circ = 0.92 - 1.66 = -0.74$ V

18.6 (a) $Cl_2(g) + 2\ e^-\ \rightarrow\ 2\ Cl^-(aq)$ $E^\circ = 1.36$ V
        $Ag^+(aq) + e^-\ \rightarrow\ Ag(s)$ $E^\circ = 0.80$ V
Cl_2 has the greater tendency to be reduced (larger E°). The species that has the greater tendency to be reduced is the stronger oxidizing agent. Cl_2 is the stronger oxidizing agent.

(b) $Fe^{2+}(aq) + 2\ e^- \rightarrow Fe(s)$ $\quad\quad E^\circ = -0.45\ V$

$\quad\quad Mg^{2+}(aq) + 2\ e^- \rightarrow Mg(s)$ $\quad\quad E^\circ = -2.37\ V$

The second half-reaction has the lesser tendency to occur in the forward direction (more negative E°) and the greater tendency to occur in the reverse direction. Therefore, Mg is the stronger reducing agent.

18.7 (a) $\quad 2\ Fe^{3+}(aq) + 2\ I^-(aq) \rightarrow 2\ Fe^{2+}(aq) + I_2(s)$

reduction: $\quad Fe^{3+}(aq) + e^- \rightarrow Fe^{2+}(aq)$ $\quad\quad E^\circ = 0.77\ V$

oxidation: $\quad 2\ I^-(aq) \rightarrow I_2(s) + 2\ e^-$ $\quad\quad\quad\quad \underline{E^\circ = -0.54\ V}$

$\quad\quad\quad\quad\quad\quad\quad\quad\quad\quad\quad\quad\quad\quad\quad$ overall $E^\circ = 0.23\ V$

Because E° for the overall reaction is positive, this reaction can occur under standard-state conditions.

(b) $\quad 3\ Ni(s) + 2\ Al^{3+}(aq) \rightarrow 3\ Ni^{2+}(aq) + 2\ Al(s)$

oxidation: $\quad Ni(s) \rightarrow Ni^{2+}(aq) + 2\ e^-$ $\quad\quad\quad\quad E^\circ = 0.26\ V$

reduction: $\quad Al^{3+}(aq) + 3\ e^- \rightarrow Al(s)$ $\quad\quad\quad \underline{E^\circ = -1.66\ V}$

$\quad\quad\quad\quad\quad\quad\quad\quad\quad\quad\quad\quad\quad\quad\quad$ overall $E^\circ = -1.40\ V$

Because E° for the overall reaction is negative, this reaction cannot occur under standard-state conditions. This reaction can occur in the reverse direction.

18.8 $\quad Cu(s) + 2\ Fe^{3+}(aq) \rightarrow Cu^{2+}(aq) + 2\ Fe^{2+}(aq)$

$E^\circ = E^\circ_{Cu \to Cu^{2+}} + E^\circ_{Fe^{3+} \to Fe^{2+}} = -0.34\ V + 0.77\ V = 0.43\ V; \quad n = 2\ mol\ e^-$

$E = E^\circ - \dfrac{0.0592}{n} \log \dfrac{[Cu^{2+}][Fe^{2+}]^2}{[Fe^{3+}]^2} = 0.43\ V - \dfrac{(0.0592\ V)}{2} \log \dfrac{(0.25)(0.20)^2}{(1.0 \times 10^{-4})^2} = 0.25\ V$

18.9 $\quad H_2(g) + Pb^{2+}(aq) \rightarrow 2\ H^+(aq) + Pb(s)$

$E^\circ = E^\circ_{H_2 \to H^+} + E^\circ_{Pb^{2+} \to Pb} = 0\ V + (-0.13\ V) = -0.13\ V; \quad n = 2\ mol\ e^-$

$E = E^\circ - \dfrac{0.0592}{n} \log \dfrac{[H_3O^+]^2}{[Pb^{2+}](P_{H_2})}$

$0.28\ V = -0.13\ V - \dfrac{(0.0592\ V)}{2} \log \dfrac{[H_3O^+]^2}{(1)(1)} = -0.13\ V - (0.0592\ V) \log [H_3O^+]$

$pH = -\log [H_3O^+]$ therefore $0.28\ V = -0.13\ V + (0.0592\ V)\ pH$

$pH = \dfrac{(0.28\ V + 0.13\ V)}{0.0592\ V} = 6.9$

18.10 $\quad 4\ Fe^{2+}(aq) + O_2(g) + 4\ H^+(aq) \rightarrow 4\ Fe^{3+}(aq) + 2\ H_2O(l)$

$E^\circ = E^\circ_{Fe^{2+} \to Fe^{3+}} + E^\circ_{O_2 \to H_2O} = -0.77\ V + 1.23\ V = 0.46\ V; \quad n = 4\ mol\ e^-$

$E^\circ = \dfrac{0.0592}{n} \log K$

$\log K = \dfrac{nE^\circ}{0.0592} = \dfrac{(4)(0.46\ V)}{0.0592\ V} = 31; \quad\quad K = 10^{31}$ at 25°C

18.11 $E° = \dfrac{0.0592}{n} \log K = \dfrac{0.0592}{2} \log(1.8 \times 10^{-5}) = -0.140 \text{ V}$

18.12 (a) $Zn(s) + 2 MnO_2(s) + 2 NH_4^+(aq) \rightarrow Zn^{2+}(aq) + Mn_2O_3(s) + 2 NH_3(aq) + H_2O(l)$
    (b) $Zn(s) + 2 MnO_2(s) \rightarrow ZnO(s) + Mn_2O_3(s)$
    (c) $Zn(s) + HgO(s) \rightarrow ZnO(s) + Hg(l)$
    (d) $Cd(s) + 2 NiO(OH)(s) + 2 H_2O(l) \rightarrow Cd(OH)_2(s) + 2 Ni(OH)_2(s)$

18.13 (a) cathode reaction $\quad 2 H_2O(l) + 2 e^- \rightarrow H_2(g) + 2 OH^-(aq)$
        anode reaction $\quad \underline{2 Cl^-(aq) \rightarrow Cl_2(g) + 2 e^-}$
        overall reaction $\quad 2 Cl^-(aq) + 2 H_2O(l) \rightarrow Cl_2(g) + H_2(g) + 2 OH^-(aq)$
    (b) cathode reaction $\quad 2 Cu^{2+}(aq) + 4 e^- \rightarrow 2 Cu(s)$
        anode reaction $\quad \underline{2 H_2O(l) \rightarrow O_2(g) + 4 H^+(aq) + 4 e^-}$
        overall reaction $\quad 2 Cu^{2+}(aq) + 2 H_2O(l) \rightarrow 2 Cu(s) + O_2(g) + 4 H^+(aq)$

18.14 $\text{Charge} = \left(1.00 \times 10^5 \dfrac{C}{s}\right)(8.00 \text{ h})\left(\dfrac{60 \text{ min}}{h}\right)\left(\dfrac{60 \text{ s}}{\text{min}}\right) = 2.88 \times 10^9 \text{ C}$

$\text{Moles of } e^- = (2.88 \times 10^9 \text{ C})\left(\dfrac{1 \text{ mol } e^-}{96,500 \text{ C}}\right) = 2.98 \times 10^4 \text{ mol } e^-$

cathode reaction: $Al^{3+} + 3 e^- \rightarrow Al$

$\text{mass Al} = (2.98 \times 10^4 \text{ mol } e^-) \times \dfrac{1 \text{ mol Al}}{3 \text{ mol } e^-} \times \dfrac{26.98 \text{ g Al}}{1 \text{ mol Al}} \times \dfrac{1 \text{ kg}}{1000 \text{ g}} = 268 \text{ kg Al}$

18.15 $3.00 \text{ g Ag} \times \dfrac{1 \text{ mol Ag}}{107.9 \text{ g Ag}} = 0.0278 \text{ mol Ag}$

cathode reaction: $Ag^+(aq) + e^- \rightarrow Ag(s)$

$\text{Charge} = (0.0278 \text{ mol Ag})\left(\dfrac{1 \text{ mol } e^-}{1 \text{ mol Ag}}\right)\left(\dfrac{96,500 \text{ C}}{1 \text{ mol } e^-}\right) = 2682.7 \text{ C}$

$\text{Time} = \dfrac{C}{A} = \left(\dfrac{2682.7 \text{ C}}{0.100 \text{ C/s}} \times \dfrac{1 \text{ h}}{3600 \text{ s}}\right) = 7.45 \text{ h}$

Understanding Key Concepts

18.16 (a) - (d)

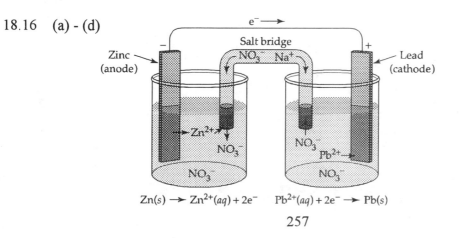

(e) anode reaction $Zn(s) \rightarrow Zn^{2+}(aq) + 2\,e^-$
    cathode reaction $\underline{Pb^{2+}(aq) + 2\,e^- \rightarrow Pb(s)}$
    overall reaction $Zn(s) + Pb^{2+}(aq) \rightarrow Zn^{2+}(aq) + Pb(s)$

18.17 (a) anode is Ni; cathode is Pt
    (b) anode reaction $3\,Ni(s) \rightarrow 3\,Ni^{2+}(aq) + 6\,e^-$
        cathode reaction $\underline{Cr_2O_7^{2-}(aq) + 14\,H^+(aq) + 6\,e^- \rightarrow 2\,Cr^{3+}(aq) + 7\,H_2O(l)}$
        overall reaction $Cr_2O_7^{2-}(aq) + 3\,Ni(s) + 14\,H^+(aq) \rightarrow$
$$2\,Cr^{3+}(aq) + 3\,Ni^{2+}(aq) + 7\,H_2O(l)$$

    (c) $Ni(s)\,|\,Ni^{2+}(aq)\,\|\,Cr_2O_7^{2-}(aq),\,Cr^{3+}\,|\,Pt(s)$

18.18 (a) The three cell reactions are the same except for cation concentrations.
    anode reaction $Cu(s) \rightarrow Cu^{2+}(aq) + 2\,e^-$ $E° = -0.34$ V
    cathode reaction $\underline{2\,Fe^{3+}(aq) + 2\,e^- \rightarrow 2\,Fe^{2+}(aq)}$ $\underline{E° = \;\;0.77}$ V
    overall reaction $Cu(s) + 2\,Fe^{3+}(aq) \rightarrow Cu^{2+}(aq) + 2\,Fe^{2+}(aq)$ $E° = \;\;0.43$ V

(b)

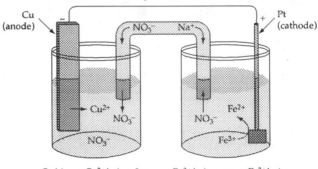

$$Cu(s) \rightarrow Cu^{2+}(aq) + 2e^- \qquad Fe^{3+}(aq) + e^- \rightarrow Fe^{2+}(aq)$$

(c) $E = E° - \dfrac{0.0592}{n} \log \dfrac{[Cu^{2+}][Fe^{2+}]^2}{[Fe^{2+}]^2}; \quad n = 2$ mol e^-

(1) $E = E° = 0.43$ V because all cation concentrations are 1 M.

(2) $E = E° - \dfrac{0.0592}{2} \log \dfrac{(1)(5)^2}{(1)^2} = 0.39$ V

(3) $E = E° - \dfrac{0.0592}{2} \log \dfrac{(0.1)(0.1)^2}{(0.1)^2} = 0.46$ V

Cell (3) has the largest potential, while cell (2) has the smallest as calculated from the Nernst equation.

18.19 (a) - (b)

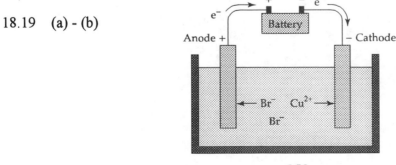

(c) anode reaction $2\,Br^-(aq) \rightarrow Br_2(aq) + 2\,e^-$

cathode reaction $\underline{Cu^{2+}(aq) + 2\,e^- \rightarrow Cu(s)}$

overall reaction $Cu^{2+}(aq) + 2\,Br^-(aq) \rightarrow Cu(s) + Br_2(aq)$

18.20   (a) oxidizing agents: PbO_2, H^+, $Cr_2O_7^{2-}$;   reducing agents: Al, Fe, Ag

(b) PbO_2 is the strongest oxidizing agent. H^+ is the weakest oxidizing agent.

(c) Al is the strongest reducing agent.   Ag is the weakest reducing agent.

(d) oxidized by Cu^{2+}: Fe and Al;   reduced by H_2O_2: PbO_2 and $Cr_2O_7^{2-}$

18.21   (a) From: $B + A^+ \rightarrow B^+ + A$, A^+ is reduced more easily than B^+

From: $C + A^+ \rightarrow C^+ + A$, A^+ is reduced more easily than C^+

From: $B + C^+ \rightarrow B^+ + C$, C^+ is reduced more easily than B^+

$A^+ + e^- \rightarrow A$

$C^+ + e^- \rightarrow C$

$B^+ + e^- \rightarrow B$

(b) A^+ is the strongest oxidizing agent;   B is the strongest reducing agent

(c) $A^+ + B \rightarrow B^+ + A$

Additional Problems
Galvanic Cells

18.22   The electrode where oxidation takes place is called the anode. For example, the lead electrode in the lead storage battery.

The electrode where reduction takes place is called the cathode. For example, the PbO_2 electrode in the lead storage battery.

18.24   The cathode of a galvanic cell is considered to be the positive electrode because electrons flow through the external circuit toward the positive electrode (the cathode).

18.26   (a) $Cd(s) + Sn^{2+}(aq) \rightarrow Cd^{2+}(aq) + Sn(s)$

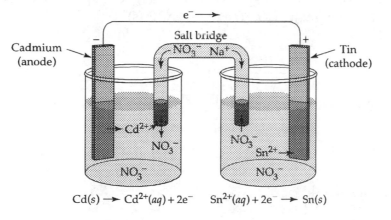

(b) $2 \, Al(s) \; + \; 3 \, Cd^{2+}(aq) \; \rightarrow \; 2 \, Al^{3+}(aq) \; + \; 3 \, Cd(s)$

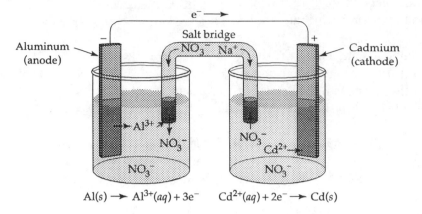

(c) $6 \, Fe^{2+}(aq) + Cr_2O_7^{2-}(aq) + 14 \, H^+(aq) \; \rightarrow \; 6 \, Fe^{3+}(aq) + 2 \, Cr^{3+}(aq) + 7 \, H_2O(l)$

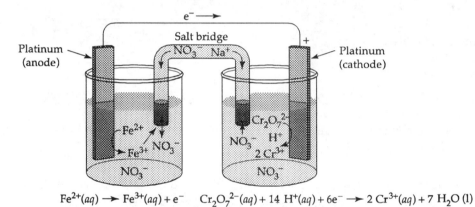

18.28   (a) $Cd(s)\,|\,Cd^{2+}(aq)\,\|\,Sn^{2+}(aq)\,|\,Sn(s)$ (b) $Al(s)\,|\,Al^{3+}(aq)\,\|\,Cd^{2+}(aq)\,|\,Cd(s)$
        (c) $Pt(s)\,|\,Fe^{2+}(aq),\, Fe^{3+}(aq)\,\|\,Cr_2O_7^{2-}(aq),\, Cr^{3+}(aq)\,|\,Pt(s)$

18.30   (a)

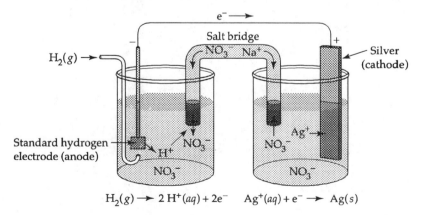

  (b) anode reaction $H_2(g) \rightarrow 2\,H^+(aq) + 2\,e^-$
    cathodes reaction $\underline{2\,Ag^+(aq) + 2\,e^- \rightarrow 2\,Ag(s)}$
    overall reaction $H_2(g) + 2\,Ag^+(aq) \rightarrow 2\,H^+(aq) + 2\,Ag(s)$
  (c) $Pt(s)|H_2(g)|H^+(aq)\|Ag^+(aq)|Ag(s)$

18.32 (a) anode reaction $Co(s) \rightarrow Co^{2+}(aq) + 2\,e^-$
    cathode reaction $\underline{Cu^{2+}(aq) + 2\,e^- \rightarrow Cu(s)}$
    overall reaction $Co(s) + Cu^{2+}(aq) \rightarrow Co^{2+}(aq) + Cu(s)$

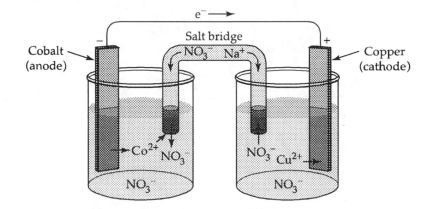

  (b) anode reaction $2\,Fe(s) \rightarrow 2\,Fe^{2+}(aq) + 4\,e^-$
    cathode reaction $\underline{O_2(g) + 4\,H^+(aq) + 4\,e^- \rightarrow 2\,H_2O(l)}$
    overall reaction $2\,Fe(s) + O_2(g) + 4\,H^+(aq) \rightarrow 2\,Fe^{2+}(aq) + 2\,H_2O(l)$

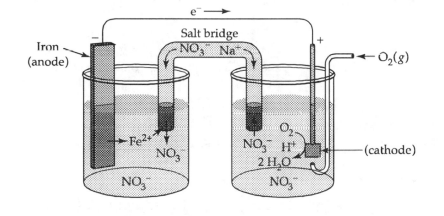

Cell Potentials and Free-Energy Changes; Standard Reduction potentials

18.34 The SI unit of electrical potential is the volt (V).
   The SI unit of charge is the coulomb (C).
   The SI unit of energy is the joule (J).
   $1\,J = 1\,C \cdot 1\,V$

18.36 E is the standard cell potential (E°) when all reactants and products are in their standard states--solutes at 1 M concentrations, gases at a partial pressure of 1 atm, solids and liquids in pure form, all at $25^\circ C$.

18.38 $Zn(s) + Ag_2O(s) \rightarrow ZnO(s) + 2\ Ag(s)$; $n = 2$ mol e^- and 1 J = 1 C x 1 V

$$\Delta G = -nFE = -(2 \text{ mol } e^-)\left(\frac{96,500 \text{ C}}{1 \text{ mol } e^-}\right)(1.60 \text{ V}) = -308,800 \text{ J} = -309 \text{ kJ}$$

18.40 $2\ H_2(g) + O_2(g) \rightarrow 2\ H_2O(l)$; $n = 4$ mol e^- and 1 V = 1 J/C
$\Delta G^\circ = 2\ \Delta G^\circ_f(H_2O(l)) = (2 \text{ mol})(-237.2 \text{ kJ/mol}) = -474.4 \text{ kJ}$
$\Delta G^\circ = -nFE^\circ$

$$E^\circ = \frac{-\Delta G^\circ}{nF} = \frac{-(-474,400 \text{ J})}{(4 \text{ mol } e^-)\left(\dfrac{96,500 \text{ C}}{1 \text{ mol } e^-}\right)} = +1.23 \text{ J/C} = +1.23 \text{ V}$$

18.42
oxidation: $Zn(s) \rightarrow Zn^{2+}(aq) + 2\ e^-$ $E^\circ = 0.76$ V
reduction: $Eu^{3+}(aq) + e^- \rightarrow Eu^{2+}(aq)$ $E^\circ = ?$
overall $Zn(s) + 2\ Eu^{3+}(aq) \rightarrow Zn^{2+}(aq) + 2\ Eu^{2+}(aq)$ $E^\circ = 0.40$ V
The standard reduction potential for the Eu^{3+}/Eu^{2+} half cell is: $E^\circ = 0.40 - 0.76 = -0.36$ V

18.44 From Table 18.1, the following oxidizing agents are listed in order of increasing strength: $Cu^{2+} < O_2 < Cl_2$

18.46 From Table 18.1:
MnO_4^- is the strongest oxidizing and Fe^{2+} is the weakest oxidizing agent.

18.48 (a) $Cd(s) + Sn^{2+}(aq) \rightarrow Cd^{2+}(aq) + Sn(s)$
oxidation: $Cd(s) \rightarrow Cd^{2+}(aq) + 2\ e^-$ $E^\circ = 0.40$ V
reduction: $Sn^{2+}(aq) + 2\ e^- \rightarrow Sn(s)$ $E^\circ = -0.14$ V
 overall $E^\circ = 0.26$ V
$n = 2$ mol e^- and 1 J = 1 C x 1 V

$$\Delta G^\circ = -nFE^\circ = -(2 \text{ mol } e^-)\left(\frac{96,500 \text{ C}}{1 \text{ mol } e^-}\right)(0.26 \text{ V}) = -50,180 \text{ J} = -50 \text{ kJ}$$

(b) $2\ Al(s) + 3\ Cd^{2+}(aq) \rightarrow 2\ Al^{3+}(aq) + 3\ Cd(s)$
oxidation: $2\ Al(s) \rightarrow 2\ Al^{3+}(aq) + 6\ e^-$ $E^\circ = 1.66$ V
reduction: $3\ Cd^{2+}(aq) + 6\ e^- \rightarrow 3\ Cd(s)$ $E^\circ = -0.40$ V
 overall $E^\circ = 1.26$ V
$n = 6$ mol e^- and 1 J = 1 C x 1 V

$$\Delta G^\circ = -nFE^\circ = -(6 \text{ mol } e^-)\left(\frac{96,500 \text{ C}}{1 \text{ mol } e^-}\right)(1.26 \text{ V}) = -729,540 \text{ J} = -730 \text{ kJ}$$

(c) $6 Fe^{2+}(aq) + Cr_2O_7^{2-}(aq) + 14 H^+(aq) \rightarrow 6 Fe^{3+}(aq) + 2 Cr^{3+}(aq) + 7 H_2O(l)$

oxidation:	$6 Fe^{2+}(aq) \rightarrow 6 Fe^{3+}(aq) + 6 e^-$	$E^\circ = -0.77$ V
reduction:	$Cr_2O_7^{2-}(aq) + 14 H^+(aq) + 6 e^- + \rightarrow 2 Cr^{3+}(aq) + 7 H_2O(l)$	$\underline{E^\circ = 1.33\ V}$
		overall $E^\circ = 0.56$ V

$n = 6$ mol e⁻ and $1 J = 1 C \times 1 V$

$$\Delta G^\circ = -nFE^\circ = -(6\ \text{mol e}^-)\left(\frac{96,500\ C}{1\ \text{mol e}^-}\right)(0.56\ V) = -324,240\ J = -324\ kJ$$

18.50   (a) $2 Fe^{2+}(aq) + Pb^{2+}(aq) \rightarrow 2 Fe^{3+}(aq) + Pb(s)$

oxidation:	$2 Fe^{2+}(aq) \rightarrow 2 Fe^{3+}(aq) + 2 e^-$	$E^\circ = -0.77$ V
reduction:	$Pb^{2+}(aq) + 2 e^- \rightarrow Pb(s)$	$\underline{E^\circ = -0.13\ V}$
		overall $E^\circ = -0.90$ V

Because E° is negative, this reaction is nonspontaneous.

(b) $Mg(s) + Ni^{2+}(aq) \rightarrow Mg^{2+}(aq) + Ni(s)$

oxidation:	$Mg(s) \rightarrow Mg^{2+}(aq) + 2 e^-$	$E^\circ = 2.37$ V
reduction:	$Ni^{2+}(aq) + 2 e^- \rightarrow Ni(s)$	$\underline{E^\circ = -0.26\ V}$
		overall $E^\circ = 2.11$ V

Because E° is positive, this reaction is spontaneous.

18.52   (a) oxidation: $Sn^{2+}(aq) \rightarrow Sn^{4+}(aq) + 2 e^-$ $E^\circ = -0.15$ V

       reduction: $Br_2(l) + 2 e^- \rightarrow 2 Br^-(aq)$ $\underline{E^\circ = 1.09\ V}$

                       overall $E^\circ = +0.94$

Because the overall E° is positive, $Sn^{2+}(aq)$ can be oxidized by $Br_2(l)$.

(b) oxidation: $Sn^{2+}(aq) \rightarrow Sn^{4+}(aq) + 2 e^-$ $E^\circ = -0.15$ V

    reduction: $Ni^{2+}(aq) + 2 e^- \rightarrow Ni(s)$ $\underline{E^\circ = -0.26\ V}$

                    overall $E^\circ = -0.41$ V

Because the overall E° is negative, $Ni^{2+}(aq)$ cannot be reduced by $Sn^{2+}(aq)$.

(c) oxidation: $2 Ag(s) \rightarrow 2 Ag^+(aq) + 2 e^-$ $E^\circ = -0.80$ V

    reduction: $Pb^{2+}(aq) + 2 e^- \rightarrow Pb(s)$ $\underline{E^\circ = -0.13\ V}$

                    overall $E^\circ = -0.93$ V

Because the overall E° is negative, $Ag(s)$ cannot be oxidized by $Pb^{2+}(aq)$.

(d) oxidation: $Cu(s) \rightarrow Cu^{2+}(aq) + 2 e^-$ $E^\circ = -0.34$ V

    reduction: $I_2(s) + 2 e^- \rightarrow 2 I^-(aq)$ $\underline{E^\circ = 0.54\ V}$

                    overall $E^\circ = +0.20$ V

Because the overall E° positive, $I_2(s)$ can be reduced by $Cu(s)$.

The Nernst Equation

18.54 $2 Ag^+(aq) + Sn(s) \rightarrow 2 Ag(s) + Sn^{2+}(aq)$

oxidation:	$Sn(s) \rightarrow Sn^{2+}(aq) + 2 e^-$	$E^\circ = 0.14$ V
reduction:	$2 Ag^+(aq) + 2 e^- \rightarrow 2 Ag(s)$	$\underline{E^\circ = 0.80\ V}$
		overall $E^\circ = 0.94$ V

$$E = E° - \frac{0.0592}{n} \log \frac{[Sn^{2+}]}{[Ag^+]^2} = 0.94\ V - \frac{(0.0592\ V)}{2} \log \frac{(0.020)}{(0.010)^2} = 0.87\ V$$

18.56 $Pb(s) + Cu^{2+}(aq) \rightarrow Pb^{2+}(aq) + Cu(s)$
oxidation: $Pb(s) \rightarrow Pb^{2+}(aq) + 2\ e^-$ $E° = 0.13\ V$
reduction: $Cu^{2+}(aq) + 2\ e^- \rightarrow Cu(s)$ $\underline{E° = 0.34\ V}$
$\qquad\qquad\qquad\qquad\qquad\qquad$ overall $E° = 0.47\ V$

$$E = E° - \frac{0.0592}{n} \log \frac{[Pb^{2+}]}{[Cu^{2+}]} = 0.47\ V - \frac{(0.0592\ V)}{2} \log \frac{1.0}{(1.0 \times 10^{-4})} = 0.35\ V$$

When $E = 0$, $0 = E° - \frac{0.0592}{n} \log \frac{[Pb^{2+}]}{[Cu^{2+}]} = 0.47\ V - \frac{(0.0592\ V)}{2} \log \frac{1.0}{[Cu^{2+}]}$

$$0 = 0.47\ V + \frac{(0.0592\ V)}{2} \log [Cu^{2+}]$$

$$\log [Cu^{2+}] = (-0.47\ V)\left(\frac{2}{0.0592\ V}\right) = -15.88; \qquad [Cu^{2+}] = 10^{-15.88} = 1 \times 10^{-16}\ M$$

18.58 (a) $E = E° - \frac{0.0592}{n} \log [I^-]^2 = 0.54\ V - \frac{(0.0592\ V)}{2} \log (0.020)^2 = 0.64\ V$

(b) $E = E° - \frac{0.0592}{n} \log \frac{[Fe^{2+}]}{[Fe^{3+}]} = 0.77\ V - \frac{(0.0592\ V)}{1} \log\left(\frac{0.10}{0.10}\right) = 0.77\ V$

(c) $E = E° - \frac{0.0592}{n} \log \frac{[Sn^{4+}]}{[Sn^{2+}]} = -0.15\ V - \frac{(0.0592\ V)}{2} \log\left(\frac{0.40}{0.0010}\right) = -0.23\ V$

(d) $E = E° - \frac{0.0592}{n} \log \frac{[Cr_2O_7^{2-}][H^+]^{14}}{[Cr^{3+}]^2} = -1.33\ V - \frac{(0.0592\ V)}{6} \log\left(\frac{(1.0)(0.010)^{14}}{1.0}\right)$

$E = -1.33\ V - \frac{(0.0592\ V)}{6}(14) \log (0.010) = -1.05\ V$

18.60 $H_2(g) + Ni^{2+}(aq) \rightarrow 2\ H^+(aq) + Ni(s)$
$E° = E°_{H_2 \rightarrow H^+} + E°_{Ni^{2+} \rightarrow Ni} = 0\ V + (-0.26\ V) = -0.26\ V$

$$E = E° - \frac{0.0592}{n} \log \frac{[H_3O^+]^2}{[Ni^{2+}](P_{H_2})}$$

$$0.27\ V = -0.26\ V - \frac{(0.0592\ V)}{2} \log \frac{[H_3O^+]^2}{(1)(1)}$$

$0.27\ V = -0.26\ V - (0.0592\ V) \log [H_3O^+]$

$pH = -\log [H_3O^+]$ therefore $0.27\ V = -0.26\ V + (0.0592\ V)\ pH$
$pH = \frac{(0.27\ V + 0.26\ V)}{0.0592\ V} = 9.0$

Standard Cell Potentials and Equilibrium Constants

18.62 $\Delta G° = -nFE°$

Because n and F are always positive, $\Delta G°$ is negative when $E°$ is positive because of the negative sign in the equation.

$$E° = \frac{0.0592}{n} \log K; \quad \log K = \frac{nE°}{0.0592}; \quad K = 10^{\frac{nE°}{0.0592}}$$

If $E°$ is positive, the exponent is positive (because n is positive), and K is greater than 1.

18.64 $Ni(s) + 2\ Ag^+(aq) \rightarrow Ni^{2+}(aq) + 2\ Ag(s)$
  oxidation: $Ni(s) \rightarrow Ni^{2+}(aq) + 2\ e^-$ $E° = 0.26\ V$
  reduction: $2\ Ag^+(aq) + 2\ e^- \rightarrow 2\ Ag(s)$ $\underline{E° = 0.80\ V}$
                            overall $E° = 1.06\ V$

$$E° = \frac{0.0592}{n} \log K; \quad \log K = \frac{nE°}{0.0592} = \frac{(2)(1.06)}{0.0592} = 35.8; \quad K = 10^{35.8} = 6 \times 10^{35}$$

18.66 $E°$ and n are from Problem 18.48.

$$E° = \frac{0.0592}{n} \log K; \quad \log K = \frac{nE°}{0.0592}$$

(a) $Cd(s) + Sn^{2+}(aq) \rightarrow Cd^{2+}(aq) + Sn(s);$ $E° = 0.26\ V$ and $n = 2\ mol\ e^-$

$$\log K = \frac{(2)(0.26)}{0.0592} = 8.8; \qquad K = 10^{8.8} = 6 \times 10^8$$

(b) $2\ Al(s) + 3\ Cd^{2+}(aq) \rightarrow 2\ Al^{3+}(aq) + 3\ Cd(s);$ $E° = 1.26\ V$ and $n = 6\ mol\ e^-$

$$\log K = \frac{(6)(1.26)}{0.0592} = 128; \qquad K = 10^{128}$$

(c) $6\ Fe^{2+}(aq) + Cr_2O_7^{2-}(aq) + 14\ H^+(aq) \rightarrow 6\ Fe^{3+}(aq) + 2\ Cr^{3+}(aq) + 7\ H_2O(l)$
$E° = 0.56\ V$ and $n = 6\ mol\ e^-$

$$\log K = \frac{(6)(0.56)}{0.0592} = 57; \qquad K = 10^{57}$$

18.68 $Hg_2^{2+}(aq) \rightarrow Hg(l) + Hg^{2+}(aq)$
  oxidation: $\frac{1}{2}[Hg_2^{2+}(aq) \rightarrow 2\ Hg^{2+}(aq) + 2\ e^-]$ $E° = -0.92\ V$
  reduction: $\frac{1}{2}[Hg_2^{2+}(aq) + 2\ e^- \rightarrow 2\ Hg(l)]$ $\underline{E° = \ \ 0.80\ V}$
                            overall $E° = -0.12\ V$

$$E° = \frac{0.0592}{n} \log K; \quad \log K = \frac{nE°}{0.0592} = \frac{(1)(-0.12)}{0.0592} = -2.027; \quad K = 10^{-2.027} = 9 \times 10^{-3}$$

Batteries; Corrosion

18.70 Rust is a hydrated form of iron(III) oxide ($Fe_2O_3 \cdot H_2O$). Rust forms from the oxidation of Fe in the presence of O_2 and H_2O. Rust can be prevented by coating Fe with Zn (galvanizing).

18.72   Cathodic protection is the attachment of a more easily oxidized metal to the metal you want to protect. This forces the metal you want to protect to be the cathode, hence the name, cathodic protection.
Zn and Al can offer cathodic protection to Fe (Ni and Sn cannot).

18.74   (a)

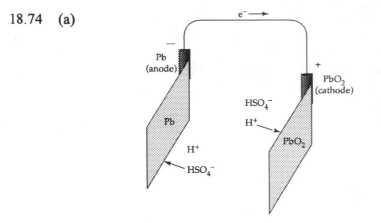

(b)   Anode: $Pb(s) + HSO_4^-(aq) \rightarrow PbSO_4(s) + H^+(aq) + 2\ e^-$ $E° = 0.296$ V
        Cathode: $\underline{PbO_2(s) + 3\ H^+(aq) + HSO_4^-(aq) + 2\ e^- \rightarrow PbSO_4(s) + 2\ H_2O(l)}$ $\underline{E° = 1.628\ V}$
    Overall $Pb(s) + PbO_2(s) + 2\ H^+(aq) + 2\ HSO_4^-(aq) \rightarrow 2\ PbSO_4(s) + 2\ H_2O(l)$ $E° = 1.924$ V

    (c) $E° = \dfrac{0.0592}{n} \log K$; $\log K = \dfrac{nE°}{0.0592} = \dfrac{(2)(1.924)}{0.0592} = 65.0$; $K = 10^{65}$

    (d) When the cell reaction reaches equilibrium the cell voltage = 0.

18.76 $Zn(s) + HgO(s) \rightarrow ZnO(s) + Hg(l)$;   Zn, 65.39 amu; HgO, 216.59 amu
    mass HgO = 2.00 g Zn x $\dfrac{1\ mol\ Zn}{65.39\ g\ Zn}$ x $\dfrac{1\ mol\ HgO}{1\ mol\ Zn}$ x $\dfrac{216.59\ g\ HgO}{1\ mol\ HgO}$ = 6.62 g HgO

Electrolysis

18.78   (a)

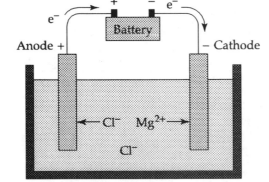

    (b) anode: $2\ Cl^-(l) \rightarrow Cl_2(g) + 2\ e^-$
        cathode: $\underline{Mg^{2+}(l) + 2\ e^- \rightarrow Mg(l)}$
        overall: $Mg^{2+}(l) + 2\ Cl^-(l) \rightarrow Mg(l) + Cl_2(g)$

18.80   possible anode reactions:

$$2 \text{ Cl}^-(aq) \rightarrow \text{Cl}_2(g) + 2 \text{ e}^-$$
$$2 \text{ H}_2\text{O}(l) \rightarrow \text{O}_2(g) + 4 \text{ H}^+(aq) + 4 \text{ e}^-$$

possible cathode reactions:

$$2 \text{ H}_2\text{O}(l) + 2 \text{ e}^- \rightarrow \text{H}_2(g) + 2 \text{ OH}^-(aq)$$
$$\text{Mg}^{2+}(aq) + 2 \text{ e}^- \rightarrow \text{Mg}(s)$$

actual reactions:

anode: $2 \text{ Cl}^-(aq) \rightarrow \text{Cl}_2(g) + 2 \text{ e}^-$

cathode: $2 \text{ H}_2\text{O}(l) + 2 \text{ e}^- \rightarrow \text{H}_2(g) + 2 \text{ OH}^-(aq)$

This anode reaction takes place instead of $2 \text{ H}_2\text{O}(l) \rightarrow \text{O}_2(g) + 4 \text{ H}^+(aq) + 4 \text{ e}^-$ because of a high overvoltage for formation of gaseous O_2.
This cathode reaction takes place instead of $\text{Mg}^{2+}(aq) + 2 \text{ e}^- \rightarrow \text{Mg}(s)$ because H_2O is easier to reduce than Mg^{2+}.

18.82   (a) NaBr

anode: $2 \text{ Br}^-(aq) \rightarrow \text{Br}_2(l) + 2 \text{ e}^-$

cathode: $\underline{2 \text{ H}_2\text{O}(l) + 2 \text{ e}^- \rightarrow \text{H}_2(g) + 2 \text{ OH}^-(aq)}$

overall: $2 \text{ H}_2\text{O}(l) + 2 \text{ Br}^-(aq) \rightarrow \text{Br}_2(l) + \text{H}_2(g) + 2 \text{ OH}^-(aq)$

(b) CuCl_2

anode: $2 \text{ Cl}^-(aq) \rightarrow \text{Cl}_2(g) + 2 \text{ e}^-$

cathode: $\underline{\text{Cu}^{2+}(aq) + 2 \text{ e}^- \rightarrow \text{Cu}(s)}$

overall: $\text{Cu}^{2+}(aq) + 2 \text{ Cl}^-(aq) \rightarrow \text{Cu}(s) + \text{Cl}_2(g)$

(c) LiOH

anode: $4 \text{ OH}^-(aq) \rightarrow \text{O}_2(g) + 2 \text{ H}_2\text{O}(l) + 4 \text{ e}^-$

cathode: $\underline{4 \text{ H}_2\text{O}(l) + 4 \text{ e}^- \rightarrow 2 \text{ H}_2(g) + 4 \text{ OH}^-(aq)}$

overall: $2 \text{ H}_2\text{O}(l) \rightarrow \text{O}_2(g) + 2 \text{ H}_2(g)$

18.84 $\text{Ag}^+(aq) + \text{e}^- \rightarrow \text{Ag}(s);$ $1 \text{ A} = 1 \text{ C/s}$

$$\text{mass Ag} = 2.40 \frac{\text{C}}{\text{s}} \times 20.0 \text{ min} \times \frac{60 \text{ s}}{1 \text{ min}} \times \frac{1 \text{ mol e}^-}{96,500 \text{ C}} \times \frac{1 \text{ mol Ag}}{1 \text{ mol e}^-} \times \frac{107.87 \text{ g Ag}}{1 \text{ mol Ag}} = 3.22 \text{ g}$$

18.86 $2 \text{ Na}^+(l) + 2 \text{ Cl}^-(l) \rightarrow 2 \text{ Na}(l) + \text{Cl}_2(g)$

$\text{Na}^+(l) + \text{e}^- \rightarrow \text{Na}(l);$ $1 \text{ A} = 1 \text{ C/s};$ $1.00 \times 10^3 \text{ kg} = 1.00 \times 10^6 \text{ g}$

$$\text{Charge} = 1.00 \times 10^6 \text{ g Na} \times \frac{1 \text{ mol Na}}{22.99 \text{ g Na}} \times \frac{1 \text{ mol e}^-}{1 \text{ mol Na}} \times \frac{96,500 \text{ C}}{1 \text{ mol e}^-} = 4.20 \times 10^9 \text{ C}$$

$$\text{Time} = \frac{4.20 \times 10^9 \text{ C}}{30,000 \text{ C/s}} \times \frac{1 \text{ h}}{3600 \text{ s}} = 38.9 \text{ h}$$

$$1.00 \times 10^6 \text{ g Na} \times \frac{1 \text{ mol Na}}{22.99 \text{ g Na}} \times \frac{1 \text{ mol Cl}_2}{2 \text{ mol Na}} = 21,748.6 \text{ mol Cl}_2$$

$PV = nRT$

$$V = \frac{nRT}{P} = \frac{(21,748.6 \text{ mol})\left(0.08206 \dfrac{\text{L} \cdot \text{atm}}{\text{mol} \cdot \text{K}}\right)(273.15 \text{ K})}{1.00 \text{ atm}} = 4.87 \times 10^5 \text{ L Cl}_2$$

267

18.88 $PbSO_4(s) + H^+(aq) + 2 e^- \rightarrow Pb(s) + HSO_4^-(aq)$

$$\text{mass PbSO}_4 = 10.0 \frac{C}{s} \times 1.50 \text{ h} \times \frac{3600 \text{ s}}{1 \text{ h}} \times \frac{1 \text{ mol e}^-}{96,500 \text{ C}} \times \frac{1 \text{ mol PbSO}_4}{2 \text{ mol e}^-} \times \frac{303.3 \text{ g PbSO}_4}{1 \text{ mol PbSO}_4}$$

mass $PbSO_4$ = 84.9 g $PbSO_4$

General Problems

18.90 (a) oxidizing agent – a substance that causes an oxidation by gaining electrons.
reducing agent – a substance that causes a reduction by losing electrons.
(b) galvanic cell – an electrochemical cell in which a spontaneous chemical reaction generates an electric current.
electrolytic cell – an electrochemical cell in which an electric current is used to drive a nonspontaneous reaction.
(c) battery – a self-contained galvanic cell.
fuel cell – a galvanic cell in which one of the reactants is a traditional fuel such as CH_4 or H_2.
(d) anode – electrode where oxidation occurs.
cathode – electrode where reduction occurs.

18.92 mercury battery $Zn(s) + HgO(s) \rightarrow ZnO(s) + Hg(l)$

$$E = E^\circ - \frac{0.0592}{2} \log(1)$$

In the Nernst equation for the mercury battery, the Q term is 1 (and log Q = 0) because all reactants and products are solids and a liquid. Consequently, the potential is independent of concentration and is stable until the reactants are consumed.

18.94 (a) $2 MnO_4^-(aq) + 16 H^+(aq) + 5 Sn^{2+}(aq) \rightarrow 2 Mn^{2+}(aq) + 5 Sn^{4+}(aq) + 8 H_2O(l)$
(b) MnO_4^- is the oxidizing agent; Sn^{2+} is the reducing agent.
(c) E° = 1.51 V + (–0.15 V) = 1.36 V

18.96 (a) Ag^+ is the strongest oxidizing agent because Ag^+ has the most positive standard reduction potential.
Pb is the strongest reducing agent because Pb has the most positive standard oxidation potential.
(b)
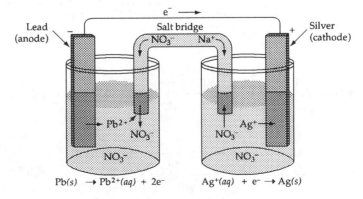

(c) $Pb(s) + 2 Ag^+(aq) \rightarrow Pb^{2+}(aq) + 2 Ag(s);$ $n = 2$ mol e^- and $1 J = 1 C \times 1 V$
$E^\circ = E^\circ_{ox} + E^\circ_{red} = 0.13 V + 0.80 V = 0.93 V$

$$\Delta G^\circ = -nFE^\circ = -(2 \text{ mol } e^-) \left(\frac{96,500 C}{1 \text{ mol } e^-} \right) (0.93 V) = -179,490 J = -180 \text{ kJ}$$

$$E^\circ = \frac{0.0592}{n} \log K; \quad \log K = \frac{nE^\circ}{0.0592} = \frac{(2)(0.93)}{0.0592} = 31; \quad K = 10^{31}$$

(d) $E = E^\circ - \dfrac{0.0592}{n} \log \dfrac{[Pb^{2+}]}{[Ag^+]^2} = 0.93 V - \dfrac{0.0592}{2} \log \left(\dfrac{0.01}{(0.01)^2} \right) = 0.87 V$

18.98 $\underline{E^\circ (V)}$
$MnO_4^-(aq) + 8 H^+(aq) + 5 e^- \rightarrow Mn^{2+}(aq) + 4 H_2O(l)$ 1.51
$Br_2(l) + 2 e^- \rightarrow 2 Br^-(aq)$ 1.09
$Fe^{3+}(aq) + e^- \rightarrow Fe^{2+}(aq)$ 0.77
$I_2(s) + 2 e^- \rightarrow 2 I^-(aq)$ 0.54
$Cu^{2+}(aq) + 2 e^- \rightarrow Cu(s)$ 0.34
$Pb^{2+}(aq) + 2 e^- \rightarrow Pb(s)$ −0.13
$Fe^{2+}(aq) + 2 e^- \rightarrow Fe(s)$ −0.45
$Al^{3+}(aq) + 3 e^- \rightarrow Al(s)$ −1.66

Fe^{2+} can oxidize any substance with a standard reduction potential more negative than −0.45 V. Fe^{2+} can oxidize Al.
Fe^{2+} can reduce any substance with a standard reduction potential more positive than 0.77 V. Fe^{2+} can reduce $Br_2(l)$ and MnO_4^-.

18.100 (a)

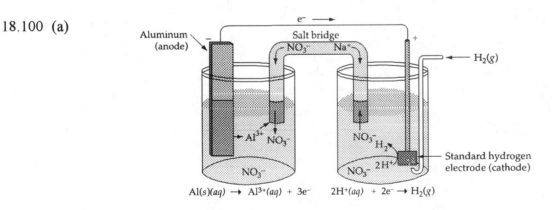

(b) $2 Al(s) + 6 H^+(aq) \rightarrow 2 Al^{3+}(aq) + 3 H_2(g)$
$E^\circ = E^\circ_{ox} + E^\circ_{red} = 1.66 V + 0.00 V = 1.66 V$
(c)

$$E = E^\circ - \frac{0.0592}{n} \log \frac{[Al^{3+}]^2 (P_{H_2})^3}{[H^+]^6} = 1.66 V - \frac{(0.0592 V)}{6} \log \left(\frac{(0.10)^2 (10.0)^3}{(0.10)^6} \right) = 1.59 V$$

(d) $1 J = 1 C \times 1 V$

$$\Delta G^\circ = -nFE^\circ = -(6 \text{ mol } e^-)\left(\frac{96,500 \text{ C}}{1 \text{ mol } e^-}\right)(1.66 \text{ V}) = -961,140 \text{ J} = -961 \text{ kJ}$$

$$E^\circ = \frac{0.0592}{n} \log K; \quad \log K = \frac{nE^\circ}{0.0592} = \frac{(6)(1.66)}{0.0592} = 168; \quad K = 10^{168}$$

(e) mass Al $= 10.0 \dfrac{C}{s} \times 25.0 \text{ min} \times \dfrac{60 \text{ s}}{1 \text{ min}} \times \dfrac{1 \text{ mol } e^-}{96,500 \text{ C}} \times \dfrac{1 \text{ mol Al}}{3 \text{ mol } e^-} \times \dfrac{26.98 \text{ g Al}}{1 \text{ mol Al}} = 1.40 \text{ g}$

18.102 $2 \text{ Cl}^-(aq) \rightarrow \text{Cl}_2(g) + 2 e^-$

13 million tons = 13×10^6 tons; Cl_2, 70.91 amu

$$13 \times 10^6 \text{ tons} \times \frac{907,200 \text{ g}}{1 \text{ ton}} \times \frac{1 \text{ mol Cl}_2}{70.91 \text{ g Cl}_2} = 1.66 \times 10^{11} \text{ mol Cl}_2$$

$$\text{Charge} = 1.66 \times 10^{11} \text{ mol Cl}_2 \times \frac{2 \text{ mol } e^-}{1 \text{ mol Cl}_2} \times \frac{96,500 \text{ C}}{1 \text{ mol } e^-} = 3.20 \times 10^{16} \text{ C}$$

$1 J = 1 C \times 1 V$; Energy $= (3.20 \times 10^{16} \text{ C})(4.5 \text{ V}) = 1.44 \times 10^{17} \text{ J}$

$$\text{kWh} = (1.44 \times 10^{17} \text{ J})\left(\frac{1 \text{ kWh}}{3.6 \times 10^6 \text{ J}}\right) = 4.0 \times 10^{10} \text{ kWh}$$

18.104 (a) $3 \text{ CH}_3\text{CH}_2\text{OH}(aq) + 2 \text{ Cr}_2\text{O}_7^{2-}(aq) + 16 \text{ H}^+(aq) \rightarrow$

$\qquad\qquad\qquad\qquad 3 \text{ CH}_3\text{COOH}(aq) + 4 \text{ Cr}^{3+}(aq) + 11 \text{ H}_2\text{O}(l)$

oxidation:

$3 \text{ CH}_3\text{CH}_2\text{OH}(aq) + 3 \text{ H}_2\text{O}(l) \rightarrow 3 \text{ CH}_3\text{COOH}(aq) + 12 \text{ H}^+(aq) + 12 e^- \qquad E^\circ = -0.058\text{V}$

reduction:

$2 \text{ Cr}_2\text{O}_7^{2-}(aq) + 28 \text{ H}^+(aq) + 12 e^- + \rightarrow 4 \text{ Cr}^{3+}(aq) + 14 \text{ H}_2\text{O}(l) \qquad\qquad \underline{E^\circ = \quad 1.33 \text{ V}}$

$\qquad\qquad\qquad\qquad\qquad\qquad\qquad\qquad\qquad\qquad\qquad$ overall $E^\circ = \quad 1.27 \text{ V}$

(b) $E = E^\circ - \dfrac{0.0592}{n} \log \dfrac{[\text{CH}_3\text{COOH}]^3[\text{Cr}^{3+}]^4}{[\text{CH}_3\text{CH}_2\text{OH}]^3[\text{Cr}_2\text{O}_7^{2-}]^2[\text{H}^+]^{16}}$

pH = 4.00, $[\text{H}^+] = 0.000\ 10$ M

$$E = 1.27 \text{ V} - \frac{(0.0592 \text{ V})}{12} \log\left(\frac{(1.0)^3(1.0)^4}{(1.0)^3(1.0)^2(0.000\ 10)^{16}}\right)$$

$$E = 1.27 \text{ V} - \frac{(0.0592 \text{ V})}{12} \log \frac{1}{(0.000\ 10)^{16}} = 0.95 \text{ V}$$

18.106 (a) $4 \text{ CH}_2=\text{CHCN} + 2 \text{ H}_2\text{O} \rightarrow 2 \text{ NC(CH}_2)_4\text{CN} + \text{O}_2$

(b) mol $e^- = 3000$ C/s $\times 10.0$ h $\times \dfrac{3600 \text{ s}}{1 \text{ h}} \times \dfrac{1 \text{ mol } e^-}{96,500 \text{ C}} = 1119.2$ mol e^-

mass adiponitrile =

1119.2 mol $e^- \times \dfrac{1 \text{ mol adiponitrile}}{2 \text{ mol } e^-} \times \dfrac{108.14 \text{ g adiponitrile}}{1 \text{ mol adiponitrile}} \times \dfrac{1.0 \text{ kg}}{1000 \text{ g}} = 60.5 \text{ kg}$

(c) $1119.2 \text{ mol e}^- \times \dfrac{1 \text{ mol O}_2}{4 \text{ mol e}^-} = 279.8 \text{ mol O}_2$

$PV = nRT$

$$V = \frac{nRT}{P} = \frac{(279.8 \text{ mol})\left(0.08206 \dfrac{L \cdot atm}{mol \cdot K}\right)(298 \text{ K})}{\left(740 \text{ mm Hg} \times \dfrac{1 \text{ atm}}{760 \text{ mm Hg}}\right)} = 7030 \text{ L O}_2$$

18.108 anode: $\quad Ag(s) + Cl^-(aq) \rightarrow AgCl(s) + e^-$
   cathode: $\quad \underline{Ag^+(aq) + e^- \rightarrow Ag(s)}$
   overall: $\quad Ag^+(aq) + Cl^-(aq) \rightarrow AgCl(s) \qquad\qquad E^\circ = 0.578 \text{ V}$
   For $AgCl(s) \rightleftarrows Ag^+(aq) + Cl^-(aq) \qquad\qquad\qquad E^\circ = -0.578 \text{ V}$

$E^\circ = \dfrac{0.0592}{n} \log K; \quad \log K = \dfrac{nE^\circ}{0.0592} = \dfrac{(1)(-0.578)}{0.0592} = -9.76$

$K = K_{sp} = 10^{-9.76} = 1.7 \times 10^{-10}$

18.110 First calculate E° for the galvanic cell in order to determine E°_1.
   anode: $\quad 5 [2 \text{ Hg}(l) + 2 \text{ Br}^-(aq) \rightarrow \text{Hg}_2\text{Br}_2(s) + 2 e^-] \qquad E^\circ_1 = ?$
   cathode: $\quad 2 [\text{MnO}_4^-(aq) + 8 \text{ H}^+(aq) + 5 e^- \rightarrow \text{Mn}^{2+}(aq) + 4 \text{ H}_2\text{O}(l)] \quad E^\circ_2 = 1.51 \text{ V}$
   overall: $\quad 2 \text{ MnO}_4^-(aq) + 10 \text{ Hg}(l) + 10 \text{ Br}^-(aq) + 16 \text{ H}^+(aq) \rightarrow$
   $\qquad\qquad\qquad\qquad 2 \text{ Mn}^{2+}(aq) + 5 \text{ Hg}_2\text{Br}_2(s) + 8 \text{ H}_2\text{O}(l)$

$n = 10 \text{ mol e}^-$

$$E = E^\circ - \frac{0.0592}{n} \log \frac{[\text{Mn}^{2+}]^2}{[\text{Br}^-]^{10}[\text{MnO}_4^-]^2[\text{H}^+]^{16}}$$

$$1.214 \text{ V} = E^\circ - \frac{(0.0592 \text{ V})}{10} \log\left(\frac{(0.10)^2}{(0.10)^{10}(0.10)^2(0.10)^{16}}\right)$$

$$1.214 \text{ V} = E^\circ - \frac{(0.0592 \text{ V})}{10} \log \frac{1}{(0.10)^{26}} = E^\circ - 0.154 \text{ V}$$

$E^\circ = 1.214 + 0.154 = 1.368 \text{ V}$
$E^\circ_1 + E^\circ_2 = 1.368 \text{ V}; \quad E^\circ_1 + 1.51 \text{ V} = 1.368 \text{ V}; \quad E^\circ_1 = 1.368 \text{ V} - 1.51 \text{ V} = -0.142 \text{ V}$
   oxidation: $\quad 2 \text{ Hg}(l) \rightarrow \text{Hg}_2^{2+}(aq) + 2 e^- \qquad\qquad E^\circ = -0.80 \text{ V (Appendix D)}$
   reduction: $\quad \underline{\text{Hg}_2\text{Br}_2(s) + 2 e^- \rightarrow 2 \text{ Hg}(l) + 2 \text{ Br}^-(aq)} \quad E^\circ = -0.142 \text{ V (from } E^\circ_1)$
   overall: $\quad \text{Hg}_2\text{Br}_2(s) \rightarrow \text{Hg}_2^{2+}(aq) + 2 \text{ Br}^-(aq) \qquad E^\circ = -0.658 \text{ V}$

$E^\circ = \dfrac{0.0592}{n} \log K; \quad \log K = \dfrac{nE^\circ}{0.0592} = \dfrac{(2)(-0.658)}{0.0592} = -22.2$

$K = K_{sp} = 10^{-22.2} = 6 \times 10^{-23}$

18.112 (a) anode: $\quad 4[\text{Al}(s) \rightarrow \text{Al}^{3+}(aq) + 3 e^-] \qquad\qquad\qquad E^\circ = 1.66 \text{ V}$
   cathode: $\quad \underline{3[\text{O}_2(g) + 4 \text{ H}^+(aq) + 4 e^- \rightarrow 2 \text{ H}_2\text{O}(l)]} \qquad E^\circ = 1.23 \text{ V}$
   overall: $\quad 4 \text{ Al}(s) + 3 \text{ O}_2(g) + 12 \text{ H}^+(aq) \rightarrow 4 \text{ Al}^{3+}(aq) + 6 \text{ H}_2\text{O}(l) \qquad E^\circ = 2.89 \text{ V}$

(b) & (c) $E = E° - \dfrac{2.303\,RT}{nF} \log \dfrac{[Al^{3+}]^4}{(P_{O_2})^3[H^+]^{12}}$

$E = 2.89\text{ V} - \dfrac{(2.303)\left(8.314 \times 10^{-3}\,\dfrac{kJ}{mol \cdot K}\right)(310\text{ K})}{(12\text{ mol }e^-)(96,500\text{ C/mol }e^-)} \log\left(\dfrac{(1.0 \times 10^{-9})^4}{(0.20)^3(1.0 \times 10^{-7})^{12}}\right)$

$E = 2.89\text{ V} - 0.257\text{ V} = 2.63\text{ V}$

18.114 $Cr_2O_7^{2-}(aq) + 6\,Fe^{2+}(aq) + 14\,H^+(aq) \rightarrow 2\,Cr^{3+}(aq) + 6\,Fe^{3+}(aq) + 7\,H_2O(l)$
The two half reactions are:
oxidation: $Fe^{2+}(aq) \rightarrow Fe^{3+}(aq) + e^-$ $E° = -0.77$ V
reduction: $Cr_2O_7^{2-}(aq) + 14\,H^+(aq) + 6\,e^- \rightarrow 2\,Cr^{3+}(aq) + 7\,H_2O(l)$ $E° = 1.33$ V
At the equivalence point the potential is given by either of the following expressions:

(1) $E = 1.33\text{ V} - \dfrac{0.0592\text{ V}}{6} \log \dfrac{[Cr^{3+}]^2}{[Cr_2O_7^{2-}][H^+]^{14}}$

(2) $E = 0.77\text{ V} - \dfrac{0.0592\text{ V}}{1} \log \dfrac{[Fe^{2+}]}{[Fe^{3+}]}$

where E is the same in both because equilibrium is reached and the solution can have only one potential. Multiplying (1) by 6, adding it to (2), and using some stoichiometric relationships at the equivalence point will simplify the log term.

$7E = [(6 \times 1.33\text{ V}) + 0.77\text{ V}] - (0.0592\text{ V})\log \dfrac{[Fe^{2+}][Cr^{3+}]^2}{[Fe^{3+}][Cr_2O_7^{2-}][H^+]^{14}}$

At the equivalence point, $[Fe^{2+}] = 6[Cr_2O_7^{2-}]$ and $[Fe^{3+}] = 3[Cr^{3+}]$. Substitute these equivalencies into the previous equation.

$7E = [(6 \times 1.33\text{ V}) + 0.77\text{ V}] - (0.0592\text{ V})\log \dfrac{6[Cr_2O_7^{2-}][Cr^{3+}]^2}{3[Cr^{3+}][Cr_2O_7^{2-}][H^+]^{14}}$

Cancel equivalent terms.

$7E = [(6 \times 1.33\text{ V}) + 0.77\text{ V}] - (0.0592\text{ V})\log \dfrac{6[Cr^{3+}]}{3[H^+]^{14}}$

mol Fe^{2+} = (0.120 L)(0.100 mol/L) = 0.0120 mol Fe^{2+}

mol $Cr_2O_7^{2-}$ = 0.0120 mol Fe^{2+} x $\dfrac{1\text{ mol }Cr_2O_7^{2-}}{6\text{ mol }Fe^{2+}}$ = 0.00200 mol $Cr_2O_7^{2-}$

volume $Cr_2O_7^{2-}$ = 0.00200 mol x $\dfrac{1\text{ L}}{0.120\text{ mol}}$ = 0.0167 L

At the equivalence point assume mol Fe^{3+} = initial mol Fe^{2+} = 0.0120 mol
Total volume at the equivalence point is 0.120 L + 0.0167 L = 0.1367 L

$[Fe^{3+}] = \dfrac{0.0120\text{ mol}}{0.1367\text{ L}} = 0.0878\text{ M};\quad [Cr^{3+}] = [Fe^{3+}]/3 = (0.0878\text{ M})/3 = 0.0293\text{ M}$

$[H^+] = 10^{-pH} = 10^{-2.00} = 0.010\text{ M}$

$$7E = [(6 \times 1.33 \text{ V}) + 0.77 \text{ V}] - (0.0592 \text{ V})\log \frac{6(0.0293)}{3(0.010)^{14}} = [8.75] - 1.585 = 7.165 \text{ V}$$

$$E = \frac{7.165 \text{ V}}{7} = 1.02 \text{ V at the equivalence point.}$$

19 — The Main-Group Elements

19.1 (a) B lies above Al in group 3A, and therefore B is more nonmetallic than Al.
(b) Ge and Br are in the same row of the periodic table, but Br (group 7A) lies to the right of Ge (group 4A). Therefore, Br is more nonmetallic.
(c) Se (group 6A) has more nonmetallic than In because it lies above and to the right of In (group 3A).

19.2 (a) HNO_3 H_3PO_4

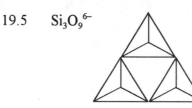

Nitrogen can form very strong pπ - pπ bonds. Phosphorus forms weaker pπ - pπ bonds, so it tends to form more single bonds.
(b) Sulfur can use empty 3d orbitals to form octahedral sp^3d^2 hybrid orbitals to bond with six fluorines in SF_6. With just 2s and 2p valence orbitals, oxygen cannot expand its valence orbitals beyond four. OF_2 has two single bonds and two lone pairs of electrons.

19.3 $H—C\equiv N\colon$ The carbon is sp hybridized.

19.4 $Hb–O_2 + CO \rightleftarrows Hb–CO + O_2$
Mild cases of carbon monoxide poisoning can be treated with O_2. Le Châtelier's principle says that adding a product (O_2) will cause the reaction to proceed in the reverse direction, back to $Hb–O_2$.

19.5 $Si_3O_9^{6-}$

19.6 H_3PO_4 $H_6P_4O_{13}$

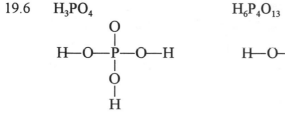
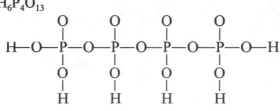

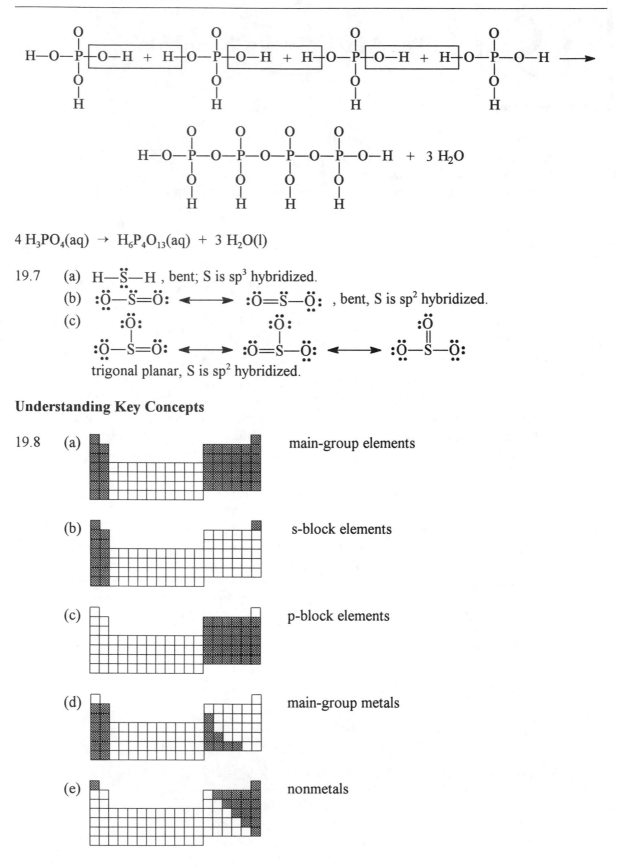

$$4\,H_3PO_4(aq)\ \rightarrow\ H_6P_4O_{13}(aq)\ +\ 3\,H_2O(l)$$

19.7 (a) H—S̈—H , bent; S is sp^3 hybridized.

(b) :Ö—S̈=Ö: ⟷ :Ö=S̈—Ö: , bent, S is sp^2 hybridized.

(c)

:Ö: :Ö: :Ö

:Ö—S=Ö: ⟷ :Ö=S—Ö: ⟷ :Ö—S—Ö:

trigonal planar, S is sp^2 hybridized.

Understanding Key Concepts

19.8 (a) main-group elements

(b) s-block elements

(c) p-block elements

(d) main-group metals

(e) nonmetals

(f) 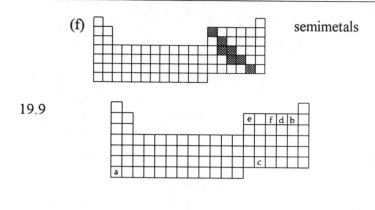 semimetals

19.9

19.10 (a) N_2, O_2, F_2, P_4, S_8, Cl_2
(b)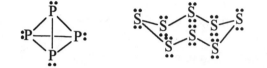

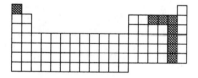

(c) The smaller N and O can form effective π bonds, whereas P and S cannot. In both F_2 and Cl_2, the atoms are joined by a single bond.

19.11 (a) gases, (b) liquid, and (c) solids

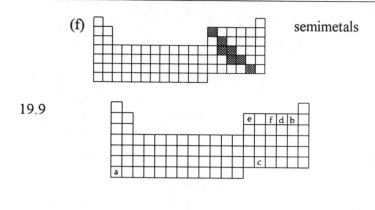

(a) ▨
(b) ▱
(c) ⊠

(d) diatomic molecules

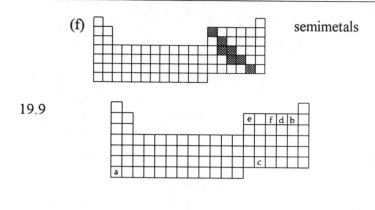

19.12 (a) H_2O, CH_4, HF, B_2H_6, NH_3
(b)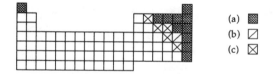

There is a problem in drawing an electron-dot structure for B_2H_6 because this molecule is electron deficient and has two three-center, two-electron bonds.

19.13 (a) SiO_4^{4-} (b) $Si_3O_{10}^{8-}$ (c) $Si_4O_{12}^{8-}$

Additional Problems
General Properties and Periodic Trends

19.14   (a) Cl (group 7A) lies to the right of S (group 6A) in the same row of the periodic table. Cl has the higher ionization energy.
(b) Si lies above Ge in group 4A. Si has the higher ionization energy.
(c) O (group 6A) lies above and to the right of In (group 3A) in the periodic table. O has the higher ionization energy.

19.16   (a) Al lies below B in group 3A. Al has the larger atomic radius.
(b) P (group 5A) lies to the left of S (group 6A) in the same row of the periodic table. P has the larger atomic radius.
(c) Pb (group 4A) lies below and to the left of Br (group 7A) in the periodic table. Pb has the larger atomic radius.

19.18   (a) I (group 7A) lies to the right of Te (group 6A) in the same row of the periodic table. I has the higher electronegativity.
(b) N lies above P in group 5A. N has the higher electronegativity.
(c) F (group 7A) lies above and to the right of In (group 3A) in the periodic table. F has the higher electronegativity.

19.20   (a) Sn lies below Si in group 4A. Sn has more metallic character.
(b) Ge (group 4A) lies to the left of Se (group 6A) in the same row of the periodic table. Ge has more metallic character.
(c) Bi (group 5A) lies below and to the left of I (group 7A) in the periodic table. Bi has more metallic character.

19.22   (a) CaH_2 (b) Ga_2O_3 (c) KCl
In each case the more ionic compound is the one formed between a metal and nonmetal.

19.24   Molecular   (a) B_2H_6 (c) SO_3 (d) $GeCl_4$
Extended three-dimensional structure   (b) $KAlSi_3O_8$

19.26   Nonmetal oxides are acidic. Metal oxides are basic or amphoteric.
(a) P_4O_{10} (b) B_2O_3 (c) SO_2

19.28   (a) Sn   (b) Cl   (c) Sn   (d) Se   (e) B

19.30   Boron has only 2s and 2p valence orbitals and can form a maximum of four bonds. Aluminum has available 3d orbitals and can use octahedral sp^3d^2 hybrid orbitals to bond to six F^- ions.

19.32   In O_2 a π bond is formed by 2p orbitals on each O. S does not form strong π bonds with its 3p orbitals, which leads to the S_8 ring structure.

The Group 3A Elements

19.34   +3 for B, Al, Ga and In;   +1 for Tl

19.36   Boron is a hard semiconductor with a high melting point. Boron forms only molecular compounds and does not form an aqueous B^{3+} ion. $B(OH)_3$ is an acid.

19.38 $$2\ BBr_3(g) + 3\ H_2(g) \xrightarrow[1200°C]{Ta\ wire} 2\ B(s) + 6\ HBr(g)$$

This reaction is used to produce crystalline boron.

19.40   (a) An electron deficient molecule is a molecule that doesn't have enough electrons to form a two center-two electron bond between each pair of bonded atoms. B_2H_6 is an electron deficient molecule.
(b) A three-center two-electron bond has three atoms bonded together using just two electrons. The B-H-B bridging bond in B_2H_6 is a three-center two- electron bond.

19.42   (a) Al   (b) Tl   (c) B   (d) B

The Group 4A Elements

19.44   (a) Pb   (b) C   (c) Si   (d) C

19.46   (a) $GeBr_4$, tetrahedral; Ge is sp^3 hybridized.
(b) CO_2, linear; C is sp hybridized.
(c) CO_3^{2-}, trigonal planar; C is sp^2 hybridized.
(d) $Sn(OH)_6^{2-}$, octahedral; Sn is sp^3d^2 hybridized.

19.48   Diamond is a very hard, high melting solid. It is an electrical insulator.
Diamond has a covalent network structure in which each C atom uses sp^3 hybrid orbitals to form a tetrahedral array of σ bonds. The interlocking, three-dimensional network of strong bonds makes diamond the hardest known substance with the highest melting point for an element. Because the valence electrons are localized in the σ bonds, diamond is an electrical insulator.

19.50   Graphite has a two-dimensional sheetlike structure in which each C atom uses sp^2 hybrid orbital to form trigonal planar σ bonds to three neighboring C atoms. In addition, each C atom uses its remaining p orbital, perpendicular to the plane of the sheet, to form a π bond. Because each C atom must share its π bond with its three neighbors, the π electrons are delocalized and are free to move in the plane of the sheet. As a result, the electrical conductivity of graphite in a direction parallel to the sheets is about 10^{20} times greater than the conductivity of diamond. The conductivity of graphite perpendicular to the sheets of C atoms is lower because electrons must hop from one sheet to the next.

19.52   (a) carbon tetrachloride, CCl_4 (b) carbon monoxide, CO   (c) methane, CH_4

19.54 Some uses for CO_2 are:
(1) To provide the "bite" in soft drinks; $CO_2(aq) + H_2O(l) \rightleftharpoons H_2CO_3(aq)$
(2) CO_2 fire extinguishers; CO_2 is nonflammable and 1.5 times more dense than air.
(3) Refrigerant; dry ice, sublimes at $-78°C$.

19.56 $SiO_2(l) + 2\,C(s) \rightarrow Si(l) + 2\,CO(g)$
(sand)

Purification of silicon for semiconductor devices:
$Si(s) + 2\,Cl_2(g) \rightarrow SiCl_4(l)$; $SiCl_4$ is purified by distillation.
$SiCl_4(g) + 2\,H_2(g) \xrightarrow{\text{heat}} Si(s) + 4\,HCl(g)$; Si is purified by zone refining.

19.58 (a) SiO_4^{4-} (b) $Si_4O_{13}^{10-}$

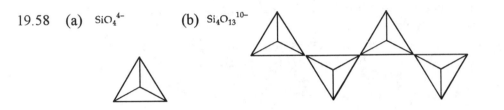

The charge on the anion is equal to the number of terminal O atoms.

19.60 (a) $Si_3O_{10}^{8-}$
(b) The charge on the anion is 8–. Because the Ca^{2+} to Cu^{2+} ratio is 1:1, there must be 2 Ca^{2+} and 2 Cu^{2+} ions in the formula for the mineral. There are also 2 waters. The formula of the mineral is: $Ca_2Cu_2Si_3O_{10} \cdot 2\,H_2O$

The Group 5A Elements

19.62 (a) P   (b) Sb and Bi   (c) N   (d) Bi

19.64 (a) N_2O, +1   (b) N_2H_4, –2   (c) Ca_3P_2, –3
(d) H_3PO_3, +3   (e) H_3AsO_4, +5

19.66 $:N{\equiv}N:$; N_2 is unreactive because of the large amount of energy necessary to break the $N{\equiv}N$ triple bond.

19.68 (a) NO_2^-, bent; N is sp^2 hybridized.
(b) PH_3, trigonal pyramidal; P is sp^3 hybridized.
(c) PF_5, trigonal bipyramidal; P is sp^3d hybridized.
(d) PCl_4^+, tetrahedral; P is sp^3 hybridized.

19.70   White phosphorus

Red phosphorus

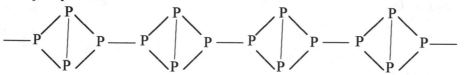

White phosphorus is reactive due to the considerable strain in the P_4 molecule.

19.72   (a) The structure for phosphorous acid is

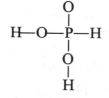

Only the two hydrogens bonded to oxygen are acidic.
(b) Nitrogen forms strong π bonds, and in N_2 the nitrogen atoms are triple bonded to each other. Phosphorus does not form strong $p\pi - p\pi$ bonds, and so the P atoms are single bonded to each other in P_4

19.74   (a) $2\ NO(g)\ +\ O_2(g)\ \rightarrow\ 2\ NO_2(g)$
        (b) $4\ HNO_3(aq)\ \rightarrow\ 4\ NO_2(aq)\ +\ O_2(g)\ +\ 2\ H_2O(l)$
        (c) $3\ Ag(s)\ +\ 4\ H^+(aq)\ +\ NO_3^-(aq)\ \rightarrow\ 3\ Ag^+(aq)\ +\ NO(g)\ +\ 2\ H_2O(l)$
        (d) $N_2H_4(aq)\ +\ 2\ I_2(aq)\ \rightarrow\ N_2(g)\ +\ 4\ H^+(aq)\ +\ 4\ I^-(aq)$

The Group 6A Elements

19.76   (a) O   (b) Te   (c) Po   (d) O

19.78   (a) rhombic sulfur – yellow crystalline solid (mp 113°C) that contains crown-shaped S_8 rings.
        (b) monoclinic sulfur – an allotrope of sulfur in which the S_8 rings pack differently in the crystal.
        (c) plastic sulfur – when sulfur is cooled rapidly, the sulfur forms disordered, tangled chains, yielding an amorphous, rubbery material called plastic sulfur.
        (d) Liquid sulfur between 160 and 195°C becomes dark reddish-brown and very viscous forming long polymer chains (S_n, n > 200,000).

19.80   (a) hydrogen sulfide, H_2S (b) sulfur dioxide, SO_2 (c) sulfur trioxide, SO_3
              lead(II) sulfide, PbS   sulfurous acid, H_2SO_3 sulfur hexafluoride, SF_6

19.82 Sulfuric acid (H_2SO_4) is manufactured by the contact process, a three-step reaction sequence in which (1) sulfur burns in air to give SO_2, (2) SO_2 is oxidized to SO_3 in the presence of a vanadium(V) oxide catalyst, and (3) SO_3 reacts with water to give H_2SO_4.

(1) $S(s) + O_2(g) \rightarrow SO_2(g)$

(2) $2\,SO_2(g) + O_2(g) \xrightarrow{\text{heat}}_{\text{V_2O_5 catalyst}} 2\,SO_3(g)$

(3) $SO_3(g) + H_2O \text{ (in conc } H_2SO_4) \rightarrow H_2SO_4(l)$

19.84 (a) $Zn(s) + 2\,H_3O^+(aq) \rightarrow Zn^{2+}(aq) + H_2(g) + 2\,H_2O(l)$
(b) $BaSO_3(s) + 2\,H_3O^+(aq) \rightarrow H_2SO_3(aq) + Ba^{2+}(aq) + 2\,H_2O(l)$
(c) $Cu(s) + 2\,H_2SO_4(l) \rightarrow Cu^{2+}(aq) + SO_4^{2-}(aq) + SO_2(g) + 2\,H_2O(l)$
(d) $H_2S(aq) + I_2(aq) \rightarrow S(s) + 2\,H^+(aq) + 2\,I^-(aq)$

19.86 (a) Acid strength increases as the number of O atoms increases.
(b) In comparison with S, O is much too electronegative to form compounds of O in the +4 oxidation state. Also, S uses sp^3d hybrid orbitals for bonding in SF_4, but O doesn't have valence d orbitals and so it can't form four bonds to F.
(c) Each S is sp^3 hybridized with two lone pairs of electrons. The bond angles are therefore 109.5°. A planar ring would require bond angles of 135°.

Halogen Oxoacids and Oxoacid Salts

19.88 (a) $HBrO_3$, +5   (b) HIO, +1   (c) $NaClO_2$, +3   (d) $NaIO_4$, +7

19.90 (a) iodic acid   (b) chlorous acid   (c) sodium hypobromite
(d) lithium perchlorate

19.92 (a) HIO_3

:Ö—Ï—Ö—H   trigonal pyramidal
        |
       :Ö:

(b) ClO_2^-

$\left[\begin{array}{c} \text{:Ö—Ċl—Ö:} \end{array}\right]^-$ bent

(c) HOCl

H—Ö—Ċl: bent

(d) IO_6^{5-}

$\begin{bmatrix} & \text{:Ö:} & \\ \text{:Ö} & | & \text{Ö:} \\ & \text{I} & \\ \text{:Ö} & | & \text{Ö:} \\ & \text{:Ö:} & \end{bmatrix}^{5-}$ octahedral

19.94 Oxygen atoms are highly electronegative. Increasing the number of oxygen atoms increases the polarity of the O–H bond and increases the acid strength.

19.96   (a) $Br_2(l) + 2\ OH^-(aq) \rightarrow OBr^-(aq) + Br^-(aq) + H_2O(l)$
      (b) $Cl_2(g) + H_2O(l) \rightarrow HOCl(aq) + H^+(aq) + Cl^-(aq)$
      (c) $3\ Cl_2(g) + 6\ OH^-(aq) \rightarrow ClO_3^-(aq) + 5\ Cl^-(aq) + 3\ H_2O(l)$

General Problems

19.98   (a) LiCl is an ionic compound. PCl_3 is a covalent molecular compound.
The ionic compound, LiCl, has the higher melting point.
(b) Carbon forms strong π bonds with oxygen, and CO_2 is a covalent molecular
compound with a low melting point. Silicon prefers to form single bonds with oxygen.
SiO_2 is three dimensional extended structure with alternating silicon and oxygen singly
bonded to each other. SiO_2 is a high melting solid.
(c) Nitrogen forms strong π bonds with oxygen and NO_2 is a covalent molecular
compound with a low melting point. Phosphorus prefers to form single bonds with
oxygen. P_4O_{10} is a larger covalent molecular compound than NO_2, with a higher metling
point.

19.100   (a) Zone refining is a purification technique in which a heater melts a narrow zone at the
top of a rod of some material and then sweeps slowly down the rod, bringing impurities
with it.
(b) The silicate chain anion, $Si_2O_6^{4-}$, is an anion in which tetrahedral SiO_4 units share two
O atoms to give an extended silicate chain with $Si_2O_6^{4-}$ as the repeating unit.
(c) Partial substitution of the Si^{4+} in SiO_2 with Al^{3+} gives aluminosilicates called feldspars,
the most abundant of all minerals.

19.102   In silicate and phosphate anions, both Si and P are surrounded by tetrahedra of O atoms,
which can link together to form chains and rings.

19.104   (a) $Na_2B_4O_7 \cdot 10\ H_2O$ (b) $Ca_3(PO_4)_2$
      (c) elemental sulfur, FeS_2, PbS, HgS, $CaSO_4 \cdot 2\ H_2O$

19.106   Earth's crust: O, Si, Al, Fe;   Human body: O, C, H, N

19.108   C, Si, Ge and Sn have allotropes with the diamond structure.
Sn and Pb have metallic allotropes.
C (nonmetal), Si (semimetal), Ge (semimetal), Sn (semimetal and metal), Pb (metal)

19.110   (a) $H_3PO_4(aq) + H_2O(l) \rightleftharpoons H_3O^+(aq) + H_2PO_4^-(aq)$
H_3PO_4 is a Brønsted-Lowry acid.
(b) $B(OH)_3(aq) + 2\ H_2O(l) \rightleftharpoons B(OH)_4^-(aq) + H_3O^+(aq)$
$B(OH)_3$ is a Lewis acid.

19.112   (a) In diamond each C is covalently bonded to four additional C atoms in a rigid three-
dimensional network solid. Graphite is a two-dimensional covalent network solid of
carbon sheets that can slide over each other. Both are high melting because melting

requires the breaking of C–C bonds.

(b) Chlorine does not form perhalic acids of the type H_5XO_6 because its smaller size favors a tetrahedral structure over an octahedral one.

19.114 The angle required by P_4 is 60°. The strain would not be reduced by using sp^3 hybrid orbitals because their angle is ~109°.

19.116 Carbon is a versatile element that can form millions of very stable compounds with elements such as N, O, and H. Biomolecules contain chains and rings with many C–C bonds. Si–Si bonds are much less stable and chains of Si atoms are uncommon. In addition, carbon can form very stable $p\pi$-$p\pi$ multiple bonds. On the other hand, the chemistry of silicon (which cannot form stable $p\pi$-$p\pi$ bonds) is dominated by structures based on the SiO_4^{4-} anion.

19.118

$$2\ In^+(aq) + 2\ e^- \rightarrow 2\ In(s) \qquad\qquad E° = -0.14\ V$$
$$\underline{In^+(aq) \rightarrow In^{3+}(aq) + 2\ e^- \qquad\qquad E° =\ \ 0.44\ V}$$
$$3\ In^+(aq) \rightarrow In^{3+}(aq) + 2\ In(s) \qquad E° =\ \ 0.30\ V$$

$1\ J = 1\ V\cdot C$

$\Delta G° = -nFE° = -(2)(96,500\ C)(0.30\ V) = -5.8 \times 10^4\ J = -58\ kJ$

Because $\Delta G° < 0$, the disproportionation is spontaneous.

$$2\ Tl^+(aq) + 2\ e^- \rightarrow 2\ Tl(s) \qquad\qquad E° = -0.34\ V$$
$$\underline{Tl^+(aq) \rightarrow Tl^{3+}(aq) + 2\ e^- \qquad\qquad E° = -1.25\ V}$$
$$3\ Tl^+(aq) \rightarrow Tl^{3+}(aq) + 2\ Tl(s) \qquad E° = -1.59\ V$$

$\Delta G° = -nFE° = -(2)(96,500\ C)(-1.59\ V) = +3.07 \times 10^5\ J = +307\ kJ$

Because $\Delta G° > 0$, the disproportionation is nonspontaneous.

19.120 NH_4NO_3, 80.04 amu; $(NH_4)_2HPO_4$, 132.06 amu

(a) % N in $NH_4NO_3 = \dfrac{2 \times (14.007\ amu\ N)}{80.04\ amu\ NH_4NO_3} \times 100\% = 35.0\%\ N$

% N in $(NH_4)_2HPO_4 = \dfrac{2 \times (14.007\ amu\ N)}{132.06\ amu\ (NH_4)_2HPO_4} \times 100\% = 21.2\%\ N$

Let x equal the fraction of NH_4NO_3 and $(1 - x)$ equal the fraction of $(NH_4)_2HPO_4$ in the mixture.

$0.3381 = x(0.350) + (1 - x)(0.212) = 0.138x + 0.212$

$0.1261 = 0.138x \qquad x = \dfrac{0.1261}{0.138} = 0.9138$

$(1 - x) = (1 - 0.9138) = 0.0862$

There is 8.62 % $(NH_4)_2HPO_4$ and 91.38% NH_4NO_3 in the mixture.

(b) sample mass = 0.965 g

mass $NH_4NO_3 = (0.965\ g)(0.9138) = 0.8818\ g$

mass $(NH_4)_2HPO_4 = 0.965\ g - 0.8818\ g = 0.0832\ g$

mol $NH_4NO_3 = 0.8818\ g \times \dfrac{1\ mol\ NH_4NO_3}{80.04\ g\ NH_4NO_3} = 0.0110\ mol\ NH_4NO_3$

$$\text{mol } (NH_4)_2HPO_4 = 0.0832 \text{ g} \times \frac{1 \text{ mol } (NH_4)_2HPO_4}{132.06 \text{ g } (NH_4)_2HPO_4} = 6.30 \times 10^{-4} \text{ mol } (NH_4)_2HPO_4$$

$$[NH_4^+] = \frac{0.0110 \text{ mol} + 2(6.30 \times 10^{-4} \text{ mol})}{0.0500 \text{ L}} = 0.245 \text{ M}$$

$$[HPO_4^{2-}] = \frac{6.30 \times 10^{-4} \text{ mol}}{0.0500 \text{ L}} = 0.0126 \text{ M}$$

The equilibrium constant for the transfer of a proton from the cation of a salt to the anion of the salt is equal to $\dfrac{K_a K_b}{K_w}$.

$K_a(NH_4^+) = 5.56 \times 10^{-10}$ and $K_b(HPO_4^{2-}) = 1.61 \times 10^{-7}$

$$K = \frac{K_a K_b}{K_w} = \frac{(5.56 \times 10^{-10})(1.61 \times 10^{-7})}{1.0 \times 10^{-14}} = 9.12 \times 10^{-3}$$

	$NH_4^+(aq)$	+ $HPO_4^{2-}(aq)$	$\rightleftarrows$	$H_2PO_4^-(aq)$	+ $NH_3(aq)$
initial (M)	0.245	0.0126		0	0
change (M)	$-x$	$-x$		$+x$	$+x$
equil (M)	$0.245 - x$	$0.0126 - x$		x	x

$$K = \frac{[H_2PO_4^-][NH_3]}{[NH_4^+][HPO_4^{2-}]} = 9.12 \times 10^{-3} = \frac{x^2}{(0.245 - x)(0.0126 - x)}$$

$0.991x^2 + 0.00235x - (2.82 \times 10^{-5}) = 0$

Use the quadratic formula to solve for x.

$$x = \frac{(-0.00235) \pm \sqrt{(0.00235)^2 - (4)(0.991)(-2.82 \times 10^{-5})}}{2(0.991)} = \frac{(-0.00235) \pm 0.01083}{2(0.991)}$$

$x = 0.00428$ and -0.00665

Of the two solutions for x, only the positive value of x has physical meaning because x is equal to both the $[H_2PO_4^-]$ and $[NH_3]$.

$$K_a(NH_4^+) = \frac{[H_3O^+][NH_3]}{[NH_4^+]} = 5.56 \times 10^{-10} = \frac{[H_3O^+](0.00428)}{(0.245 - 0.00428)}$$

$$[H_3O^+] = \frac{(5.56 \times 10^{-10})(0.245 - 0.00428)}{0.00428} = 3.127 \times 10^{-8} \text{ M}$$

$pH = -\log[H_3O^+] = -\log(3.127 \times 10^{-8}) = 7.50$

Transition Elements and Coordination Chemistry

20.1   (a) V, [Ar] $3d^3 4s^2$ (b) Co^{2+}, [Ar] $3d^7$ (c) Mn^{4+} in MnO_2, [Ar] $3d^3$

20.2 $[Cr(NH_3)_2(SCN)_4]^-$

20.3   In $Na_4[Fe(CN)_6]$ each sodium is in the +1 oxidation state (+4 total); each cyanide (CN^-) has a −1 charge (−6 total). The compound is neutral; therefore, the oxidation state of the iron is +2.

20.4   (a) tetraamminecopper(II) sulfate   (b) sodium tetrahydroxochromate(III)
        (c) triglycinatocobalt(III)   (d) pentaaquaisothiocyanatoiron(III) ion

20.5   (a) $[Zn(NH_3)_4](NO_3)_2$ (b) $Ni(CO)_4$ (c) $K[Pt(NH_3)Cl_3]$ (d) $[Au(CN)_2]^-$

20.6   (a) Two diastereoisomers are possible.

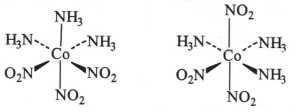

cis   trans

(b) No isomers are possible for a tetrahedral complex of the type MA_2B_2.

(c) Two diastereoisomers are possible.

$$
\begin{array}{cc}
\underset{O_2N}{\underset{H_3N\cdots}{}}\overset{NH_3}{\overset{|}{\underset{|}{Co}}}\overset{\cdots NH_3}{\underset{NO_2}{\overset{}{}}} &
\underset{O_2N}{\underset{H_3N\cdots}{}}\overset{NO_2}{\overset{|}{\underset{|}{Co}}}\overset{\cdots NH_3}{\underset{NO_2}{\overset{}{}}}
\end{array}
$$

(d) No isomers are possible for a complex of this type.

(e) Two diastereoisomers are possible.

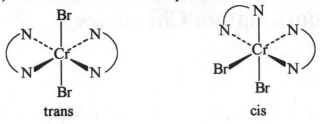

trans   cis

(f) No diastereoisomers are possible for a complex of this type.

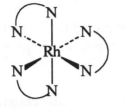

20.7   (a) chair, no   (b) foot, yes   (c) pencil, no   (d) corkscrew, yes   (e) banana, no

20.8   (a) $[Fe(C_2O_4)_3]^{3-}$ can exist as enantiomers.

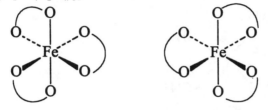

(b) $[Co(NH_3)_4en]^{3+}$ cannot exist as enantiomers.
(c) $[Co(NH_3)_2(en)_2]^{3+}$ can exist as enantiomers.

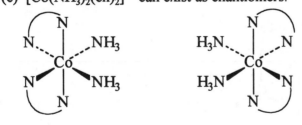

(d) $[Cr(H_2O)_4Cl_2]^+$ cannot exist as enantiomers.

20.9   (a) Fe^{3+} [Ar] ↑ ↑ ↑ ↑ ↑ ___ __ __ __
                                    3d   4s   4p

         $[Fe(CN)_6]^{3-}$ [Ar] ↑↓ ↑↓ ↑ ↑↓ ↑↓   ↑↓   ↑↓ ↑↓ ↑↓
                                    3d   4s   4p

                                 d^2sp^3 1 unpaired e⁻

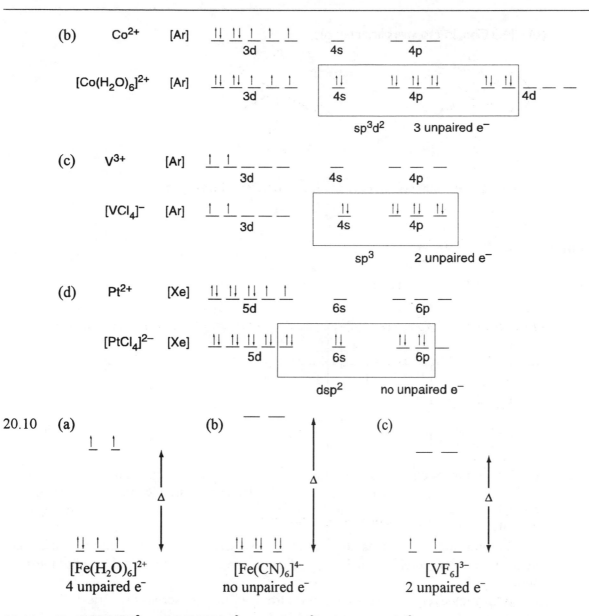

20.11   Both $[NiCl_4]^{2-}$ and $[Ni(CN)_4]^{2-}$ contain Ni^{2+} with a [Ar] $3d^8$ electron configuration.
(a) $[NiCl_4]^{2-}$ (tetrahedral)

(b) $[Ni(CN)_4]^{2-}$ (square planar)

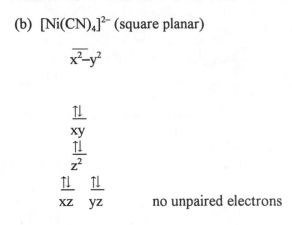

no unpaired electrons

Understanding Key Concepts

20.12   (a) Co   (b) Cr   (c) Zr   (d) Pr

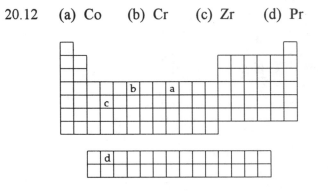

20.13   (a) The atomic radii decrease, at first markedly and then more gradually. Toward the end of the series, the radii increase again. The decrease in atomic radii is a result of an increase in Z_{eff}. The increase is due to electron-electron repulsions in doubly occupied d orbitals.

(b) The densities of the transition metals are inversely related to their atomic radii. The densities initially increase from left to right and then decrease toward the end of the series.

(c) Ionization energies generally increase from left to right across the series. The general trend correlates with an increase in Z_{eff} and a decrease in atomic radii.

(d) The standard oxidation potentials generally decrease from left to right across the first transition series. This correlates with the general trend in ionization energies.

20.14   (a) $NH_2-CH_2-CH_2-NH_2$ is a bidentate ligand. It can form a chelate ring using the atoms indicated in bold.

(b) $CH_3-CH_2-CH_2-NH_2$ is a monodentate ligand.

(c) $NH_2-CH_2-CH_2-NH-CH_2-CO_2^-$ is a tridentate ligand. It can form chelate rings using the atoms indicated in bold.

(d) $NH_2-CH_2-CH_2-NH_3^+$ is a monodentate ligand. The first N can coordinate to a metal.

20.15 (a) Na[Au(CN)$_2$]
1 Na$^+$ 2 CN$^-$
The oxidation state of the Au is +1.   Coordination number = 2; Linear
(b) [Co(NH$_3$)$_5$Br]SO$_4$
1 Br$^-$ 1 SO$_4^{2-}$ 5 NH$_3$ (no charge)
The oxidation state of the Co is +3.   Coordination number = 6; Octahedral
(c) Pt(en)Cl$_2$
2 Cl$^-$ en = NH$_2$–CH$_2$–CH$_2$–NH$_2$ (no charge)
The oxidation state of the Pt is +2.   Coordination number = 4; Square planar
(d) (NH$_4$)$_2$[PtCl$_2$(C$_2$O$_4$)$_2$]
2 NH$_4^+$ 2 Cl$^-$ 2 C$_2$O$_4^{2-}$
The oxidation state of the Pt is +4.   Coordination number = 6; Octahedral

20.16. (a) (1) cis; (2) trans; (3) trans; (4) cis
(b) (1) and (4) are the same.   (2) and (3) are the same.
(c) None of the isomers exist as enantiomers because their mirror images are identical.

20.17 (a) (1) chiral; (2) achiral; (3) chiral; (4) chiral

(b)

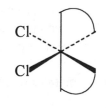

(1)   enantiomer of (1)

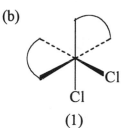
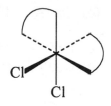

(3)   enantiomer of (3)

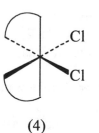

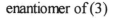

(4)   enantiomer of (4)

(c) (1) and (4) are enantiomers.

20.18

20.19   The tetrahedral complex is chiral because it is not identical to its mirror image. The square planar complex is achiral because it has a symmetry plane (the plane of the molecule).

Additional Problems
Electron Configurations and Properties of Transition Elements

20.20   (a) Ni, [Ar] $3d^8 4s^2$ (b) Cr, [Ar] $3d^5 4s^1$
          (c) Zr, [Kr] $4d^2 5s^2$ (d) Zn, [Ar] $3d^{10} 4s^2$

20.22   (a) Co^{2+}, [Ar] $3d^7$ (b) Fe^{3+}, [Ar] $3d^5$
          (c) Mo^{3+}, [Kr] $4d^3$ (d) Cr(VI), [Ar] $3d^0$

20.24   (a) Cu^{2+},   [Ar] $3d^9$ ↿⇂ ↿⇂ ↿⇂ ↿⇂ ↿   1 unpaired e^-
                             3d

          (b) Ti^{2+},   [Ar] $3d^2$ ↿ ↿ _ _ _   2 unpaired e^-
                             3d

          (c) Zn^{2+},   [Ar] $3d^{10}$ ↿⇂ ↿⇂ ↿⇂ ↿⇂ ↿⇂   0 unpaired e^-
                             3d

          (d) Cr^{3+},   [Ar] $3d^3$ ↿ ↿ ↿ _ _   3 unpaired e^-
                             3d

20.26   Ti is harder than K and Ca largely because the sharing of d, as well as s, electrons results in stronger metallic bonding.

20.28   (a) The decrease in radii with increasing atomic number is expected because the added d electrons only partially shield the added nuclear charge. As a result, Z_{eff} increases. With increasing Z_{eff}, the electrons are more strongly attracted to the nucleus, and atomic size decreases.
          (b) The densities of the transition metals are inversely related to their atomic radii.

20.30   The smaller than expected sizes of the third-transition series atoms are associated with what is called the lanthanide contraction, the general decrease in atomic radii of the f-block lanthanide elements.
          The lanthanide contraction is due to the increase in Z_{eff} as the 4f subshell is filled.

20.32   Sc $(631 + 1235) = 1866$ kJ/mol
           Ti $(658 + 1310) = 1968$ kJ/mol
           V $(650 + 1414) = 2064$ kJ/mol
           Cr $(653 + 1592) = 2225$ kJ/mol
           Mn $(717 + 1509) = 2226$ kJ/mol
           Fe $(759 + 1561) = 2320$ kJ/mol
           Co $(758 + 1646) = 2404$ kJ/mol
           Ni $(737 + 1753) = 2490$ kJ/mol
           Cu $(745 + 1958) = 2703$ kJ/mol
           Zn $(906 + 1733) = 2639$ kJ/mol

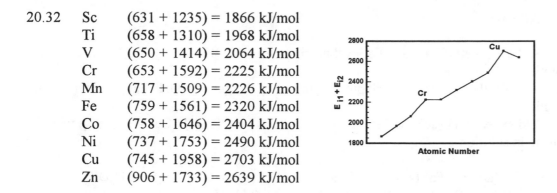

Across the first transition element series, Z_{eff} increases and there is an almost linear increase in the sum of the first two ionization energies. This is what is expected if the two electrons are removed from the 4s orbital. Higher than expected values for the sum of the first two ionization energies are observed for Cr and Cu because of their anomalous electron configurations (Cr $3d^5 4s^1$; Cu $3d^{10} 4s^1$). An increasing Z_{eff} affects 3d orbitals more than the 4s orbital and the second ionization energy for an electron from the 3d orbital is higher than expected.

20.34   (a) $Cr(s) + 2 H^+(aq) \rightarrow Cr^{2+}(aq) + H_2(g)$
           (b) $Zn(s) + 2 H^+(aq) \rightarrow Zn^{2+}(aq) + H_2(g)$
           (c) N.R.
           (d) $Fe(s) + 2 H^+(aq) \rightarrow Fe^{2+}(aq) + H_2(g)$

Oxidation States

20.36   (b) Mn   (d) Cu

20.38   Sc(III), Ti(IV), V(V), Cr(VI), Mn(VII), Fe(VI), Co(III), Ni(II), Cu(II), Zn(II)

20.40 Cu^{2+} is a stronger oxidizing agent than Cr^{2+} because of a higher Z_{eff}.

20.42 Cr^{2+} is more easily oxidized than Ni^{2+} because of a smaller Z_{eff}.

20.44 $Mn^{2+} < MnO_2 < MnO_4^-$ because of increasing oxidation state of Mn.

Chemistry of Selected Transition Elements

20.46   (a) $Cr_2O_3(s) + 2 Al(s) \rightarrow 2 Cr(s) + Al_2O_3(s)$
           (b) $Cu_2S(l) + O_2(g) \rightarrow 2 Cu(l) + SO_2(g)$

20.48   (a) $Cr_2O_7^{2-}$ (b) Cr^{3+} (c) Cr^{2+} (d) Fe^{2+} (e) Cu^{2+}

20.50 $Cr(OH)_2 < Cr(OH)_3 < CrO_2(OH)_2$
           Acid strength increases with polarity of the O–H bond, which increases in turn with the oxidation state of Cr.

20.52   (c) $Cr(OH)_3$

20.54   (a) Add excess KOH(aq) and Fe^{3+} will precipitate as $Fe(OH)_3$(s). Na^+(aq) will remain in solution.
       (b) Add NaOH(aq) and Fe^{3+} will precipitate as $Fe(OH)_3$(s). $Cr(OH)_4^-$(aq) will remain in solution.
       (c) Add excess NH_3(aq) and Fe^{3+} will precipitate as $Fe(OH)_3$(s). $Cu(NH_3)_4^{2+}$(aq) will remain in solution.

20.56   (a) $Cr_2O_7^{2-}(aq) + 6\ Fe^{2+}(aq) + 14\ H^+(aq) \rightarrow 2\ Cr^{3+}(aq) + 6\ Fe^{3+}(aq) + 7\ H_2O(l)$
       (b) $4\ Fe^{2+}(aq) + O_2(g) + 4\ H^+(aq) \rightarrow 4\ Fe^{3+}(aq) + 2\ H_2O(l)$
       (c) $Cu_2O(s) + 2\ H^+(aq) \rightarrow Cu(s) + Cu^{2+}(aq) + H_2O(l)$
       (d) $Fe(s) + 2\ H^+(aq) \rightarrow Fe^{2+}(aq) + H_2(g)$

20.58   (a) $2\ CrO_4^{2-}(aq) + 2\ H_3O^+(aq) \rightarrow Cr_2O_7^{2-}(aq) + 3\ H_2O(l)$
          (yellow)   (orange)
       (b) $[Fe(H_2O)_6]^{3+}(aq) + SCN^-(aq) \rightarrow [Fe(H_2O)_5(SCN)]^{2+}(aq) + H_2O(l)$
                                          (red)
       (c) $3\ Cu(s) + 2\ NO_3^-(aq) + 8\ H^+(aq) \rightarrow 3\ Cu^{2+}(aq) + 2\ NO(g) + 4\ H_2O(l)$
                               (blue)
       (d) $Cr(OH)_3(s) + OH^-(aq) \rightarrow Cr(OH)_4^-(aq)$
          $2\ Cr(OH)_4^-(aq) + 3\ HO_2^-(aq) \rightarrow 2\ CrO_4^{2-}(aq) + 5\ H_2O(l) + OH^-(aq)$
                                (yellow)

Coordination Compounds; Ligands

20.60 Ni^{2+} accepts six pairs of electrons, two each from the three ethylenediamine ligands. Ni^{2+} is an electron pair acceptor, a Lewis acid. The two nitrogens in each ethylenediamine donate a pair of electrons to the Ni^{2+}. The ethylenediamine is an electron pair donor, a Lewis base. The formation of $[Ni(en)_3]^{2+}$ is a Lewis acid-base reaction.

20.62   (a) $[Ag(NH_3)_2]^+$ (b) $[Ni(CN)_4]^{2-}$ (c) $[Cr(H_2O)_6]^{3+}$

20.64   (a) $[HgCl_4]^{2-}$
       $4\ Cl^-$
       The oxidation state of the Hg is +2.
       (b) $[Cr(H_2O)_4Cl_2]^+$
       $4\ H_2O$ (no charge) $2\ Cl^-$
       The oxidation state of the Cr is +3.
       (c) $[Au(CN)_2]^-$
       $2\ CN^-$
       The oxidation state of the Au is +1.
       (d) $[ZrF_8]^{4-}$
       $8\ F^-$
       The oxidation state of the Zr is +4.

(e) $[Mo(CO)_5Br]^-$
5 CO (no charge)   1 Br$^-$
The oxidation state of the Mo is 0.

20.66   (a) $Ni(CO)_4$ (b) $[Ag(NH_3)_2]^+$ (c) $[Fe(CN)_6]^{3-}$ (d) $[Ni(CN)_4]^{2-}$

20.68

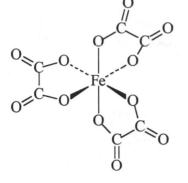

The iron is in the +3 oxidation state, and the coordination number is six. The geometry about the Fe is octahedral. The oxalate ligand is behaving as a bidentate chelating ligand. There are three chelate rings, one formed by each oxalate ligand.

20.70   (a) $Co(NH_3)_3(NO_2)_3$
3 NH$_3$ (no charge)   3 NO$_2^-$
The oxidation state of the Co is +3.
(b) $[Ag(NH_3)_2]NO_3$
2 NH$_3$ (no charge)   1 NO$_3^-$
The oxidation state of the Ag is +1
(c) $K_3[Cr(C_2O_4)_2Cl_2]$
3 K$^+$ 2 C$_2$O$_4^{2-}$ 2 Cl$^-$
The oxidation state of the Cr is +3.
(d) $Cs[CuCl_2]$
1 Cs$^+$ 2 Cl$^-$
The oxidation state of the Cu is +1

Naming Coordination Compounds

20.72   (a) tetrachloromanganate(II)   (b) hexaamminenickel(II)
(c) tricarbonatocobaltate(III)   (d) bis(ethylenediamine)dithiocyanatoplatinum(IV)

20.74   (a) cesium tetrachloroferrate(III)   (b) hexaaquavanadium(III) nitrate
(c) tetraamminedibromocobalt(III) bromide   (d) diglycinatocopper(II)

20.76   (a) $[Pt(NH_3)_4]Cl_2$ (b) $Na_3[Fe(CN)_6]$
(c) $[Pt(en)_3](SO_4)_2$ (d) $Rh(NH_3)_3(SCN)_3$

Isomers

20.78

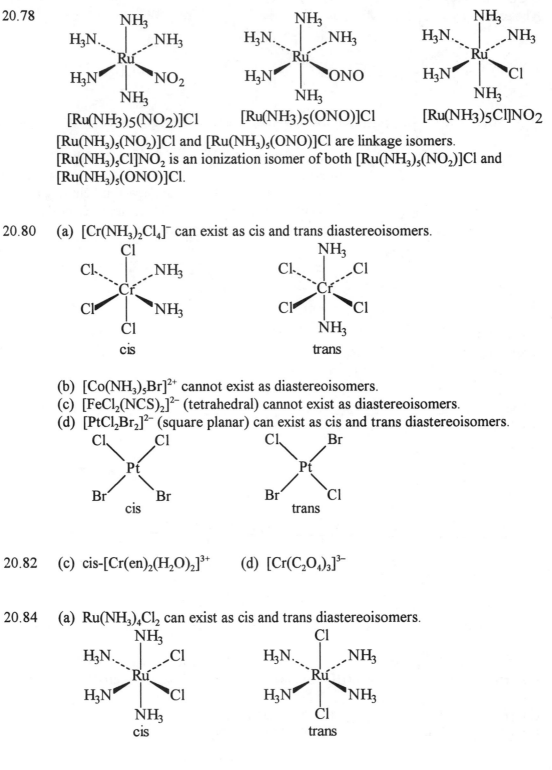

$[Ru(NH_3)_5(NO_2)]Cl$ $[Ru(NH_3)_5(ONO)]Cl$ $[Ru(NH_3)_5Cl]NO_2$

$[Ru(NH_3)_5(NO_2)]Cl$ and $[Ru(NH_3)_5(ONO)]Cl$ are linkage isomers.
$[Ru(NH_3)_5Cl]NO_2$ is an ionization isomer of both $[Ru(NH_3)_5(NO_2)]Cl$ and $[Ru(NH_3)_5(ONO)]Cl$.

20.80   (a) $[Cr(NH_3)_2Cl_4]^-$ can exist as cis and trans diastereoisomers.

cis   trans

(b) $[Co(NH_3)_5Br]^{2+}$ cannot exist as diastereoisomers.
(c) $[FeCl_2(NCS)_2]^{2-}$ (tetrahedral) cannot exist as diastereoisomers.
(d) $[PtCl_2Br_2]^{2-}$ (square planar) can exist as cis and trans diastereoisomers.

cis   trans

20.82   (c) cis-$[Cr(en)_2(H_2O)_2]^{3+}$ (d) $[Cr(C_2O_4)_3]^{3-}$

20.84   (a) $Ru(NH_3)_4Cl_2$ can exist as cis and trans diastereoisomers.

cis   trans

(b) [Pt(en)₃]⁴⁺ can exist as enantiomers.

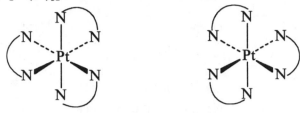

(c) [Pt(en)₂ClBr]²⁺ can exist as both diastereoisomers and enantiomers.

diastereoisomers

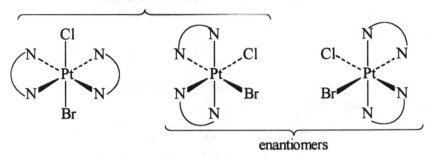

enantiomers

20.86 Plane-polarized light is light in which the electric vibrations of the light wave are restricted to a single plane. The following chromium complex can rotate the plane of plane-polarized light.

[Cr(en)₃]³⁺

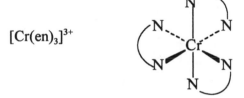

Color of Complexes; Valence Bond and Crystal Field Theories

20.88 The measure of the amount of light absorbed by a substance is called the absorbance, and a graph of absorbance versus wavelength is called an absorption spectrum.
If a complex absorbs at 455 nm, its color is orange (use the color wheel in Figure 20.24).

20.90 (a) $[Ti(H_2O)_6]^{3+}$
Ti^{3+} [Ar]   ↑ __ __ __ __   __   __ __ __
                         3d   4s   4p

$[Ti(H_2O)_6]^{3+}$ [Ar]   ↑ __ __ | ↑↓ ↑↓   ↑↓   ↑↓ ↑↓ ↑↓ |
                             3d   4s   4p

                         d^2sp^3 1 unpaired e⁻

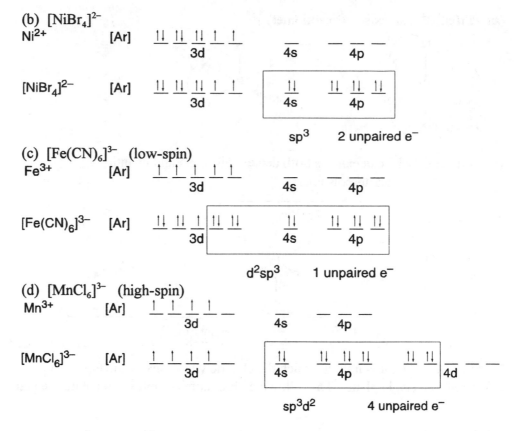

(b) $[NiBr_4]^{2-}$

Ni^{2+} [Ar]   3d   4s   4p

$[NiBr_4]^{2-}$ [Ar]   3d   4s   4p

sp^3 2 unpaired e$^-$

(c) $[Fe(CN)_6]^{3-}$ (low-spin)

Fe^{3+} [Ar]   3d   4s   4p

$[Fe(CN)_6]^{3-}$ [Ar]   3d   4s   4p

d^{2}sp^3 1 unpaired e$^-$

(d) $[MnCl_6]^{3-}$ (high-spin)

Mn^{3+} [Ar]   3d   4s   4p

$[MnCl_6]^{3-}$ [Ar]   3d   4s   4p   4d

sp^{3}d^2 4 unpaired e$^-$

20.92 $[Ti(H_2O)_6]^{3+}$ Ti^{3+} 3d^1

E

z^2 x^2–y^2

Δ   crystal field splitting

xz   yz   xy

$[Ti(H_2O)_6]^{3+}$ is colored because it can absorb light in the visible region, exciting the electron to the higher-energy set of orbitals

20.94 $\lambda = 544$ nm $= 544 \times 10^{-9}$ m

$$\Delta = \frac{hc}{\lambda} = \frac{(6.626 \times 10^{-34} \text{ J} \cdot \text{s})(3.00 \times 10^8 \text{ m/s})}{(544 \times 10^{-9} \text{ m})} = 3.65 \times 10^{-19} \text{ J}$$

$\Delta = (3.65 \times 10^{-19}$ J/ion$)(6.022 \times 10^{23}$ ion/mol$) = 219,803$ J/mol $= 220$ kJ/mol

For $[Ti(H_2O)_6]^{3+}$, $\Delta = 240$ kJ/mol

Because $\Delta_{NCS^-} < \Delta_{H_2O}$ for the Ti complex, NCS$^-$ is a weaker-field ligand than H_2O. If $[Ti(NCS)_6]^{3-}$ absorbs at 544 nm, its color should be red (use the color wheel in Figure 20.24).

20.96 (a) $[CrF_6]^{3-}$ (b) $[V(H_2O)_6]^{3+}$ (c) $[Fe(CN)_6]^{3-}$

    — —   — —   — —

↿ ↿ ↿
3 unpaired e⁻

               ↿ ↿ _
               2 unpaired e⁻

                                ⇅ ⇅ ↿
                                  1 unpaired e⁻

20.98 $Ni^{2+}(aq)$ $Zn^{2+}(aq)$

    ↿ ↿   ⇅ ⇅

    ⇅ ⇅ ⇅   ⇅ ⇅ ⇅

$Ni^{2+}(aq)$ is green because the Ni^{2+} ion can absorb light, which promotes electrons from the filled d orbitals to the higher energy half-filled d orbitals. $Zn^{2+}(aq)$ is colorless because the d orbitals are completely filled and no electrons can be promoted, so no light is absorbed.

20.100 Weak-field ligands produce a small Δ. Strong-field ligands produce a large Δ. For a metal complex with weak-field ligands, Δ < P, where P is the pairing energy, and it is easier to place an electron in either d_{z^2} or $d_{x^2-y^2}$ than to pair up electrons; high-spin complexes result. For a metal complex with strong-field ligands, Δ > P and it is easier to pair up electrons than to place them in either d_{z^2} or $d_{x^2-y^2}$; low-spin complexes result.

20.102   __ x^2-y^2

             ⇅ xy

             ⇅ z^2

          ⇅ ⇅ xz, yz

Square planar geometry is most common for metal ions with d^8 configurations because this configuration favors low-spin complexes in which all four lower energy d orbitals are filled and the higher energy $d_{x^2-y^2}$ orbital is vacant.

General Problems

20.104 (a) $[Mn(CN)_6]^{3-}$ Mn^{3+} $[Ar]\ 3d^4$
CN$^-$ is a strong-field ligand. The Mn^{3+} complex is low-spin.

— —

$\underline{\uparrow\downarrow}\ \underline{\uparrow}\ \underline{\uparrow}$ 2 unpaired e$^-$, paramagnetic

(b) $[Zn(NH_3)_4]^{2+}$ Zn^{2+} $[Ar]\ 3d^{10}$
$[Zn(NH_3)_4]^{2+}$ is tetrahedral.

$\underline{\uparrow\downarrow}\ \underline{\uparrow\downarrow}\ \underline{\uparrow\downarrow}$

$\underline{\uparrow\downarrow}\ \underline{\uparrow\downarrow}$ 0 unpaired e$^-$, diamagnetic

(c) $[Fe(CN)_6]^{4-}$ Fe^{2+} $[Ar]\ 3d^6$
CN$^-$ is a strong-field ligand. The Fe^{2+} complex is low-spin.

— —

$\underline{\uparrow\downarrow}\ \underline{\uparrow\downarrow}\ \underline{\uparrow\downarrow}$ 0 unpaired e$^-$, diamagnetic

(d) $[FeF_6]^{4-}$ Fe^{2+} $[Ar]\ 3d^6$
F$^-$ is a weak-field ligand. The Fe^{2+} complex is high-spin.

$\underline{\uparrow}\ \underline{\uparrow}$

$\underline{\uparrow\downarrow}\ \underline{\uparrow}\ \underline{\uparrow}$ 4 unpaired e$^-$, paramagnetic

20.106 (a) $4\ [Co^{3+}(aq) + e^- \rightarrow Co^{2+}(aq)]$
$\underline{2\ H_2O(l) \rightarrow O_2(g) + 4\ H^+(aq) + 4\ e^-}$
$4\ Co^{3+}(aq) + 2\ H_2O(l) \rightarrow 4\ Co^{2+}(aq) + O_2(g) + 4\ H^+(aq)$
(b) $4\ Cr^{2+}(aq) + O_2(g) + 4\ H^+(aq) \rightarrow 4\ Cr^{3+}(aq) + 2\ H_2O(l)$
(c) $3\ [Cu(s) \rightarrow Cu^{2+}(aq) + 2\ e^-]$
$\underline{Cr_2O_7^{2-}(aq) + 14\ H^+(aq) + 6\ e^- \rightarrow 2\ Cr^{3+}(aq) + 7\ H_2O(l)}$
$3\ Cu(s) + Cr_2O_7^{2-}(aq) + 14\ H^+(aq) \rightarrow 3\ Cu^{2+}(aq) + 2\ Cr^{3+}(aq) + 7\ H_2O(l)$
(d) $2\ CrO_4^{2-}(aq) + 2\ H^+(aq) \rightarrow Cr_2O_7^{2-}(aq) + H_2O(l)$

20.108 mol Fe^{2+} = (0.1000 L)(0.400 mol/L) = 0.0400 mol Fe^{2+}
mol $Cr_2O_7^{2-}$ = (0.1000 L)(0.100 mol/L) = 0.0100 mol $Cr_2O_7^{2-}$
      $6\ [Fe^{2+}(aq) \rightarrow Fe^{3+}(aq) + e^-]$
$\underline{Cr_2O_7^{2-}\ (aq) + 14\ H^+(aq) + 6\ e^- \rightarrow 2\ Cr^{3+}(aq) + 7\ H_2O(l)}$
$6\ Fe^{2+}(aq) + Cr_2O_7^{2-}(aq) + 14\ H^+(aq) \rightarrow 6\ Fe^{3+}(aq) + 2\ Cr^{3+}(aq) + 7\ H_2O(l)$

	$\underline{Fe^{2+}}$	$\underline{Cr_2O_7^{2-}}$	$\underline{Fe^{3+}}$	$\underline{Cr^{3+}}$
initial (mol)	0.0400	0.0100	0	0
change (mol)	−0.0400	$-\dfrac{0.0400}{6}$	+0.0400	$+\dfrac{0.0400}{3}$
after (mol)	0	0.00333	0.0400	0.0133

$$[Fe^{3+}] = \frac{0.0400 \text{ mol}}{0.2000 \text{ L}} = 0.200 \text{ M} \qquad [Cr^{3+}] = \frac{0.0133 \text{ mol}}{0.200 \text{ L}} = 0.0665 \text{ M}$$

$$[Cr_2O_7^{2-}] = \frac{0.00333 \text{ mol}}{0.200 \text{ L}} = 0.0166 \text{ M}$$

20.110 (a) sodium aquabromodioxalatoplatinate(IV)
(b) hexaamminechromium(III) trioxalatocobaltate(III)
(c) hexaamminecobalt(III) trioxalatochromate(III)
(d) diamminebis(ethylenediamine)rhodium(III) sulfate

20.112 Cl^- is a weak-field ligand, whereas CN^- is a strong-field ligand. Δ for $[Fe(CN)_6]^{3-}$ is larger than the pairing energy P; Δ for $[FeCl_6]^{3-}$ is smaller than P. Fe^{3+} has a $3d^5$ electron configuration.

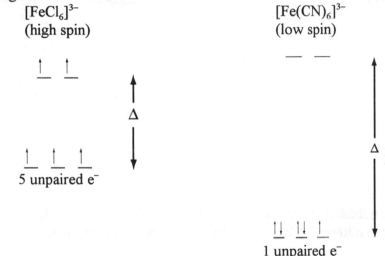

$[FeCl_6]^{3-}$
(high spin)

$[Fe(CN)_6]^{3-}$
(low spin)

5 unpaired e⁻

1 unpaired e⁻

Because of the difference in Δ, $[FeCl_6]^{3-}$ is high-spin with five unpaired electrons, whereas $[Fe(CN)_6]^{3-}$ is low-spin with only one unpaired electron.

20.114 A choice between high-spin and low-spin electron configurations arises only for complexes of metal ions with four to seven d electrons, the so-called d^4–d^7 complexes. For d^1–d^3 and d^8–d^{10} complexes, only one ground-state electron configuration is possible. In d^1 - d^3 complexes, all the electrons occupy the lower-energy d orbitals, independent of the value of Δ. In d^8 – d^{10} complexes, the lower-energy set of d orbitals is filled with three pairs of electrons, while the higher-energy set contains two, three, or four electrons, again independent of the value of Δ.

↑ _ _   ↑ ↑ _   ↑ ↑ ↑

d^1 d^2 d^3

↑ _   _ _   ↑ ↑   _ _

↑ ↑ ↑   ↑↓ ↑ ↑   ↑ ↑ ↑   ↑↓ ↑↓ ↑

d^4 high-spin d^4 low-spin d^5 high-spin d^5 low-spin

↑ ↑   _ _   ↑ ↑   ↑ _

↑↓ ↑ ↑   ↑↓ ↑↓ ↑↓   ↑↓ ↑↓ ↑   ↑↓ ↑↓ ↑↓

d^6 high-spin d^6 low-spin d^7 high-spin d^7 low-spin

↑ ↑   ↑↓ ↑   ↑↓ ↑↓

↑↓ ↑↓ ↑↓   ↑↓ ↑↓ ↑↓   ↑↓ ↑↓ ↑↓

d^8 d^9 d^{10}

20.116 $[CoCl_4]^{2-}$ is tetrahedral. $[Co(H_2O)_6]^{2+}$ is octahedral. Because $\Delta_{tet} < \Delta_{oct}$, these complexes have different colors. $[CoCl_4]^{2-}$ has absorption bands at longer wavelengths.

20.118 $Co(gly)_3$

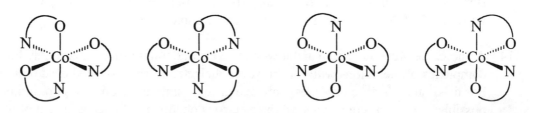

20.120

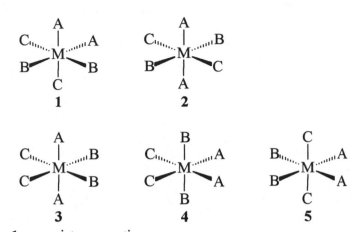

1 can exist as enantiomers.

20.122 (1) $Ni(H_2O)_6^{2+}(aq) + 6 NH_3(aq) \rightleftharpoons Ni(NH_3)_6^{2+}(aq) + 6 H_2O(l)$ $K_f = 5.6 \times 10^8$

(2) $Ni(H_2O)_6^{2+}(aq) + 3 en(aq) \rightleftharpoons Ni(en)_3^{2+}(aq) + 6 H_2O(l)$ $K_f = 1.4 \times 10^{18}$

(a) Reaction (2) should have the larger entropy change because three bidentate en ligands displace six water molecules.

(b) $\Delta G° = \Delta H° - T\Delta S°$

Because $\Delta H°_1$ and $\Delta H°_2$ are almost the same, the difference in $\Delta G°$ is determined by the difference in $\Delta S°$. Because $\Delta S°_2$ is larger than $\Delta S°_1$, $\Delta G°_2$ is more negative than $\Delta G°_1$ which is consistent with the greater stability of $Ni(en)_3^{2+}$

(c) $\Delta H° - T\Delta S° = \Delta G° = -2.303RT \log K_f$

$\Delta H°_1 - T\Delta S°_1 - (\Delta H°_2 - T\Delta S°_2) = -2.303RT \log K_f(1) - [-2.303RT \log K_f(2)]$

$T\Delta S°_2 - T\Delta S°_1 = 2.303RT \log K_f(2) - 2.303RT \log K_f(1) = 2.303RT \log \dfrac{K_f(2)}{K_f(1)}$

$\Delta S°_2 - \Delta S°_1 = 2.303 R \log \dfrac{K_f(2)}{K_f(1)} = (2.303)[8.314 \text{ J/(K} \cdot \text{mol)}] \log \dfrac{1.4 \times 10^{18}}{5.6 \times 10^8}$

$\Delta S°_2 - \Delta S°_1 = 180 \text{ J/(K} \cdot \text{mol)}$

Metals and Solid-State Materials

21.1
(a) $Cr_2O_3(s) + 2\ Al(s) \rightarrow 2\ Cr(s) + Al_2O_3(s)$
(b) $Cu_2S(s) + O_2(g) \rightarrow 2\ Cu(s) + SO_2(g)$
(c) $PbO(s) + C(s) \rightarrow Pb(s) + CO(g)$

$$2\ K^+(l) + 2\ Cl^-(l) \xrightarrow{\text{electrolysis}} 2\ K(l) + Cl_2(g)$$
(d)

21.2
The electron configuration for Hg is $[Xe]\ 4f^{14}5d^{10}6s^2$. Assuming the 5d and 6s bands overlap, the composite band can accomodate 12 valence electrons per metal atom. Weak bonding and a low melting point are expected for Hg because both the bonding and antibonding MOs are occupied.

21.3
Ge doped with As is an n-type semiconductor because As has an additional valence electron. The extra electrons are in the conduction band. The number of electrons in the conduction band of the doped Ge is much higher than for pure Ge, and the conductivity of the doped semiconductor is higher.

21.4
8 Cu at corners $8 \times 1/8 = 1$ Cu
8 Cu on edges $8 \times 1/4 = \underline{2\ Cu}$
                                      Total $= 3$ Cu

12 O on edges $12 \times 1/4 = 3$ O
8 O on faces $8 \times 1/2 = \underline{4\ O}$
                                      Total $= 7$ O

21.5
$Si(OCH_3)_4 + 4\ H_2O \rightarrow Si(OH)_4 + 4\ HOCH_3$

21.6
$Ba[OCH(CH_3)_2]_2 + Ti[OCH(CH_3)_2]_4 + 6\ H_2O \rightarrow BaTi(OH)_6(s) + 6\ HOCH(CH_3)_2$

$$BaTi(OH)_6(s) \xrightarrow{\text{heat}} BaTiO_3(s)$$

21.7
(a) cobalt/tungsten carbide is a ceramic-metal composite.
(b) silicon carbide/zirconia is a ceramic-ceramic composite.
(c) boron nitride/epoxy is a ceramic-polymer composite.
(d) boron carbide/titanium is a ceramic-metal composite.

Understanding Key Concepts

21.8
A – metal oxide;   B – metal sulfide;   C – metal carbonate;   D – free metal

21.9
(a) (iii) electrolysis   (b) (i) roasting a metal sulfide

21.10   (a) (1) and (4) are semiconductors; (2) is a metal; (3) is an insulator

(b) (3) < (1) < (4) < (2). The conductivity increases with decreasing band gap.

(c) (1) and (4) increases; (2) decreases; (3) not much change.

21.11   (a) (2), bonding MO's are filled.

(b) (3), bonding and antibonding MO's are filled.

(c) (3) < (1) < (2). Hardness increases with increasing MO bond order.

21.12

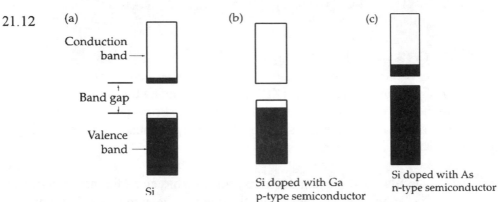

(d) The electrical conductivity of the doped silicon in both cases is higher than for pure silicon. Si doped with Ga is a p-type semiconductor with many more positive holes in the conduction band than in pure Si. This results in a higher conductivity. Si doped with As is an n-type semiconductor with many more electrons in the conduction band than in pure Si. This also results in a higher conductivity.

21.13

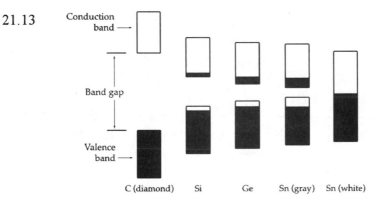

Additional Problems
Sources of the Metallic Elements

21.14 TiO_2, MnO_2, and Fe_2O_3

21.16   (a) Cu is found in nature as a sulfide.   (b) Zr is found in nature as an oxide.

(c) Pd is found in nature uncombined.   (d) Bi is found in nature as a sulfide.

21.18   The less electronegative early transition metals tend to form ionic compounds by losing electrons to highly electronegative nonmetals such as oxygen. The more electronegative late transition metals tend to form compounds with more covalent character by bonding to the less electronegative nonmetals such as sulfur.

21.20   (a) Fe_2O_3, hematite   (b) PbS, galena
         (c) TiO_2, rutile   (d) $CuFeS_2$, chalcopyrite

Metallurgy

21.22   The flotation process exploits the differences in the ability of water and oil to wet the surfaces of the mineral and the gangue. The gangue, which contains ionic silicates, is moistened by the polar water molecules and sinks to the bottom of the tank. The mineral particles, which contain the less polar metal sulfide, are coated by the oil and become attached to the soapy air bubbles created by the detergent. The metal sulfide particles are carried to the surface in the soapy froth, which is skimmed off at the top of the tank. This process would not work well for a metal oxide because it is too polar and will be wet by the water and sink with the gangue.

21.24   Since $E° < 0$ for Zn^{2+}, the reduction of Zn^{2+} is not favored.
         Since $E° > 0$ for Hg^{2+}, the reduction of Hg^{2+} is favored.
         The roasting of CdS should yield CdO because, like Zn^{2+}, $E° < 0$ for the reduction of Cd^{2+}.

21.26   (a) $V_2O_5(s) + 5 Ca(s) \rightarrow 2 V(s) + 5 CaO(s)$
         (b) $2 PbS(s) + 3 O_2(g) \rightarrow 2 PbO(s) + 2 SO_2(g)$
         (c) $MoO_3(s) + 3 H_2(g) \rightarrow Mo(s) + 3 H_2O(g)$
         (d) $3 MnO_2(s) + 4 Al(s) \rightarrow 3 Mn(s) + 2 Al_2O_3(s)$

         (e) $MgCl_2(l) \xrightarrow{\text{electrolysis}} Mg(l) + Cl_2(g)$

21.28 $2 ZnS(s) + 3 O_2(g) \rightarrow 2 ZnO(s) + 2 SO_2(g)$
         $\Delta H° = [2 \Delta H°_f(ZnO) + 2 \Delta H°_f(SO_2)] - [2 \Delta H°_f(ZnS)]$
         $\Delta H° = [(2 \text{ mol})(-348.3 \text{ kJ/mol}) + (2 \text{ mol})(-296.8 \text{ kJ/mol})]$
                   $- (2 \text{ mol})(-206.0 \text{ kJ/mol}) = -878.2 \text{ kJ}$
         $\Delta G° = [2 \Delta G°_f(ZnO) + 2 \Delta G°_f(SO_2)] - [2 \Delta G°_f(ZnS)]$
         $\Delta G° = [(2 \text{ mol})(-318.3 \text{ kJ/mol}) + (2 \text{ mol})(-300.2 \text{ kJ/mol})]$
                   $- (2 \text{ mol})(-201.3 \text{ kJ/mol}) = -834.4 \text{ kJ}$
         $\Delta H°$ and $\Delta G°$ are different because of the entropy change associated with the reaction. The minus sign for $(\Delta H° - \Delta G°)$ indicates that the entropy is negative, which is consistent with a decrease in the number of moles of gas from 3 mol to 2 mol.

21.30 $FeCr_2O_4(s) + 4 C(s) \rightarrow Fe(s) + 2 Cr(s) + 4 CO(g)$
                          ferrochrome
         (a) $FeCr_2O_4$, 223.84 amu; Cr, 52.00 amu;   236 kg = 236 x 10^3 g

$$\text{mass Cr} = 236 \times 10^3 \text{ g} \times \frac{1 \text{ mol FeCr}_2\text{O}_4}{223.84 \text{ g}} \times \frac{2 \text{ mol Cr}}{1 \text{ mol FeCr}_2\text{O}_4} \times \frac{52.00 \text{ g Cr}}{1 \text{ mol Cr}} \times \frac{1.00 \text{ kg}}{1000 \text{ g}} = 110 \text{ kg Cr}$$

(b) $\text{mol CO} = 236 \times 10^3 \text{ g} \times \dfrac{1 \text{ mol FeCr}_2\text{O}_4}{223.84 \text{ g}} \times \dfrac{4 \text{ mol CO}}{1 \text{ mol FeCr}_2\text{O}_4} = 4217.3 \text{ mol CO}$

$$PV = nRT; \quad V = \frac{nRT}{P} = \frac{(4217.3 \text{ mol})\left(0.08206 \dfrac{\text{L} \cdot \text{atm}}{\text{mol} \cdot \text{K}}\right)(298 \text{ K})}{\left(740 \text{ mm Hg} \times \dfrac{1 \text{ atm}}{760 \text{ mm Hg}}\right)} = 1.06 \times 10^5 \text{ L CO}$$

21.32 $Ni^{2+}(aq) + 2 e^- \rightarrow Ni(s); 1 \text{ A} = 1 \text{ C/s}$

$$\text{mass Ni} = 52.5 \frac{C}{s} \times 8 \text{ h} \times \frac{3600 \text{ s}}{1 \text{ h}} \times \frac{1 \text{ mol e}^-}{96,500 \text{ C}} \times \frac{1 \text{ mol Ni}}{2 \text{ mol e}^-} \times \frac{58.69 \text{ g Ni}}{1 \text{ mol Ni}} \times \frac{1.00 \text{ kg}}{1000 \text{ g}}$$

mass Ni = 0.460 kg Ni

Iron and Steel

21.34 $Fe_2O_3(s) + 3 CO(g) \rightarrow 2 Fe(l) + 3 CO_2(g)$
Fe_2O_3 is the oxidizing agent. CO is the reducing agent.

21.36   Slag is a byproduct of iron production, consisting mainly of $CaSiO_3$. It is produced from the gangue in iron ore.

21.38   Molten iron from a blast furnace is exposed to a jet of pure oxygen gas for about 20 minutes. The impurities are oxidized to yield a molten slag that can be poured off.
$P_4(l) + 5 O_2(g) \rightarrow P_4O_{10}(l)$
$6 CaO(s) + P_4O_{10}(l) \rightarrow 2 Ca_3(PO_4)_2(l)$ (slag)

$2 Mn(l) + O_2(g) \rightarrow 2 MnO(s)$
$MnO(s) + SiO_2(s) \rightarrow MnSiO_3(l)$ (slag)

21.40 $SiO_2(s) + 2 C(s) \rightarrow Si(s) + 2 CO(g)$
$Si(s) + O_2(g) \rightarrow SiO_2(s)$
$CaO(s) + SiO_2(s) \rightarrow CaSiO_3(l)$ (slag)

Bonding in Metals

21.42

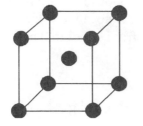

Each K has a single valence electron and has eight nearest neighbor K atoms. The valence electrons can't be localized in an electron-pair bond between any particular pair of K atoms.

21.44   Malleability and ductility of metals follows from the fact that the delocalized bonding extends in all directions. When a metallic crystal is deformed, no localized bonds are broken. Instead, the electron sea simply adjusts to the new distribution of cations, and the energy of the deformed structure is similar to that of the original. Thus, the energy required to deform a metal is relatively small.

21.46   The energy required to deform a transition metal like W is greater than that for Cs because W has more valence electrons and hence more electrostatic "glue".

21.48   The difference in energy between successive MOs in a metal decreases as the number of metal atoms increases so that the MOs merge into an almost continuous band of energy levels. Consequently, MO theory for metals is often called band theory.

21.50   The energy levels within a band occur in degenerate pairs; one set of energy levels applies to electrons moving to the right, and the other set applies to electrons moving to the left. In the absence of an electrical potential, the two sets of levels are equally populated. As a result there is no net electric current. In the presence of an electrical potential those electrons moving to the right are accelerated, those moving to the left are slowed down, and some change direction. Thus, the two sets of energy levels are now unequally populated. The number of electrons moving to the right is now greater than the number moving to the left, and so there is a net electric current.

21.52   (a)   (b)

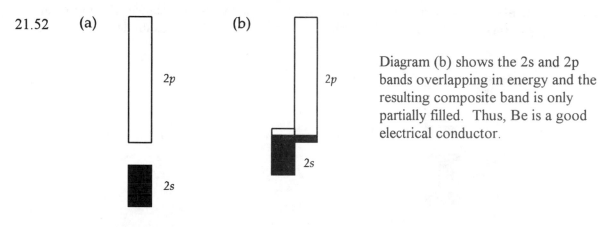

Diagram (b) shows the 2s and 2p bands overlapping in energy and the resulting composite band is only partially filled. Thus, Be is a good electrical conductor.

21.54   Transition metals have a d band that can overlap the s band to give a composite band consisting of six MOs per metal atom. Half of the MOs are bonding and half are antibonding, and thus one expects maximum bonding for metals that have six valence electrons per metal atom. Accordingly, the melting points of the transition metals go through a maximum at or near group 6B.

Semiconductors

21.56   A semiconductor is a material that has an electrical conductivity intermediate between that of a metal and that of an insulator. Si, Ge, and Sn (gray) are semiconductors.

21.58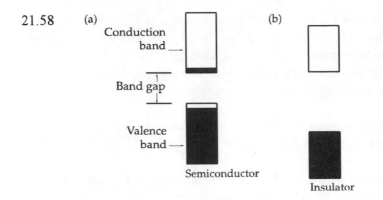

The MOs of a semiconductor are similar to those of an insulator, but the band gap in a semiconductor is smaller. As a result, a few electrons have enough energy to jump the gap and occupy the higher-energy, conduction band. The conduction band is thus partially filled, and the valence band is partially empty. When an electrical potential is applied to a semiconductor, it conducts a small amount of current because the potential can accelerate the electrons in the partially filled bands.

21.60 As the band gap increases, the number of electrons able to jump the gap and occupy the higher-energy conduction band decreases, and thus the conductivity decreases.

21.62 An n-type semiconductor is a semiconductor doped with a substance with more valence electrons than the semiconductor itself. Si doped with P is an example.

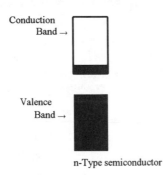

n-Type semiconductor

21.64 In the MO picture, the extra electrons occupy the conduction band. The number of electrons in the conduction band of the doped Ge is much greater than for pure Ge, and the conductivity of the doped semiconductor is correspondingly higher.

21.66 (a) p−type (In is electron deficient with respect to Si)
(b) n−type (Sb is electron rich with respect to Ge)
(c) n−type (As is electron rich with respect to gray Sn)

21.68 Al_2O_3 < Ge < Ge doped with In < Fe < Cu

Superconductors

21.70   (1) A superconductor is able to levitate a magnet.
(2) In a superconductor, once an electric current is started, it flows indefinitely without loss of energy. A superconductor has no electrical resistance.

21.72   The fullerides are three-dimensional superconductors, whereas the copper oxide ceramics are two-dimensional superconductors.

Ceramics and Composites

21.74   Ceramics are inorganic, nonmetallic, nonmolecular solids, including both crystalline and amorphous materials. Ceramics have higher melting points, and they are stiffer, harder, and more resistant to wear and corrosion than are metals.

21.76   Ceramics have higher melting points, and they are stiffer, harder, and more wear resistant than metals because they have stronger bonding. They maintain much of their strength at high temperatures, where metals either melt or corrode because of oxidation.

21.78   The brittleness of ceramics is due to strong chemical bonding. In silicon nitride each Si atom is bonded to four N atoms and each N atom is bonded to three Si atoms. The strong, highly directional covalent bonds prevent the planes of atoms from sliding over one another when the solid is subjected to a stress. As a result, the solid can't deform to relieve the stress. It maintains its shape up to a point, but then the bonds give way suddenly and the material fails catastrophically when the stress exceeds a certain threshold value. By contrast, metals are able to deform under stress because their planes of metal cations can slide easily in the electron sea.

21.80   Ceramic processing is the series of steps that leads from raw material to the finished ceramic object.

21.82   A sol is a colloidal dispersion of tiny particles. A gel is a more rigid gelatin-like material consisting of larger particles.

21.84 $Zr[OCH(CH_3)_2]_4 + 4 H_2O \rightarrow Zr(OH)_4 + 4 HOCH(CH_3)_2$

21.86 $(HO)_3Si-O-H + H-O-Si(OH)_3 \rightarrow (HO)_3Si-O-Si(OH)_3 + H_2O$
Further reactions of this sort give a three-dimensional network of Si–O–Si bridges. On heating, SiO_2 is obtained.

21.88 $3 SiCl_4(g) + 4 NH_3(g) \rightarrow Si_3N_4(s) + 12 HCl(g)$

21.90   Graphite/epoxy composites are good materials for making tennis rackets and golf clubs because of their high strength–to–weight ratios.

General Problems

21.92 The chemical composition of the alkaline earth minerals is that of metal sulfates and sulfites, MSO_4 and MSO_3.

21.94 Band theory better explains how the number of valence electrons affects properties such as melting point and hardness.

21.96 V [Ar] $3d^3 4s^2$ Zn [Ar] $3d^{10} 4s^2$

Transition metals have a d band that can overlap the s band to give a composite band consisting of six MOs per metal atom. Half of the MOs are bonding and half are antibonding. Strong bonding and a high enthalpy of vaporization are expected for V because almost all of the bonding MOs are occupied and all of the antibonding MOs are empty. Weak bonding and a low enthalpy of vaporization are expected for Zn because both the bonding and the antibonding MOs are occupied.

21.98 With a band gap of 130 kJ/mol, GaAs is a semiconductor. Because Ge lies between Ga and As in the periodic table, GaAs is isoelectronic with Ge.

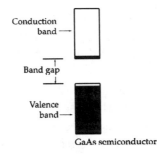

21.100 $YBa_2Cu_3O_7$, 666.20 amu; $Cu(OCH_2CH_3)_2$, 153.67 amu

$Y(OCH_2CH_3)_3$, 224.09 amu; $Ba(OCH_2CH_3)_2$, 227.45 amu

$$\text{mol } Cu(OCH_2CH_3)_2 = 75.4 \text{ g} \times \frac{1 \text{ mol}}{153.67 \text{ g}} = 0.4907 \text{ mol } Cu(OCH_2CH_3)_2$$

mass $Y(OCH_2CH_3)_3 = 0.4907$ mol $Cu(OCH_2CH_3)_2$ x

$$\frac{1 \text{ mol } Y(OCH_2CH_3)_3}{3 \text{ mol } Cu(OCH_2CH_3)_2} \text{ x } \frac{224.09 \text{ g } Y(OCH_2CH_3)_3}{1 \text{ mol } Y(OCH_2CH_3)_3} = 36.7 \text{ g } Y(OCH_2CH_3)_3$$

mass $Ba(OCH_2CH_3)_2 = 0.4907$ mol $Cu(OCH_2CH_3)_2$ x

$$\frac{2 \text{ mol } Ba(OCH_2CH_3)_2}{3 \text{ mol } Cu(OCH_2CH_3)_2} \text{ x } \frac{227.45 \text{ g } Ba(OCH_2CH_3)_2}{1 \text{ mol } Ba(OCH_2CH_3)_2} = 74.4 \text{ g } Ba(OCH_2CH_3)_2$$

mass $YBa_2Cu_3O_7 = 0.4907$ mol $Cu(OCH_2CH_3)_2$ x

$$\frac{1 \text{ mol } YBa_2Cu_3O_7}{3 \text{ mol } Cu(OCH_2CH_3)_2} \text{ x } \frac{666.20 \text{ g } YBa_2Cu_3O_7}{1 \text{ mol } YBa_2Cu_3O_7} = 109 \text{ g } YBa_2Cu_3O_7$$

21.102   (a) $6 \ Al(OCH_2CH_3)_3 + 2 \ Si(OCH_2CH_3)_4 + 26 \ H_2O \rightarrow$

$\qquad\qquad 6 \ Al(OH)_3(s) + 2 \ Si(OH)_4(s) + 26 \ HOCH_2CH_3$

$\qquad\qquad\qquad\qquad$ sol

(b) H_2O is eliminated from the sol through a series of reactions linking the sol particles together through a three-dimensional network of O bridges to form the gel.

$(HO)_2Al-O-H + H-O-Si(OH)_3 \rightarrow (HO)_2Al-O-Si(OH)_3 + H_2O$

(c) The remaining H_2O and solvent are removed from the gel by heating to produce the ceramic, $3 \ Al_2O_3 \cdot 2 \ SiO_2$.

21.104   (a)

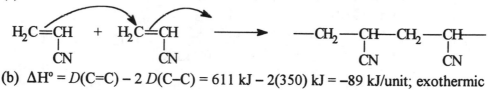

(b) $\Delta H° = D(C=C) - 2 \ D(C-C) = 611 \ kJ - 2(350) \ kJ = -89 \ kJ/unit$; exothermic

21.106 $Ni(s) + 4 \ CO(g) \rightleftharpoons Ni(CO)_4(g)$

$\Delta H° = -160.8 \ kJ; \ \Delta S° = -410 \ J/K = -410 \times 10^{-3} \ kJ/K$

(a) $150°C = 423 \ K$

$\Delta G° = \Delta H° - T\Delta S° = -160.8 \ kJ - (423 \ K)(-410 \times 10^{-3} \ kJ/K) = +12.6 \ kJ$

$\Delta G° = -2.303RT \log K$

$\log K = \dfrac{-\Delta G°}{2.303 \, RT} = \dfrac{-12.6 \ kJ}{(2.303)(8.314 \times 10^{-3} \ kJ/K)(423 \ K)} = -1.56$

$K = K_p = 10^{-1.56} = 0.028$

(b) $230°C = 503 \ K$

$\Delta G° = \Delta H° - T\Delta S° = -160.8 \ kJ - (503 \ K)(-410 \times 10^{-3} \ kJ/K) = +45.4 \ kJ$

$\Delta G° = -2.303RT \log K$

$\log K = \dfrac{-\Delta G°}{2.303 \, RT} = \dfrac{-45.4 \ kJ}{(2.303)(8.314 \times 10^{-3} \ kJ/K)(503 \ K)} = -4.71$

$K = K_p = 10^{-4.71} = 1.9 \times 10^{-5}$

(c) $\Delta S°$ is large and negative because as the reaction proceeds in the forward direction, the number of moles of gas decrease from four to one.

Because $\Delta S°$ is negative, $-T\Delta S°$ is positive, and as T increases, $\Delta G°$ becomes more positive because $\Delta G° = \Delta H° - T\Delta S°$.

(d) The reaction is exothermic because $\Delta H°$ is negative.

$Ni(s) + 4 \ CO(g) \rightleftharpoons Ni(CO)_4(g) + heat$

Heat is added as the temperature is raised and the reaction proceeds in the reverse direction to relieve this stress, as predicted by Le Châtelier's principle. As the reverse reaction proceeds, the partial pressure of CO increases and the partial pressure of

$Ni(CO)_4$ decreases. K_p decreases as calculated because $K_p = \dfrac{P_{Ni(CO)_4}}{(P_{CO})^4}$.

21.108 $SiO_2(s) + 2 \ C(s) \rightarrow Si(s) + 2 \ CO(g)$

(a) $\Delta H° = 2 \ \Delta H°_f(CO) - \Delta H°_f(SiO_2)$

$\Delta H° = (2 \ mol)(-110.5 \ kJ/mol) - (1 \ mol)(-910.9 \ kJ/mol) = 689.9 \ kJ$

$\Delta S^\circ = [S^\circ(Si) + 2\ S^\circ(CO)] - [S^\circ(SiO_2) + 2\ S^\circ(C)]$

$\Delta S^\circ = [(1\ mol)(18.8\ J/(K \cdot mol)) + (2\ mol)(197.6\ J/(K \cdot mol))]$
$\qquad\qquad - [(1\ mol)(41.8\ J/(K \cdot mol)) + (2\ mol)(5.7\ J/(K \cdot mol))]$

$\Delta S^\circ = 360.8\ J/K = 360.8 \times 10^{-3}\ kJ/K$

$\Delta G^\circ = \Delta H^\circ - T\Delta S^\circ = 689.9\ kJ - (298.15\ K)(360.8 \times 10^{-3}\ kJ/K) = 582.3\ kJ$

(b) The reaction is endothermic because $\Delta H^\circ > 0$.

(c) The number of moles of gas increases from 0 to 2 mol, therefore, $\Delta S^\circ > 0$.

(d) Because $\Delta G^\circ > 0$, the reaction is nonspontaneous at 25°C and 1 atm pressure of CO.

(e) To determine the crossover temperature, set $\Delta G^\circ = 0$ and solve for T.

$\Delta G^\circ = 0 = \Delta H^\circ - T\Delta S^\circ$

$\Delta H^\circ = T\Delta S^\circ; \qquad T = \dfrac{\Delta H^\circ}{\Delta S^\circ} = \dfrac{689.9\ kJ}{360.8 \times 10^{-3}\ kJ/K} = 1912\ K = 1639°C$

22

Nuclear Chemistry

22.1 (a) In beta emission, the mass number is unchanged, and the atomic number increases by one. $\quad ^{106}_{44}\text{Ru} \rightarrow {}^{0}_{-1}\text{e} + {}^{106}_{45}\text{Rh}$

(b) In alpha emission, the mass number decreases by four, and the atomic number decreases by two. $\quad ^{189}_{83}\text{Bi} \rightarrow {}^{4}_{2}\text{He} + {}^{185}_{81}\text{Tl}$

(c) In electron capture, the mass number is unchanged, and the atomic number decreases by one. $\quad ^{204}_{84}\text{Po} + {}^{0}_{-1}\text{e} \rightarrow {}^{204}_{83}\text{Bi}$

22.2 The mass number decreases by four, and the atomic number decreases by two. This is characteristic of alpha emission. $\quad ^{214}_{90}\text{Th} \rightarrow {}^{210}_{88}\text{Ra} + {}^{4}_{2}\text{He}$

22.3 $\quad t_{1/2} = \dfrac{0.693}{k} = \dfrac{0.693}{1.08 \times 10^{-2}\,\text{h}^{-1}} = 64.2\,\text{h}$

22.4 $\quad k = \dfrac{0.693}{t_{1/2}} = \dfrac{0.693}{5715\,\text{y}} = 1.21 \times 10^{-4}\,\text{y}^{-1}$

22.5 $\quad \ln\!\left(\dfrac{N}{N_0}\right) = -0.693\left(\dfrac{t}{t_{1/2}}\right) = -0.693\left(\dfrac{16{,}230\,\text{y}}{5715\,\text{y}}\right) = -1.968$

$\dfrac{N}{N_0} = e^{-1.968} = 0.140; \qquad \dfrac{N}{100\%} = 0.140; \qquad N = 14.0\%$

22.6 $\quad \ln\!\left(\dfrac{N}{N_0}\right) = (-0.693)\left(\dfrac{t}{t_{1/2}}\right); \qquad \dfrac{N}{N_0} = \dfrac{\text{Decay rate at time } t}{\text{Decay rate at } t = 0}$

$\ln\!\left(\dfrac{10{,}860}{16{,}800}\right) = (-0.693)\left(\dfrac{28.0\,\text{d}}{t_{1/2}}\right); \qquad -0.436 = (-0.693)\left(\dfrac{28.0\,\text{d}}{t_{1/2}}\right)$

$t_{1/2} = \dfrac{(-0.693)(28.0\,\text{d})}{(-0.436)} = 44.5\,\text{d}$

22.7 For $^{16}_{8}\text{O}$:

First, calculate the total mass of the nucleons (8 n + 8 p)
Mass of 8 neutrons = (8)(1.008 66 amu) = 8.069 28 amu
<u>Mass of 8 protons = (8)(1.007 28 amu) = 8.058 24 amu</u>
Mass of 8 n + 8 p $\qquad\qquad$ = 16.127 52 amu

Next, calculate the mass of a ^{16}O nucleus by subtracting the mass of 8 electrons from the mass of a ^{16}O atom.

Mass of ^{16}O atom	= 15.994 92 amu
–Mass of 8 electrons = –(8)(5.486 x 10^{-4} amu)	= –0.004 39 amu
Mass of ^{16}O nucleus	= 15.990 53 amu

Then subtract the mass of the ^{16}O nucleus from the mass of the nucleons to find the mass defect:

Mass defect = mass of nucleons – mass of nucleus

$$= (16.127\ 52\ \text{amu}) - (15.990\ 53\ \text{amu}) = 0.136\ 99\ \text{amu}$$

Mass defect in g/mol:

$(0.136\ 99\ \text{amu})(1.660\ 54 \times 10^{-24}\ \text{g/amu})(6.022 \times 10^{23}\ \text{mol}^{-1}) = 0.136\ 99\ \text{g/mol}$

Now, use the Einstein equation to convert the mass defect into the binding energy.

$\Delta E = \Delta mc^2 = (0.136\ 99\ \text{g/mol})(10^{-3}\ \text{kg/g})(3.00 \times 10^8\ \text{m/s})^2$

$\Delta E = 1.233 \times 10^{13}\ \text{J/mol} = 1.233 \times 10^{10}\ \text{kJ/mol}$

$$\Delta E = \frac{1.233 \times 10^{13}\ \text{J/mol}}{6.022 \times 10^{23}\ \text{nuclei/mol}} \times \frac{1\ \text{MeV}}{1.60 \times 10^{-13}\ \text{J}} \times \frac{1\ \text{nucleus}}{16\ \text{nucleons}} = 8.00\ \frac{\text{MeV}}{\text{nucleon}}$$

22.8 $\Delta E = -852\ \text{kJ/mol} = -852 \times 10^3\ \text{J/mol};$ $1\ \text{J} = 1\ \text{kg} \cdot \text{m}^2/\text{s}^2$

$$\Delta E = \Delta mc^2; \quad \Delta m = \frac{\Delta E}{c^2} = \frac{\left(-852 \times 10^3\ \dfrac{\text{kg} \cdot \text{m}^2}{\text{s}^2 \cdot \text{mol}}\right)}{\left(3.00 \times 10^8\ \text{m/s}\right)^2} = -9.47 \times 10^{-12}\ \text{kg/mol}$$

$$\Delta m = -9.47 \times 10^{-12}\ \frac{\text{kg}}{\text{mol}} \times \frac{1000\ \text{g}}{1\ \text{kg}} = -9.47 \times 10^{-9}\ \text{g/mol}$$

22.9 $^{1}_{0}n + {}^{235}_{92}U \rightarrow {}^{137}_{52}Te + {}^{97}_{40}Zr + 2\ {}^{1}_{0}n$

mass $^{235}_{92}U$	235.0439	amu
mass $^{1}_{0}n$	1.008 66	amu
–mass $^{137}_{52}Te$	–136.9254	amu
–mass $^{97}_{40}Zr$	–96.9110	amu
–mass $2\ {}^{1}_{0}n$	–(2)(1.008 66)	amu
mass change	0.1988	amu

$(0.1988\ \text{amu})(1.660\ 54 \times 10^{-24}\ \text{g/amu})(6.022 \times 10^{23}\ \text{mol}^{-1}) = 0.1988\ \text{g/mol}$

$\Delta E = \Delta mc^2 = (0.1988\ \text{g/mol})(10^{-3}\ \text{kg/g})(3.00 \times 10^8\ \text{m/s})^2$

$\Delta E = 1.79 \times 10^{13}\ \text{J/mol} = 1.79 \times 10^{10}\ \text{kJ/mol}$

22.10 ^1_1H + ^2_1H → ^3_2He

mass ^1H	1.007 83 amu
mass ^2H	2.014 10 amu
−mass ^3He	−3.016 03 amu

mass change	0.005 90 amu

$(0.005\ 90\ amu)(1.660\ 54 \times 10^{-24}\ g/amu)(6.022 \times 10^{23}\ mol^{-1}) = 0.005\ 90\ g/mol$

$\Delta E = \Delta mc^2 = (0.005\ 90\ g/mol)(10^{-3}\ kg/g)(3.00 \times 10^8\ m/s)^2$

$\Delta E = 5.31 \times 10^{11}\ J/mol = 5.31 \times 10^8\ kJ/mol$

22.11 $^{40}_{18}Ar$ + ^1_1p → $^{40}_{19}K$ + ^1_0n

22.12 $^{238}_{92}U$ + ^2_1H → $^{238}_{93}Np$ + 2 ^1_0n

22.13 $\ln\left(\dfrac{N}{N_0}\right) = (-0.693)\left(\dfrac{t}{t_{1/2}}\right)$; $\dfrac{N}{N_0} = \dfrac{\text{Decay rate at time } t}{\text{Decay rate at time } t\ =\ 0}$

$\ln\left(\dfrac{2.4}{15.3}\right) = (-0.693)\left(\dfrac{t}{5715\ y}\right)$; $t = 1.53 \times 10^4\ y$

Understanding Key Concepts

22.14 $16\ ^{40}K$ → $8\ ^{40}K$ → $4\ ^{40}K$; two half-lives have passed.

22.15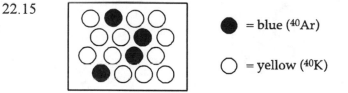

$\bullet$ = blue (^{40}Ar)

$\circ$ = yellow (^{40}K)

22.16   The isotope contains 8 neutrons and 6 protons. The isotope symbol is $^{14}_6C$.

22.17 $^{14}_6C$ would decay by beta emission because the n/p ratio is high.

22.18 $^{136}_{53}I$ is neutron rich and decays by beta emission.

$^{122}_{53}I$ is neutron poor and decays by positron emission.

22.19 ^{160}W is neutron poor and decays by alpha emission. ^{185}W is neutron rich and decays by beta emission.

Additional Problems
Nuclear Reactions and Radioactivity

22.20 Positron emission is the conversion of a proton in the nucleus into a neutron plus an ejected positron.
Electron capture is the process in which a proton in the nucleus captures an inner-shell electron and is thereby converted into a neutron.

22.22 Alpha particles move relatively slowly and can be stopped by the skin. However, inside the body, alpha particles give up their energy to the immediately surrounding tissue. Gamma rays move at the speed of light and are very penetrating. Therefore they are equally hazardous internally and externally.

22.24 There is no radioactive "neutralization" reaction like there is an acid-base neutralization reaction.

22.26 (a) $^{126}_{50}\text{Sn} \rightarrow \ ^{0}_{-1}\text{e} + \ ^{126}_{51}\text{Sb}$ (b) $^{210}_{88}\text{Ra} \rightarrow \ ^{4}_{2}\text{He} + \ ^{206}_{86}\text{Rn}$

        (c) $^{77}_{37}\text{Rb} \rightarrow \ ^{0}_{1}\text{e} + \ ^{77}_{36}\text{Kr}$ (d) $^{76}_{36}\text{Kr} + \ ^{0}_{-1}\text{e} \rightarrow \ ^{76}_{35}\text{Br}$

22.28 (a) $^{188}_{80}\text{Hg} \rightarrow \ ^{188}_{79}\text{Au} + \ ^{0}_{1}\text{e}$ (b) $^{218}_{85}\text{At} \rightarrow \ ^{214}_{83}\text{Bi} + \ ^{4}_{2}\text{He}$

        (c) $^{234}_{90}\text{Th} \rightarrow \ ^{234}_{91}\text{Pa} + \ ^{0}_{-1}\text{e}$

22.30 (a) $^{162}_{75}\text{Re} \rightarrow \ ^{158}_{73}\text{Ta} + \ ^{4}_{2}\text{He}$ (b) $^{138}_{62}\text{Sm} + \ ^{0}_{-1}\text{e} \rightarrow \ ^{138}_{61}\text{Pm}$

        (c) $^{188}_{74}\text{W} \rightarrow \ ^{188}_{75}\text{Re} + \ ^{0}_{-1}\text{e}$ (d) $^{165}_{73}\text{Ta} \rightarrow \ ^{165}_{72}\text{Hf} + \ ^{0}_{1}\text{e}$

22.32 $^{241}_{95}\text{Am} \rightarrow \ ^{237}_{93}\text{Np} + \ ^{4}_{2}\text{He}$

        $^{237}_{93}\text{Np} \rightarrow \ ^{233}_{91}\text{Pa} + \ ^{4}_{2}\text{He}$

        $^{233}_{91}\text{Pa} \rightarrow \ ^{233}_{92}\text{U} + \ ^{0}_{-1}\text{e}$

        $^{233}_{92}\text{U} \rightarrow \ ^{229}_{90}\text{Th} + \ ^{4}_{2}\text{He}$

        $^{229}_{90}\text{Th} \rightarrow \ ^{225}_{88}\text{Ra} + \ ^{4}_{2}\text{He}$

        $^{225}_{88}\text{Ra} \rightarrow \ ^{225}_{89}\text{Ac} + \ ^{0}_{-1}\text{e}$

        $^{225}_{89}\text{Ac} \rightarrow \ ^{221}_{87}\text{Fr} + \ ^{4}_{2}\text{He}$

        $^{221}_{87}\text{Fr} \rightarrow \ ^{217}_{85}\text{At} + \ ^{4}_{2}\text{He}$

        $^{217}_{85}\text{At} \rightarrow \ ^{213}_{83}\text{Bi} + \ ^{4}_{2}\text{He}$

        $^{213}_{83}\text{Bi} \rightarrow \ ^{213}_{84}\text{Po} + \ ^{0}_{-1}\text{e}$

        $^{213}_{84}\text{Po} \rightarrow \ ^{209}_{82}\text{Pb} + \ ^{4}_{2}\text{He}$

        $^{209}_{82}\text{Pb} \rightarrow \ ^{209}_{83}\text{Bi} + \ ^{0}_{-1}\text{e}$

Radioactive Decay Rates

22.34 If the half-life of ^{59}Fe is 44.5 d, it takes 44.5 days for half of the original amount of ^{59}Fe to decay.

22.36 $k = \dfrac{0.693}{t_{1/2}} = \dfrac{0.693}{2.806 \text{ d}} = 0.247 \text{ d}^{-1}$

22.38 $t_{1/2} = \dfrac{0.693}{k} = \dfrac{0.693}{0.227 \text{ d}^{-1}} = 3.05 \text{ d}$

22.40 After 65 d: $\ln\left(\dfrac{N}{N_0}\right) = -0.693\left(\dfrac{t}{t_{1/2}}\right) = -0.693\left(\dfrac{\left[\dfrac{65 \text{ d}}{365 \text{ d/y}}\right]}{432.2 \text{ y}}\right) = -0.000\ 285\ 5$

$\dfrac{N}{N_0} = e^{-0.0002855} = 0.9997; \qquad \dfrac{N}{100\%} = 0.9997; \qquad N = 99.97\%$

After 65 y: $\ln\left(\dfrac{N}{N_0}\right) = -0.693\left(\dfrac{t}{t_{1/2}}\right) = -0.693\left(\dfrac{65 \text{ y}}{432.2 \text{ y}}\right) = -0.1042$

$\dfrac{N}{N_0} = e^{-0.1042} = 0.9010; \qquad \dfrac{N}{100\%} = 0.9010; \qquad N = 90.10\%$

After 650 y: $\ln\left(\dfrac{N}{N_0}\right) = -0.693\left(\dfrac{t}{t_{1/2}}\right) = -0.693\left(\dfrac{650 \text{ y}}{432.2 \text{ y}}\right) = -1.042$

$\dfrac{N}{N_0} = e^{-1.042} = 0.3527; \qquad \dfrac{N}{100\%} = 0.3527; \qquad N = 35.27\%$

22.42 $\ln\left(\dfrac{N}{N_0}\right) = (-0.693)\left(\dfrac{t}{t_{1/2}}\right); \quad \ln(0.43) = (-0.693)\left(\dfrac{t}{5715 \text{ y}}\right); \quad t = 6960 \text{ y}$

22.44 $t_{1/2} = \dfrac{0.693}{k} = \dfrac{0.693}{7.95 \times 10^{-3} \text{ d}^{-1}} = 87.17 \text{ d}$

$\ln\left(\dfrac{N}{N_0}\right) = (-0.693)\left(\dfrac{t}{t_{1/2}}\right) = (-0.693)\left(\dfrac{185 \text{ d}}{87.17 \text{ d}}\right) = -1.4707$

$\dfrac{N}{N_0} = e^{-1.4707} = 0.2298; \qquad \dfrac{N}{100\%} = 0.2298; \qquad N = 23.0\%$

22.46 $t_{1/2} = (105 \text{ y})(365 \text{ d/y})(24 \text{ h/d})(3600 \text{ s/h}) = 3.3113 \times 10^9 \text{ s}$

$k = \dfrac{0.693}{t_{1/2}} = \dfrac{0.693}{3.3113 \times 10^9 \text{ s}} = 2.0928 \times 10^{-10} \text{ s}^{-1}$

$$N = (1.0 \times 10^{-9}\ \text{g})\left(\frac{1\ \text{mol Po}}{209\ \text{g Po}}\right)(6.022 \times 10^{23}\ \text{atoms/mol}) = 2.881 \times 10^{12}\ \text{atoms}$$

Decay rate = kN = $(2.0928 \times 10^{-10}\ \text{s}^{-1})(2.881 \times 10^{12}\ \text{atoms}) = 6.0 \times 10^2\ \text{s}^{-1}$

600 α particles are emitted in 1.0 s.

22.48   Decay rate = kN

$$N = (1.0 \times 10^{-3}\ \text{g})\left(\frac{1\ \text{mol}\ ^{79}\text{Se}}{79\ \text{g}}\right)(6.022 \times 10^{23}\ \text{atoms/mol}) = 7.6 \times 10^{18}\ \text{atoms}$$

$$k = \frac{\text{Decay rate}}{N} = \frac{2.8 \times 10^6/\text{s}}{7.6 \times 10^{18}} = 3.7 \times 10^{-13}\ \text{s}^{-1}$$

$$t_{1/2} = \frac{0.693}{k} = \frac{0.693}{3.7 \times 10^{-13}\ \text{s}^{-1}} = 1.9 \times 10^{12}\ \text{s}$$

$$t_{1/2} = (1.9 \times 10^{12}\ \text{s})\left(\frac{1\ \text{h}}{3600\ \text{s}}\right)\left(\frac{1\ \text{d}}{24\ \text{h}}\right)\left(\frac{1\ \text{y}}{365\ \text{d}}\right) = 6.0 \times 10^4\ \text{y}$$

22.50 $$\ln\left(\frac{N}{N_0}\right) = (-0.693)\left(\frac{t}{t_{1/2}}\right); \qquad \frac{N}{N_0} = \frac{\text{Decay rate at time}\ t}{\text{Decay rate at time}\ t\ =\ 0}$$

$$\ln\left(\frac{6990}{8540}\right) = (-0.693)\left(\frac{10.0\ \text{d}}{t_{1/2}}\right); \qquad t_{1/2} = 34.6\ \text{d}$$

Energy Changes During Nuclear Reactions

22.52   The loss in mass that occurs when protons and neutrons combine to form a nucleus is called the mass defect. The lost mass is converted into the binding energy that is used to hold the nucleons together.

22.54 $$E = (1.50\ \text{MeV})\left(\frac{1.60 \times 10^{-13}\ \text{J}}{1\ \text{MeV}}\right) = 2.40 \times 10^{-13}\ \text{J}$$

$$\lambda = \frac{hc}{E} = \frac{(6.626 \times 10^{-34}\ \text{J·s})(3.00 \times 10^8\ \text{m/s})}{2.40 \times 10^{-13}\ \text{J}} = 8.28 \times 10^{-13}\ \text{m} = 0.000\ 828\ \text{nm}$$

22.56   (a) For $^{52}_{26}\text{Fe}$:

First, calculate the total mass of the nucleons (26 n + 26 p)

Mass of 26 neutrons = (26)(1.008 66 amu) = 26.225 16 amu

Mass of 26 protons = (26)(1.007 28 amu) = 26.189 28 amu

Mass of 26 n + 26 p   = 52.414 44 amu

Next, calculate the mass of a ^{52}Fe nucleus by subtracting the mass of 26 electrons from the mass of a ^{52}Fe atom.

Mass of ^{52}Fe atom   = 51.948 11 amu
−Mass of 26 electrons = −(26)(5.486 x 10^{-4} amu)   = −0.014 26 amu
Mass of ^{52}Fe nucleus   = 51.933 85 amu

Then subtract the mass of the ^{52}Fe nucleus from the mass of the nucleons to find the mass defect:

Mass defect = mass of nucleons − mass of nucleus
             = (52.414 44 amu) − (51.933 85 amu) = 0.480 59 amu

Mass defect in g/mol:
(0.480 59 amu)(1.660 54 x 10^{-24} g/amu)(6.022 x 10^{23} mol^{-1}) = 0.480 59 g/mol

(b) For $^{92}_{42}$Mo :

First, calculate the total mass of the nucleons (50 n + 42 p)
Mass of 50 neutrons = (50)(1.008 66 amu) = 50.433 00 amu
Mass of 42 protons = (42)(1.007 28 amu) = 42.305 76 amu
Mass of 50 n + 42 p   = 92.738 76 amu

Next, calculate the mass of a ^{92}Mo nucleus by subtracting the mass of 42 electrons from the mass of a ^{92}Mo atom.

Mass of ^{92}Mo atom   = 91.906 81 amu
−Mass of 42 electrons = −(42)(5.486 x 10^{-4} amu)   = −0.023 04 amu
Mass of ^{92}Mo nucleus   = 91.883 77 amu

Then subtract the mass of the ^{92}Mo nucleus from the mass of the nucleons to find the mass defect:

Mass defect = mass of nucleons − mass of nucleus
             = (92.738 76 amu) − (91.883 77 amu) = 0.854 99 amu

Mass defect in g/mol:
(0.854 99 amu)(1.660 54 x 10^{-24} g/amu)(6.022 x 10^{23} mol^{-1}) = 0.854 99 g/mol

22.58   (a) For $^{58}_{28}$Ni :

First, calculate the total mass of the nucleons (30 n + 28 p)
Mass of 30 neutrons = (30)(1.008 66 amu) = 30.259 80 amu
Mass of 28 protons = (28)(1.007 28 amu) = 28.203 84 amu
Mass of 30 n + 28 p   = 58.463 64 amu

Next, calculate the mass of a ^{58}Ni nucleus by subtracting the mass of 28 electrons from the mass of a ^{58}Ni atom.

Mass of ^{58}Ni atom   = 57.935 35 amu
−Mass of 28 electrons = −(28)(5.486 x 10^{-4} amu)   = −0.015 36 amu
Mass of ^{58}Ni nucleus   = 57.919 99 amu

Then subtract the mass of the ^{58}Ni nucleus from the mass of the nucleons to find the mass defect:

Mass defect = mass of nucleons − mass of nucleus
             = (58.463 64 amu) − (57.919 99 amu) = 0.543 65 amu

Mass defect in g/mol:
(0.543 65 amu)(1.660 54 x 10^{-24} g/amu)(6.022 x 10^{23} mol^{-1}) = 0.543 65 g/mol

Now, use the Einstein equation to convert the mass defect into the binding energy.
$\Delta E = \Delta mc^2$ = (0.543 65 g/mol)(10^{-3} kg/g)(3.00 x 10^8 m/s)^2

$\Delta E = 4.893 \times 10^{13}$ J/mol $= 4.893 \times 10^{10}$ kJ/mol

$$\Delta E = \frac{4.893 \times 10^{13} \text{ J/mol}}{6.022 \times 10^{23} \text{ nuclei/mol}} \times \frac{1 \text{ MeV}}{1.60 \times 10^{-13} \text{ J}} \times \frac{1 \text{ nucleus}}{58 \text{ nucleons}} = 8.76 \frac{\text{MeV}}{\text{nucleon}}$$

(b) For $^{84}_{36}$Kr:

First, calculate the total mass of the nucleons (48 n + 36 p)
Mass of 48 neutrons = (48)(1.008 66 amu) = 48.415 68 amu
Mass of 36 protons = (36)(1.007 28 amu) = 36.262 08 amu
Mass of 48 n + 36 p   = 84.677 76 amu
Next, calculate the mass of a ^{84}Kr nucleus by subtracting the mass of 36 electrons from the mass of a ^{84}Kr atom.

Mass of ^{84}Kr atom	= 83.911 51 amu
−Mass of 36 electrons = −(36)(5.486 $\times 10^{-4}$ amu)	= −0.019 75 amu
Mass of ^{84}Kr nucleus	= 83.891 76 amu

Then subtract the mass of the ^{84}Kr nucleus from the mass of the nucleons to find the mass defect:

Mass defect = mass of nucleons − mass of nucleus
              = (84.677 76 amu) − (83.891 76 amu) = 0.786 00 amu

Mass defect in g/mol:
(0.786 00 amu)(1.660 54 $\times 10^{-24}$ g/mol)(6.022 $\times 10^{23}$ mol^{-1}) = 0.786 00 g/mol
Now, use the Einstein equation to convert the mass defect into the binding energy.
$\Delta E = \Delta mc^2 = (0.786\ 00$ g/mol)(10^{-3} kg/g)(3.00×10^8 m/s)^2
$\Delta E = 7.074 \times 10^{13}$ J/mol $= 7.074 \times 10^{10}$ kJ/mol

$$\Delta E = \frac{7.074 \times 10^{13} \text{ J/mol}}{6.022 \times 10^{23} \text{ nuclei/mol}} \times \frac{1 \text{ MeV}}{1.60 \times 10^{-13} \text{ J}} \times \frac{1 \text{ nucleus}}{84 \text{ nucleons}} = 8.74 \frac{\text{MeV}}{\text{nucleon}}$$

22.60 $^{174}_{77}$Ir   → $^{170}_{75}$Re   + $^{4}_{2}$He

mass $^{174}_{77}$Ir	173.966 66 amu
−mass $^{170}_{75}$Re	−169.958 04 amu
−mass $^{4}_{2}$He	−4.002 60 amu
mass change	0.006 02 amu

(0.006 02 amu)(1.660 54 $\times 10^{-24}$ g/amu)(6.022 $\times 10^{23}$ mol^{-1}) = 0.006 02 g/mol
$\Delta E = \Delta mc^2 = (0.006\ 02$ g/mol)(10^{-3} kg/g)(3.00×10^8 m/s)^2
$\Delta E = 5.42 \times 10^{11}$ J/mol $= 5.42 \times 10^8$ kJ/mol

22.62 $\Delta m = \dfrac{\Delta E}{c^2} = \dfrac{92.2 \times 10^3 \text{ J}}{(3.00 \times 10^8 \text{ m/s})^2} = \dfrac{92.2 \times 10^3 \text{ kg·m}^2/\text{s}^2}{(3.00 \times 10^8 \text{ m/s})^2} = 1.02 \times 10^{-12}$ kg

$\Delta m = 1.02 \times 10^{-9}$ g

22.64　Mass of positron and electron

$$= 2(9.109 \times 10^{-31} \text{ kg})(6.022 \times 10^{23} \text{ mol}^{-1}) = 1.097 \times 10^{-6} \text{ kg/mol}$$

$$\Delta E = \Delta mc^2 = (1.097 \times 10^{-6} \text{ kg/mol})(3.00 \times 10^8 \text{ m/s})^2$$

$$\Delta E = 9.87 \times 10^{10} \text{ J/mol} = 9.87 \times 10^7 \text{ kJ/mol}$$

Nuclear Transmutation

22.66　(a) $^{109}_{47}\text{Ag} + \,^{4}_{2}\text{He} \rightarrow \,^{113}_{49}\text{In}$　　(b) $^{10}_{5}\text{B} + \,^{4}_{2}\text{He} \rightarrow \,^{13}_{7}\text{N} + \,^{1}_{0}\text{n}$

22.68　$^{209}_{83}\text{Bi} + \,^{58}_{26}\text{Fe} \rightarrow \,^{266}_{109}\text{Mt} + \,^{1}_{0}\text{n}$

22.70　$^{238}_{92}\text{U} + \,^{12}_{6}\text{C} \rightarrow \,^{246}_{98}\text{Cf} + 4\,^{1}_{0}\text{n}$

General Problems

22.72　Each alpha emission decreases the mass number by four and the atomic number by two. Each beta emission increases the atomic number by one.

$$^{232}_{90}\text{Th} \rightarrow \,^{208}_{82}\text{Pb}$$

$$\text{Number of } \alpha \text{ emissions} = \frac{\text{Th mass number} - \text{Pb mass number}}{4}$$

$$= \frac{232 - 208}{4} = 6 \ \alpha \text{ emissions}$$

The atomic number decreases by 12 as a result of 6 alpha emissions. The resulting atomic number is $(90 - 12) = 78$.

Number of β emissions = Pb atomic number $- 78 = 82 - 78 = 4 \ \beta$ emissions

22.74　$\ln\left(\dfrac{N}{N_0}\right) = (-0.693)\left(\dfrac{t}{t_{1/2}}\right); \qquad \dfrac{N}{N_0} = \dfrac{\text{Decay rate at time } t}{\text{Decay rate at time } t = 0}$

$$\ln\left(\frac{100 - 99.99}{100}\right) = (-0.693)\left(\frac{t}{1.53 \text{ s}}\right); \qquad t = 20.3 \text{ s}$$

22.76　(a) For $^{50}_{24}\text{Cr}$:

First, calculate the total mass of the nucleons (26 n + 24 p)

Mass of 26 neutrons = (26)(1.008 66 amu) = 26.225 16 amu

<u>Mass of 24 protons　= (24)(1.007 28 amu)　= 24.174 72 amu</u>

Mass of 26 n + 24 p　　　　　　　　　　= 50.399 88 amu

Next, calculate the mass of a ^{50}Cr nucleus by subtracting the mass of 24 electrons from the mass of a ^{50}Cr atom.

Mass of ^{50}Cr atom	= 49.946 05 amu
−Mass of 24 electrons = −(24)(5.486 x 10^{-4} amu)	= −0.013 17 amu
Mass of ^{50}Cr nucleus	= 49.932 88 amu

Then subtract the mass of the ^{50}Cr nucleus from the mass of the nucleons to find the mass defect:

Mass defect = mass of nucleons − mass of nucleus
= (50.399 88 amu) − (49.932 88 amu) = 0.467 00 amu

Mass defect in g/mol:
(0.467 00 amu)(1.660 54 x 10^{-24} g/amu)(6.022 x 10^{23} mol^{-1}) = 0.467 00 g/mol

Now, use the Einstein equation to convert the mass defect into the binding energy.

$\Delta E = \Delta mc^2$ = (0.467 00 g/mol)(10^{-3} kg/g)(3.00 x 10^8 m/s)^2

ΔE = 4.203 x 10^{13} J/mol = 4.203 x 10^{10} kJ/mol

$$\Delta E = \frac{4.203 \times 10^{13} \text{ J/mol}}{6.022 \times 10^{23} \text{ nuclei/mol}} \times \frac{1 \text{ MeV}}{1.60 \times 10^{-13} \text{ J}} \times \frac{1 \text{ nucleus}}{50 \text{ nucleons}} = 8.72 \frac{\text{MeV}}{\text{nucleon}}$$

(b) For $^{64}_{30}$Zn:

First, calculate the total mass of the nucleons (34 n + 30 p)

Mass of 34 neutrons = (34)(1.008 66 amu)	= 34.294 44 amu
Mass of 30 protons = (30)(1.007 28 amu)	= 30.218 40 amu
Mass of 34 n + 30 p	= 64.512 84 amu

Next, calculate the mass of a ^{64}Zn nucleus by subtracting the mass of 30 electrons from the mass of a ^{64}Zn atom.

Mass of ^{64}Zn atom	= 63.929 15 amu
−Mass of 30 electrons = −(30)(5.486 x 10^{-4} amu)	= −0.016 46 amu
Mass of ^{64}Zn nucleus	= 63.912 69 amu

Then subtract the mass of the ^{64}Zn nucleus from the mass of the nucleons to find the mass defect:

Mass defect = mass of nucleons − mass of nucleus
= (64.512 84 amu) − (63.912 69 amu) = 0.600 15 amu

Mass defect in g/mol:
(0.600 15 amu)(1.660 54 x 10^{-24} g/amu)(6.022 x 10^{23} mol^{-1}) = 0.600 15 g/mol

Now, use the Einstein equation to convert the mass defect into the binding energy.

$\Delta E = \Delta mc^2$ = (0.600 15 g/mol)(10^{-3} kg/g)(3.00 x 10^8 m/s)^2

ΔE = 5.401 x 10^{13} J/mol = 5.401 x 10^{10} kJ/mol

$$\Delta E = \frac{5.401 \times 10^{13} \text{ J/mol}}{6.022 \times 10^{23} \text{ nuclei/mol}} \times \frac{1 \text{ MeV}}{1.60 \times 10^{-13} \text{ J}} \times \frac{1 \text{ nucleus}}{64 \text{ nucleons}} = 8.76 \frac{\text{MeV}}{\text{nucleon}}$$

The ^{64}Zn is more stable.

22.78 $\quad {}^2_1H + {}^3_2He \rightarrow {}^4_2He + {}^1_1H$

mass ${}^2_1H$	2.0141 amu
mass ${}^3_2He$	3.0160 amu
−mass ${}^4_2He$	−4.0026 amu
−mass ${}^1_1H$	−1.0078 amu
mass change	0.0197 amu

$(0.0197 \text{ amu})(1.660\ 54 \times 10^{-24} \text{ g/amu})(6.022 \times 10^{23} \text{ mol}^{-1}) = 0.0197 \text{ g/mol}$
$\Delta E = \Delta mc^2 = (0.0197 \text{ g/mol})(10^{-3} \text{ kg/g})(3.00 \times 10^8 \text{ m/s})^2$
$\Delta E = 1.77 \times 10^{12} \text{ J/mol} = 1.77 \times 10^9 \text{ kJ/mol}$

22.80 $\quad t_{1/2} = 1.1 \times 10^{20} \text{ y} = (1.1 \times 10^{20} \text{ y})(365 \text{ d/y}) = 4.0 \times 10^{22} \text{ d}$

$k = \dfrac{0.693}{t_{1/2}} = \dfrac{0.693}{4.0 \times 10^{22} \text{ d}} = 1.7 \times 10^{-23} \text{ d}^{-1}$

$N = 6.02 \times 10^{23}$ atoms
Decay rate $= kN = (1.7 \times 10^{-23} \text{ d}^{-1})(6.02 \times 10^{23} \text{ atoms}) = 10/\text{d}$
There are 10 disintegrations per day.

22.82 $\quad \ln\left(\dfrac{N}{N_0}\right) = (-0.693)\left(\dfrac{t}{t_{1/2}}\right); \qquad \dfrac{N}{N_0} = \dfrac{\text{Decay rate at time t}}{\text{Decay rate at time t} = 0}$

$\ln\left(\dfrac{1,350}{44,500}\right) = (-0.693)\left(\dfrac{5.00 \text{ min}}{t_{1/2}}\right); \qquad t_{1/2} = 0.991 \text{ min}$

22.84 $\quad \ln\left(\dfrac{N}{N_0}\right) = (-0.693)\left(\dfrac{t}{t_{1/2}}\right); \qquad \dfrac{N}{N_0} = \dfrac{\text{Decay rate at time t}}{\text{Decay rate at time t} = 0}$

$\ln\left(\dfrac{9.2}{15.3}\right) = (-0.693)\left(\dfrac{t}{5715 \text{ y}}\right); \qquad t = 4200 \text{ y}$

23 Organic Chemistry

23.1

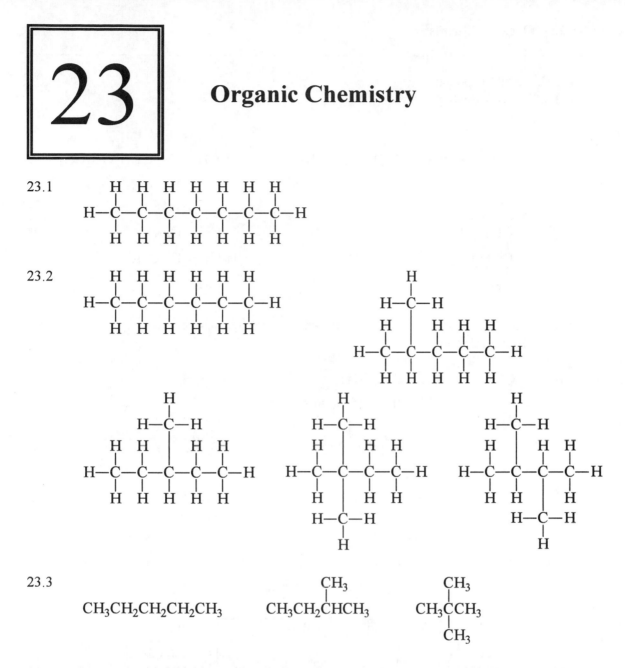

23.2

23.3

$$CH_3CH_2CH_2CH_2CH_3$$

$$CH_3CH_2\overset{\overset{\displaystyle CH_3}{|}}{C}HCH_3$$

$$CH_3\overset{\overset{\displaystyle CH_3}{|}}{\underset{\underset{\displaystyle CH_3}{|}}{C}}CH_3$$

23.4 Structures (a) and (c) are identical. They both contain a chain of six carbons with two –CH₃ branches at the fourth carbon and one –CH₃ branch at the second carbon. Structure (b) is different, having a chain of seven carbons.

23.5 (a) $CH_3CH_2CH_2CH_2CH_3$ pentane

$$CH_3CH_2\overset{\overset{\displaystyle CH_3}{|}}{C}HCH_3$$ 2-methylbutane

$$CH_3\overset{\overset{\displaystyle CH_3}{|}}{\underset{\underset{\displaystyle CH_3}{|}}{C}}CH_3$$ 2,2-dimethylpropane

(b) 3,4-dimethylhexane

(c) 2,4-dimethylpentane
(d) 2,2,5-trimethylheptane

23.6 (a)

$$CH_3CH_2\overset{\overset{\displaystyle CH_3}{|}}{CH}CHCH_2CH_2CH_2CH_2CH_3$$
$$\underset{|}{\quad\;\;}$$
$$CH_3$$

(b)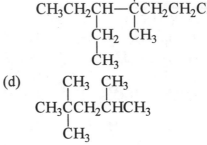

(c)

$$\overset{\overset{\displaystyle CH_3}{|}}{CH_3C}CH_2\overset{\overset{\displaystyle CH_2CH_2CH_3}{|}}{CH}CH_2CH_2CH_2CH_3$$
$$\underset{|}{\quad}$$
$$CH_3$$

(d)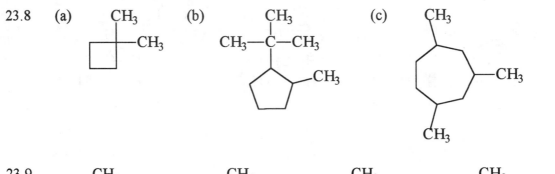

23.7 (a) 1,4-dimethylcyclohexane   (b) 1-ethyl-3-methylcyclopentane
(c) isopropylcyclobutane

23.8 (a) (b) (c)

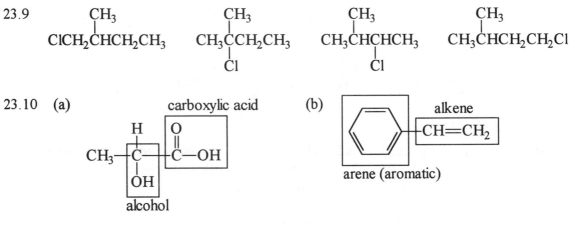

23.9

$$\overset{\overset{\displaystyle CH_3}{|}}{ClCH_2CH}CH_2CH_3 \qquad \overset{\overset{\displaystyle CH_3}{|}}{CH_3C}CH_2CH_3 \qquad \overset{\overset{\displaystyle CH_3}{|}}{CH_3CH}CHCH_3 \qquad \overset{\overset{\displaystyle CH_3}{|}}{CH_3CH}CH_2CH_2Cl$$
$$\qquad\qquad\qquad\quad\; \underset{|}{Cl} \qquad\qquad\qquad\;\; \underset{|}{Cl}$$

23.10 (a)

carboxylic acid

$$CH_3\overset{\overset{\displaystyle H}{|}}{C}\overset{\overset{\displaystyle O}{\|}}{C}OH$$
$$\underset{|}{OH}$$
alcohol

(b)

arene (aromatic) alkene CH=CH₂

23.11 (a)

$$\overset{\overset{\displaystyle O}{\|}}{CH_3CH}$$

(b)

$$\overset{\overset{\displaystyle O}{\|}}{CH_3CH_2COH}$$

23.12 (a) 3-methyl-1-butene   (b) 4-methyl-3-heptene   (c) 3-ethyl-1-hexyne

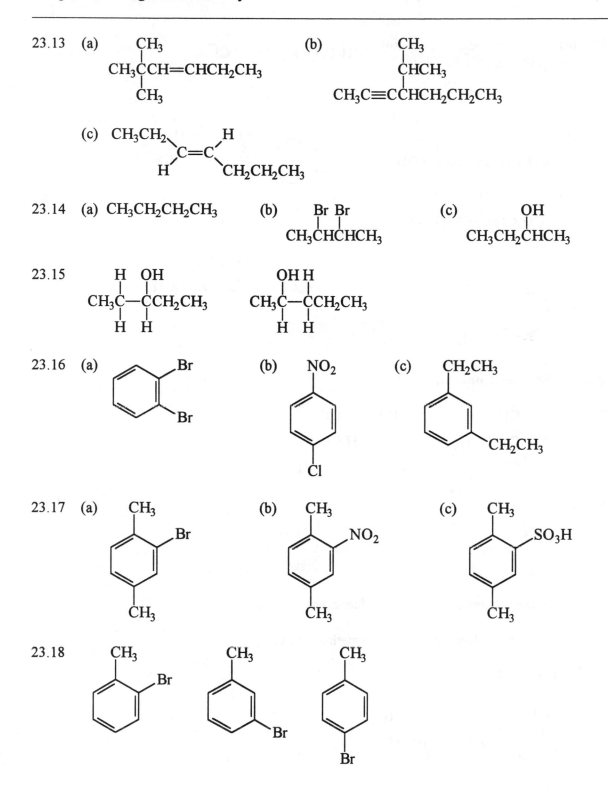

23.13 (a)

$$CH_3CCH=CHCH_2CH_3$$
with CH_3 above and CH_3 below the C

(b)

$$CH_3C\equiv CCHCH_2CH_2CH_3$$
with CH_3 and $CHCH_3$ above

(c)

$$CH_3CH_2\atop H \quad C=C \quad {H \atop CH_2CH_2CH_3}$$

23.14 (a) $CH_3CH_2CH_2CH_3$

(b)

$$CH_3CHCHCH_3$$
with Br Br above

(c)

$$CH_3CH_2CHCH_3$$
with OH above

23.15

$$CH_3C-CCH_2CH_3$$
with H OH above and H H below

$$CH_3C-CCH_2CH_3$$
with OH H above and H H below

23.16 (a) benzene with Br, Br

(b) benzene with NO_2 and Cl

(c) benzene with CH_2CH_3 and CH_2CH_3

23.17 (a) benzene with CH_3, Br, CH_3

(b) benzene with CH_3, NO_2, CH_3

(c) benzene with CH_3, SO_3H, CH_3

23.18 benzene with CH_3, Br   benzene with CH_3, Br   benzene with CH_3, Br

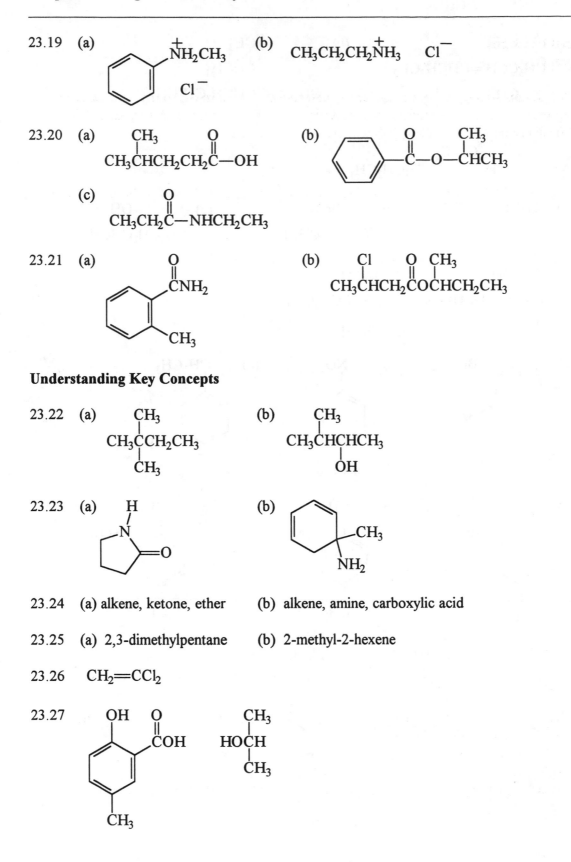

23.19 (a) Phenyl-NH₂CH₃⁺ Cl⁻

(b) CH₃CH₂CH₂NH₃⁺ Cl⁻

23.20 (a) $CH_3CHCH_2CH_2C-OH$ with CH₃ and O

(b) benzene-C(=O)-O-CHCH₃ with CH₃

(c) $CH_3CH_2C-NHCH_2CH_3$ with O

23.21 (a) benzene with C(=O)NH₂ and CH₃

(b) $CH_3CHCH_2COCHCH_2CH_3$ with Cl, O, CH₃

Understanding Key Concepts

23.22 (a) $CH_3CCH_2CH_3$ with CH₃ and CH₃

(b) $CH_3CHCHCH_3$ with CH₃ and OH

23.23 (a) pyrrolidinone ring with N-H and O

(b) cyclohexadiene with CH₃ and NH₂

23.24 (a) alkene, ketone, ether   (b) alkene, amine, carboxylic acid

23.25 (a) 2,3-dimethylpentane   (b) 2-methyl-2-hexene

23.26 $CH_2=CCl_2$

23.27 benzene with OH, COH(=O), CH₃; and HOCH with CH₃, CH₃

Additional Problems
Functional Groups and Isomers

23.28   A functional group is a part of a larger molecule and is composed of an atom or group of atoms that has a characteristic chemical behavior. They are important because their chemistry controls the chemistry in molecules that contain them.

23.30   (a)
$$CH_3CH_2\overset{\overset{\displaystyle O}{\|}}{C}CH_2CH_3$$
(b)
$$CH_3CH_2CH_2\overset{\overset{\displaystyle O}{\|}}{C}OCH_2CH_3$$
(c)
$$NH_2CH_2\overset{\overset{\displaystyle O}{\|}}{C}OH$$

23.32
$$CH_3CH_2CH_2OH \qquad CH_3\overset{\overset{\displaystyle OH}{|}}{C}HCH_3 \qquad CH_3CH_2OCH_3$$

23.34   (a) alkene and aldehyde   (b) arene, alcohol, and ketone

Alkanes

23.36   In a straight-chain alkane, all the carbons are connected in a row. In a branched-chain alkane, there are branching connections of carbons along the carbon chain.

23.38   In forming alkanes, carbon uses sp^3 hybrid orbitals.

23.40 C_3H_9 contains one more H than needed for an alkane.

23.42   (a) 4-ethyl-3-methyloctane   (b) 4-isopropyl-2-methylheptane
        (c) 2,2,6-trimethylheptane   (d) 4-ethyl-4-methyloctane

23.44   (a)
$$CH_3CH_2\overset{\overset{\displaystyle CH_2CH_3}{|}}{C}HCH_2CH_2CH_3$$
(b)
$$CH_3\overset{\overset{\displaystyle CH_3}{|}}{\underset{\underset{\displaystyle CH_3}{|}}{C}}{-}\overset{\overset{\displaystyle CH_3}{|}}{C}HCH_2CH_3$$

(c)
$$CH_3CH_2\overset{\overset{\displaystyle CH_2CH_3}{|}}{\underset{\underset{\displaystyle CH_3}{|}}{C}}{-}\overset{\underset{\displaystyle CH_3}{|}}{C}HCH_2CH_2CH_3$$
(d)
$$CH_3\overset{\displaystyle CH_3}{\underset{\displaystyle CH_3CHCH_2CH_2}{}}\overset{\overset{\displaystyle CH_3}{|}}{\underset{\underset{\displaystyle CHCH_3}{|}}{C}}HCH_2CH_2CH_3$$

23.46   (a) 1,1-dimethylcyclopentane   (b) 1-isopropyl-2-methylcyclohexane
        (c) 1,2,4-trimethylcyclooctane

23.48   The structures are shown in Problem 23.2.
        hexane, 2-methylpentane, 3-methylpentane, 2,2-dimethylbutane, and 2,3-dimethylbutane

23.50 (a)

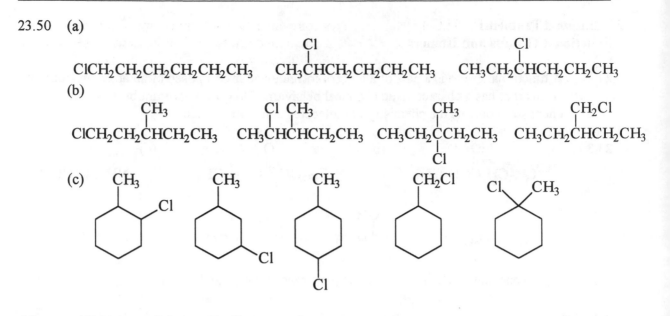

$ClCH_2CH_2CH_2CH_2CH_2CH_3$ $CH_3\overset{Cl}{\underset{|}{C}}HCH_2CH_2CH_2CH_3$ $CH_3CH_2\overset{Cl}{\underset{|}{C}}HCH_2CH_2CH_3$

(b)

$ClCH_2CH_2\overset{CH_3}{\underset{|}{C}}HCH_2CH_3$ $CH_3\overset{Cl}{\underset{|}{C}}H\overset{CH_3}{\underset{|}{C}}HCH_2CH_3$ $CH_3CH_2\overset{CH_3}{\underset{\underset{Cl}{|}}{C}}CH_2CH_3$ $CH_3CH_2\overset{CH_2Cl}{\underset{|}{C}}HCH_2CH_3$

(c)

Alkenes, Alkynes, and Aromatic Compounds

23.52   (a) sp^2 (b) sp (c) sp^2

23.54   Today the term "aromatic" refers to the class of compounds containing a six-membered ring with three double bonds, not to the fragrance of a compound.

23.56   (a) $CH_3CH=CHCH_2CH_3$ (b) $HC\equiv CCH_2CH_3$ (c)

23.58   (a) 4-methyl-2-pentene   (b) 3-methyl-1-pentene
         (c) 1,2-dichlorobenzene, or o-dichlorobenzene
         (d) 2-methyl-2-butene   (e) 7-methyl-3-octyne

23.60 $CH_2=CHCH_2CH_2CH_3$ $CH_3CH=CHCH_2CH_3$ $CH_2=\overset{CH_3}{\underset{|}{C}}CH_2CH_3$
         1-pentene   2-pentene
                                                                     2-methyl-1-butene

         $CH_3\overset{CH_3}{\underset{|}{C}}=CHCH_3$ $CH_2=CH\overset{CH_3}{\underset{|}{C}}HCH_3$ Only 2-pentene can exist as
         2-methyl-2-butene   3-methyl-1-butene   cis-trans isomers.

23.62 (a) $CH_2=CHCH_2CH_2CH_2CH_3$ This compound cannot form cis-trans isomers.
    (b) $CH_3CH=CHCH_2CH_2CH_3$ This compound can form cis-trans isomers because of the different groups on each double bond C.

    (c) $CH_3CH_2CH=CHCH_2CH_3$ This compound can form cis-trans isomers because of the different groups on each double bond C.

23.64 (a)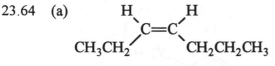

(b)

(c)

23.66 Cis-trans isomers are possible for substituted alkenes because of the lack of rotation about the carbon-carbon double bond. Alkanes and alkynes cannot form cis-trans isomers because alkanes have free rotation about carbon-carbon single bonds and alkynes are linear about the carbon-carbon triple bond.

23.68 (a)

$$CH_3C=CCH_3 \quad + \quad H_2 \quad \xrightarrow{Pd} \quad CH_3C-CCH_3$$

(b)

$$CH_3C=CCH_3 \quad + \quad Br_2 \quad \longrightarrow \quad CH_3C-CCH_3$$

(c)

$$CH_3C=CCH_3 \quad + \quad H_2O \quad \xrightarrow{H_2SO_4} \quad CH_3C-CCH_3$$

23.70 (a)

$$+ \quad Br_2 \quad \xrightarrow{FeBr_3}$$

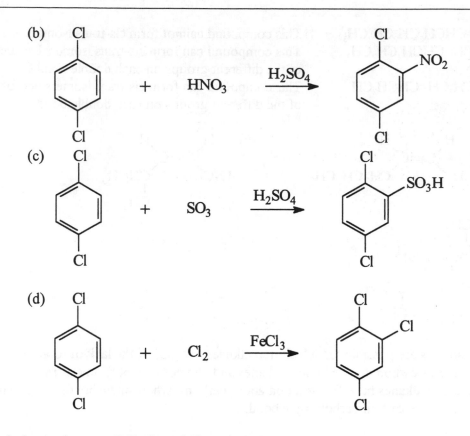

Alcohols, Amines, and Carbonyl Compounds

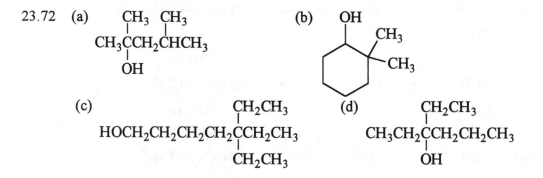

23.72 (a)

23.74 Quinine, a base will dissolve in aqueous acid, but menthol is insoluble.

23.76 An aldehyde has a terminal carbonyl group. A ketone has the carbonyl group located between two carbon atoms.

23.78 The industrial preparation of ketones and aldehydes involves the oxidation of the related alcohol.

23.80 (a) ketone   (b) aldehyde   (c) ketone   (d) amide   (e) ester

23.82

$$C_6H_5COOH(aq) + H_2O(l) \rightleftharpoons H_3O^+(aq) + C_6H_5CO_2^-(aq)$$

initial (M)	1.0	~0	0
change (M)	−x	+x	+x
equil (M)	1.0 − x	x	x

$$K_a = \frac{[H_3O^+][C_6H_5CO_2^-]}{[C_6H_5COOH]} = 6.5 \times 10^{-5} = \frac{x^2}{1.0-x} \approx \frac{x^2}{1.0}$$

$$x = [H_3O^+] = [C_6H_5COOH]_{diss} = 0.0081 \text{ M}$$

$$\% \text{ dissociation} = \frac{[C_6H_5COOH]_{diss}}{[C_6H_5COOH]_{initial}} \times 100\% = \frac{0.0081 \text{ M}}{1.0 \text{ M}} \times 100\% = 0.81\%$$

23.84 (a) methyl 4-methylpentanoate   (b) 4,4-dimethylpentanoic acid
      (c) 2-methylpentanamide

23.86 (a)

$$CH_3CH_2CH_2CH_2\overset{\overset{\displaystyle O}{\|}}{C}-OCH_3$$

(b)

$$CH_3CH_2\overset{\overset{\displaystyle O}{\|}}{C}H\overset{\overset{\displaystyle O}{\|}}{C}-O\overset{\overset{\displaystyle CH_3}{|}}{C}H$$
$$\underset{CH_3}{|} \qquad \underset{CH_3}{|}$$

(c)

$$CH_3\overset{\overset{\displaystyle O}{\|}}{C}-O-\bigcirc$$

23.88 (a)

$$CH_3CH_2CH_2CH_2\overset{\overset{\displaystyle O}{\|}}{C}OH + CH_3OH \xrightarrow{H^+} CH_3CH_2CH_2CH_2\overset{\overset{\displaystyle O}{\|}}{C}OCH_3 + H_2O$$

(b)

$$CH_3CH_2\overset{\overset{\displaystyle O}{\|}}{\underset{\underset{CH_3}{|}}{C}}HCOH + \overset{\overset{\displaystyle CH_3}{|}}{\underset{\underset{CH_3}{|}}{H}}COH \xrightarrow{H^+} CH_3CH_2\overset{\underset{\underset{CH_3}{|}}{}}{C}H\overset{\overset{\displaystyle O}{\|}}{C}O\overset{\overset{\displaystyle CH_3}{|}}{\underset{\underset{CH_3}{|}}{C}}H + H_2O$$

(c)

$$CH_3\overset{\overset{\displaystyle O}{\|}}{C}OH + HO-\bigcirc \xrightarrow{H^+} CH_3\overset{\overset{\displaystyle O}{\|}}{C}O-\bigcirc + H_2O$$

23.90 amine, arene, and ester

$$H_2N-\bigcirc-\overset{\overset{\displaystyle O}{\|}}{C}OH \qquad\qquad HOCH_2CH_2\overset{\overset{\displaystyle CH_2CH_3}{|}}{N}CH_2CH_3$$

carboxylic acid   alcohol

Polymers

23.92 Polymers are large molecules formed by the repetitive bonding together of many smaller molecules, called monomers.

23.94

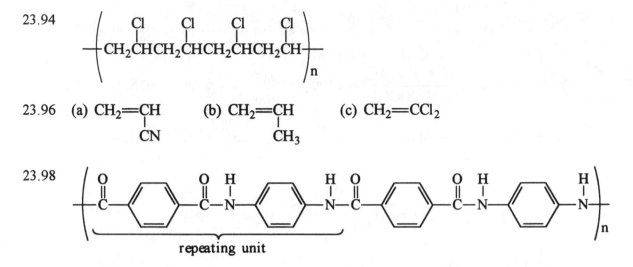

23.96 (a) $CH_2\!\!=\!\!CH$ (b) $CH_2\!\!=\!\!CH$ (c) $CH_2\!\!=\!\!CCl_2$
              $|$ $|$
              CN CH_3

23.98

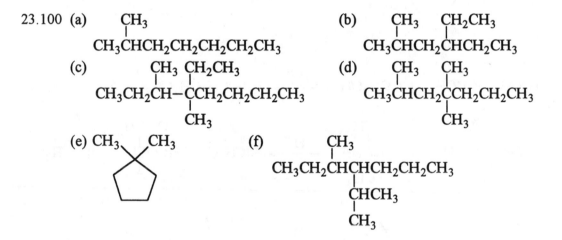

repeating unit

General Problems

23.100 (a) CH_3
                     $|$
              $CH_3CHCH_2CH_2CH_2CH_2CH_3$

(b) CH_3 CH_2CH_3
            $|$ $|$
        $CH_3CHCH_2CHCH_2CH_3$

(c) CH_3 CH_2CH_3
                  $|$ $|$
      $CH_3CH_2CH\!-\!CCH_2CH_2CH_2CH_3$
                          $|$
                         CH_3

(d) CH_3 CH_3
            $|$ $|$
        $CH_3CHCH_2CCH_2CH_2CH_3$
                      $|$
                     CH_3

(e) CH_3 CH_3

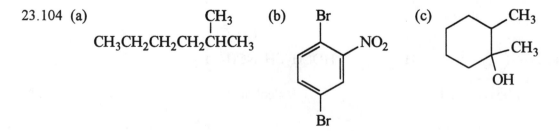

(f) CH_3
                  $|$
        $CH_3CH_2CHCHCH_2CH_2CH_3$
                   $|$
                  $CHCH_3$
                   $|$
                  CH_3

23.102 Cyclohexene will react with Br_2 and decolorize it. Cyclohexane will not react.

23.104 (a) CH_3 (b) Br (c)
                                $|$
        $CH_3CH_2CH_2CH_2CHCH_3$ NO_2 CH_3
                                                                  CH_3
                                          Br OH

23.106 (a) Calculate the empirical formula. Assume a 100.0 g sample of fumaric acid.

$$41.4 \text{ g C} \times \frac{1 \text{ mol C}}{12.01 \text{ g C}} = 3.45 \text{ mol C}; \qquad 3.5 \text{ g H} \times \frac{1 \text{ mol H}}{1.008 \text{ g H}} = 3.47 \text{ mol H}$$

$$55.1 \text{ g O} \times \frac{1 \text{ mol O}}{16.00 \text{ g O}} = 3.44 \text{ mol O}$$

Because the mol amounts for the three elements are essentially the same, the empirical formula is CHO (29 amu).

Calculate the molar mass from the osmotic pressure.

$$\Pi = MRT; \quad M = \frac{\Pi}{RT} = \frac{\left(240.3 \text{ mm Hg} \times \dfrac{1.00 \text{ atm}}{760 \text{ mm Hg}}\right)}{\left(0.082\,06 \dfrac{\text{L}\cdot\text{atm}}{\text{mol}\cdot\text{K}}\right)(298 \text{ K})} = 0.0129 \text{ M}$$

$(0.1000 \text{ L})(0.0129 \text{ mol/L}) = 1.29 \times 10^{-3} \text{ mol fumaric acid}$

$$\text{fumaric acid molar mass} = \frac{0.1500 \text{ g}}{1.29 \times 10^{-3} \text{ mol}} = 116 \text{ g/mol}$$

Determine the molecular formula. $\dfrac{\text{molar mass}}{\text{empirical formula mass}} = \dfrac{116}{29} = 4$

molecular formula $= C_{(1 \times 4)}H_{(1 \times 4)}O_{(1 \times 4)} = C_4H_4O_4$

From the titration, the number of carboxylic acid groups can be determined.

$$\text{mol } C_4H_4O_4 = 0.573 \text{ g} \times \frac{1 \text{ mol } C_4H_4O_4}{116 \text{ g}} = 0.004\,94 \text{ mol } C_4H_4O_4$$

mol NaOH used $= (0.0941 \text{ L})(0.105 \text{ mol/L}) = 0.0099 \text{ mol NaOH}$

$$\frac{\text{mol NaOH}}{\text{mol } C_4H_4O_4} = \frac{0.0099 \text{ mol}}{0.004\,94 \text{ mol}} = 2$$

Because 2 mol of NaOH are required to titrate 1 mol $C_4H_4O_4$, $C_4H_4O_4$ is a diprotic acid. Because $C_4H_4O_4$ gives an addition product with HCl and a reduction product with H_2, it contains a double bond.

(b)

(c) The correct structure is

24.1

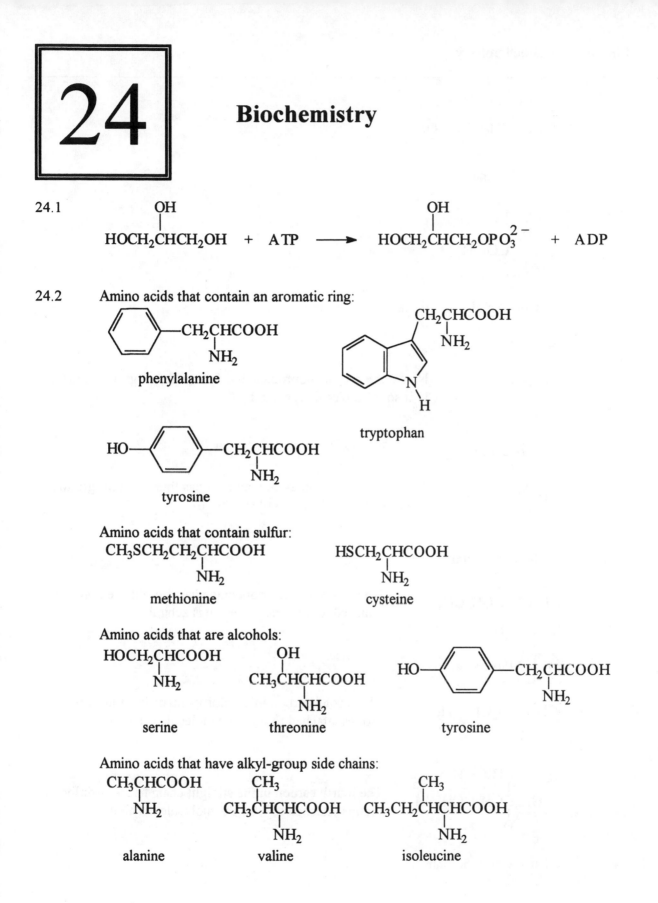

$$HOCH_2\overset{\displaystyle OH}{\underset{\displaystyle |}{C}}HCH_2OH \;+\; ATP \longrightarrow HOCH_2\overset{\displaystyle OH}{\underset{\displaystyle |}{C}}HCH_2OPO_3^{2-} \;+\; ADP$$

24.2 Amino acids that contain an aromatic ring:

phenylalanine

tryptophan

tyrosine

Amino acids that contain sulfur:

$$CH_3SCH_2CH_2\underset{\underset{\displaystyle NH_2}{|}}{C}HCOOH$$

methionine

$$HS CH_2\underset{\underset{\displaystyle NH_2}{|}}{C}HCOOH$$

cysteine

Amino acids that are alcohols:

$$HOCH_2\underset{\underset{\displaystyle NH_2}{|}}{C}HCOOH$$

serine

$$CH_3\overset{\overset{\displaystyle OH}{|}}{C}H\underset{\underset{\displaystyle NH_2}{|}}{C}HCOOH$$

threonine

tyrosine

Amino acids that have alkyl-group side chains:

$$CH_3\underset{\underset{\displaystyle NH_2}{|}}{C}HCOOH$$

alanine

$$CH_3\overset{\overset{\displaystyle CH_3}{|}}{C}H\underset{\underset{\displaystyle NH_2}{|}}{C}HCOOH$$

valine

$$CH_3CH_2\overset{\overset{\displaystyle CH_3}{|}}{C}H\underset{\underset{\displaystyle NH_2}{|}}{C}HCOOH$$

isoleucine

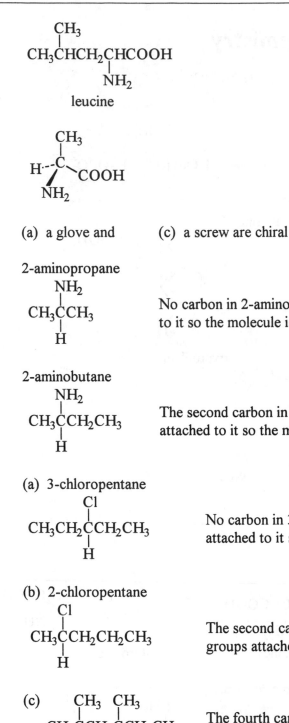

CH$_3$

CH$_3$CHCH$_2$CHCOOH

NH$_2$

leucine

24.3

CH$_3$

H---C---COOH

NH$_2$

24.4   (a) a glove and   (c) a screw are chiral

24.5   2-aminopropane

NH$_2$

CH$_3$CCH$_3$

H

No carbon in 2-aminopropane has four different groups attached to it so the molecule is achiral.

2-aminobutane

NH$_2$

CH$_3$CCH$_2$CH$_3$

H

The second carbon in 2-aminobutane has four different groups attached to it so the molecule is chiral.

24.6   (a) 3-chloropentane

Cl

CH$_3$CH$_2$CCH$_2$CH$_3$

H

No carbon in 3-chloropentane has four different groups attached to it so the molecule is achiral.

(b) 2-chloropentane

Cl

CH$_3$CCH$_2$CH$_2$CH$_3$

H

The second carbon in 2-chloropentane has four different groups attached to it so the molecule is chiral.

(c)   CH$_3$ CH$_3$

CH$_3$CCH$_2$CCH$_2$CH$_3$

H   H

The fourth carbon in the straight chain has four different groups attached to it so the molecule is chiral.

24.7   isoleucine and threonine

24.8   Val-Cys   Cys-Val

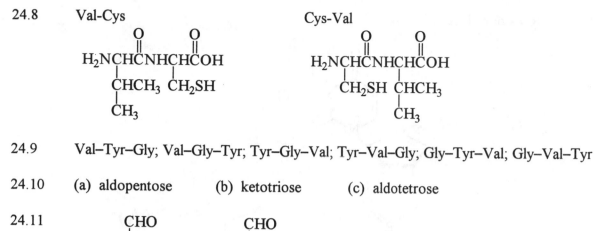

24.9   Val–Tyr–Gly; Val–Gly–Tyr; Tyr–Gly–Val; Tyr–Val–Gly; Gly–Tyr–Val; Gly–Val–Tyr

24.10   (a) aldopentose   (b) ketotriose   (c) aldotetrose

24.11

CHO   CHO

24.12   In general, a compound with n chiral carbon atoms has a maximum of 2^n possible forms. Because ribose has three chiral carbon atoms, the maximum number of ribose forms = $2^3 = 8$.

24.13

$$CH_2OC(CH_2)_7CH=CH(CH_2)_7CH_3$$

$$CHOC(CH_2)_7CH=CH(CH_2)_7CH_3$$

$$CH_2OC(CH_2)_7CH=CH(CH_2)_7CH_3$$

24.14   DNA dinucleotide A – G.

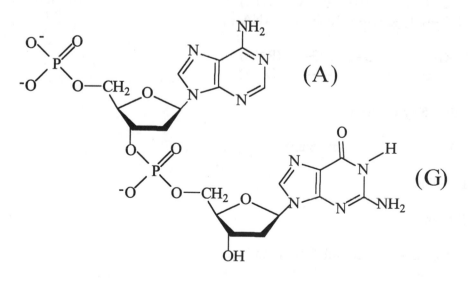

24.15   RNA dinucleotide U – A.

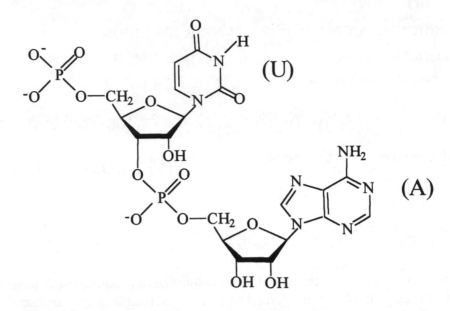

24.16   Original: G–G–C–C–C–G–T–A–A–T
        Complement: C–C–G–G–G–C–A–T–T–A

24.17

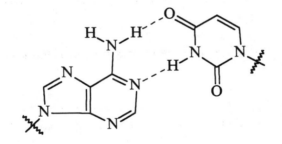

24.18   C–G–T–G–A–T–T–A–C–A   (DNA)
        G–C–A–C–U–A–A–U–G–U   (RNA)

24.19   U–G–C–A–U–C–G–A–G–U   (RNA)
        A–C–G–T–A–G–C–T–C–A   (DNA)

Understanding Key Concepts

24.20   (a) serine   (b) methionine

24.21   D-valine

24.22   β form

24.23   Ribose is an aldose with 3 chiral carbons.

24.24   Ser–Val

24.25   (a) guanine   (b) cytosine

Additional Problems
Amino Acids, Peptides, and Proteins

24.26   The "α" in α-amino acids means that the amino group in each is connected to the carbon atom alpha to (next to) the carboxylic acid group.

24.28   (a) serine   (b) threonine   (c) proline   (d) phenylalanine   (e) cysteine

24.30   (a)   O   (b)   O
                              ‖   ‖   +
                  H₂NCH₂COCH₃   HOCCH₂NH₃   Cl⁻

$$\text{(a)}\quad H_2NCH_2\overset{\displaystyle O}{\overset{\|}{C}}OCH_3 \qquad \text{(b)}\quad HO\overset{\displaystyle O}{\overset{\|}{C}}CH_2\overset{+}{N}H_3\ \ Cl^-$$

24.32   Val–Ser–Phe–Met–Thr–Ala

24.34   (a) The primary structure of a protein is the sequence in which amino acids are linked together.

(b) The secondary structure of a protein is the orientation of segments of a protein chain into a regular pattern.

(c) The tertiary structure of a protein is the way in which a protein chain folds into a specific three-dimensional shape.

24.36   A protein's tertiary structure is stabilized by amino acid side-chain interactions, which include hydrophobic interactions, covalent disulfide bridges, electrostatic salt bridges, and hydrogen bonds.

24.38   Cysteine is an important amino acid for defining the tertiary structure of proteins because the nearby cysteine residues can link together forming a disulfide bridge.

24.40   Met–Ile–Lys, Met–Lys–Ile, Ile–Met–Lys, Ile–Lys–Met, Lys–Met–Ile, Lys–Ile–Met

Molecular Handedness

24.42   (a) a shoe and   (c) a light bulb are chiral.

24.44   (a) 2,4-dimethylheptane
                 CH₃ CH₃
                  | |
     CH₃CCH₂CCH₂CH₂CH₃   The fourth carbon in the straight chain has four different
                  | | groups attached to it so the molecule is chiral.
                  H   H

(b) 5–ethyl–3,3–dimethylheptane

$$CH_3CH_2\overset{\overset{\displaystyle CH_3}{|}}{\underset{\underset{\displaystyle CH_3}{|}}{C}}CH_2\overset{\overset{\displaystyle CH_2CH_3}{|}}{\underset{\underset{\displaystyle H}{|}}{C}}CH_2CH_3$$

No carbon in 5–ethyl–3,3–dimethylheptane has four different groups attached to it so the molecule is achiral.

24.46 $CH_3CH_2CH_2CH_2CH_2OH$

$$CH_3CH_2\overset{\overset{\displaystyle OH}{|}}{C}HCH_2CH_3 \qquad CH_3\overset{\overset{\displaystyle CH_3}{|}}{\underset{\underset{\displaystyle OH}{|}}{C}}CH_2CH_3 \qquad CH_3\overset{\overset{\displaystyle CH_3}{|}}{\underset{\underset{\displaystyle CH_3}{|}}{C}}CH_2OH$$

$$CH_3\overset{\overset{\displaystyle CH_3}{|}}{C}HCH_2CH_2OH \qquad CH_3\overset{\overset{\displaystyle OH}{|}}{\underset{\underset{\displaystyle H}{|}}{C}}CH_2CH_2CH_3 \qquad CH_3CH_2\overset{\overset{\displaystyle CH_3}{|}}{\underset{\underset{\displaystyle H}{|}}{C}}CH_2OH \qquad CH_3\overset{\overset{\displaystyle CH_3}{|}}{C}H-\overset{\overset{\displaystyle OH}{|}}{\underset{\underset{\displaystyle H}{|}}{C}}CH_3$$

$$\qquad\qquad\qquad\text{chiral}\qquad\qquad\qquad\qquad\text{chiral}\qquad\qquad\qquad\text{chiral}$$

Carbohydrates

24.48   An aldose contains the aldehyde functional group while a ketose contains the ketone functional group.

24.50   Structurally, starch differs from cellulose in that it contains α- rather than β-glucose units.
Cellulose is the fibrous substance used by plants as a structural material in grasses, leaves, and stems. It consists of several thousand β-glucose molecules joined together by 1,4 links to form an immense polysaccharide.
Starch is a polymer made up of α-glucose units. Unlike cellulose, starch is edible. The starch in beans, rice, and potatoes is an essential part of the human diet.

24.52

$$HOCH_2\overset{\overset{\displaystyle O}{\|}}{C}H\overset{}{\underset{\underset{\displaystyle OH}{|}}{C}}CH_2OH$$

24.54

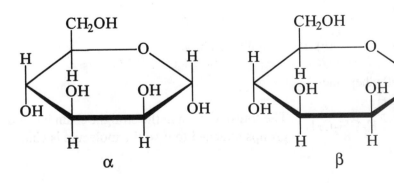

24.56

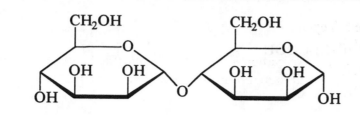

Lipids

24.58 Long-chain carboxylic acids are called fatty acids. Fatty acids are usually unbranched and have an even number of carbon atoms in the range of 12-22.

24.60

$$CH_2OC(CH_2)_{12}CH_3$$
(with C=O)

$$CHOC(CH_2)_{12}CH_3$$
(with C=O)

$$CH_2OC(CH_2)_{12}CH_3$$
(with C=O)

24.62

$$CH_2OC(CH_2)_{16}CH_3$$
$$CHOC(CH_2)_{16}CH_3$$
$$CH_2OC(CH_2)_{14}CH_3$$

$$CH_2OC(CH_2)_{16}CH_3$$
$$CHOC(CH_2)_{14}CH_3$$
$$CH_2OC(CH_2)_{16}CH_3$$

The two fat molecules differ from each other depending on where the palmitic acid chain is located. In the first fat molecule, palmitic acid is on an end and in the second it is in the middle.

24.64

$$CH_3(CH_2)_{16}CO^- \ K^+$$
potassium stearate

$$CH_3(CH_2)_7CH=CH(CH_2)_7CO^- \ K^+$$
potassium oleate

$$CH_3CH_2CH=CHCH_2CH=CHCH_2CH=CH(CH_2)_7CO^- \ K^+$$
potassium linolenate

$$HOCH_2CHCH_2OH$$
(with OH on middle carbon)
glycerol

24.66   (a)

$$CH_3(CH_2)_7\overset{\overset{\displaystyle Br}{|}}{C}H\overset{\overset{\displaystyle Br}{|}}{C}H(CH_2)_7COOH$$

(b) $CH_3(CH_2)_{16}COOH$

(c)

$$CH_3(CH_2)_7CH\!=\!CH(CH_2)_7\overset{\overset{\displaystyle O}{\|}}{C}OCH_3$$

Nucleic Acids

24.68   Just as proteins are polymers made of amino acid units, nucleic acids are polymers made up of nucleotide units linked together to form a long chain. Each nucleotide contains a phosphate group, an aldopentose sugar, and an amine base.

24.70   Most DNA of higher organisms is found in the nucleus of cells.

24.72   A chromosome is a threadlike strand of DNA in the cell nucleus. A gene is a segment of a DNA chain that contains the instructions necessary to make a specific protein.

24.74

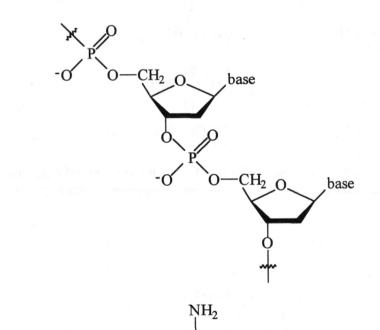

24.76

24.78   Original: T–A–C–C–G–A
         Complement: A–T–G–G–C–T

24.80   It takes three nucleotides to code for a specific amino acid. In insulin the 21 amino acid chain would require (3 x 21) = 63 nucleotides to code for it, and the 30 amino acid chain would require (3 x 30) = 90 nucleotides to code for it.

General Problems

24.82
$$CH_3(CH_2)_{18}\overset{\displaystyle O}{\overset{\|}{C}}O(CH_2)_{31}CH_3$$

24.84 $\Delta G^\circ = +\ 2870$ kJ/mol, because photosynthesis is the reverse reaction.

24.86   (a)

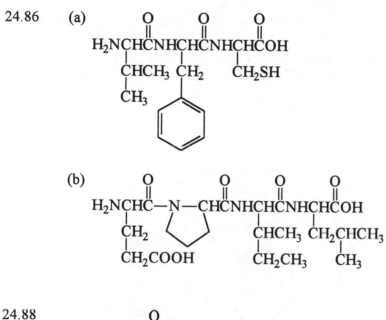

         (b)

24.88
$$CH_3(CH_2)_{16}\overset{\displaystyle O}{\overset{\|}{C}}O(CH_2)_{21}CH_3$$

24.90   Original: A–G–T–T–C–A–T–C–G
         Complement: T–C–A–A–G–T–A–G–C